AF566190

ULTRASOUND: BIOLOGICAL EFFECTS AND POTENTIAL HAZARDS

MEDICAL PHYSICS SERIES

P. N. T. Wells. Physical Principles of Ultrasonic Diagnosis. 1969
D. W. Hill and A. M. Dolan. Intensive Care Instrumentation. (2nd edn) 1982
P. N. T. Wells. Biomedical Ultrasonics. 1977
P. Rolfe. Non-invasive Physiological Measurements, Volume 1. 1979
P. Atkinson and J. P. Woodcock. Doppler Ultrasound and its use in Clinical Measurement. 1981
P. Rolfe. Non-invasive Physiological Measurements, Volume 2. 1983
G. Laszlo and M. F. Sudlow (eds). Measurement of Clinical Respiratory Physiology. 1983

ULTRASOUND: BIOLOGICAL EFFECTS AND POTENTIAL HAZARDS

A. Roy Williams

Department of Medical Biophysics, University of Manchester, Oxford Road, Manchester, England

1983

ACADEMIC PRESS

A Subsidiary of Harcourt Brace Jovanovich, Publishers

London New York

Paris San Diego San Francisco São Paulo

Sydney Tokyo Toronto

ACADEMIC PRESS INC. (LONDON) LTD.
24/28 Oval Road,
London NW1

United States Edition published by
ACADEMIC PRESS INC.
111 Fifth Avenue
New York, New York 10003

British Library Cataloguing in Publication Data

Williams, A. R.
Ultrasound.—(Medical physics, ISSN 0076–5953)
1. Ultrasonics in medicine
I. Title II. Series
610′.28 R857.U48

ISBN 0–12–756960–X

LCCCN 83–70296

Filmset in Monophoto Imprint
by Latimer Trend & Company Ltd, Plymouth
Printed in Great Britain by Galliard (Printers) Ltd, Great Yarmouth

PREFACE

This book is aimed primarily at the clinician or physical therapist who uses ultrasound in the clinic and is interested in knowing its effects on their patients. It attempts to summarize the complex field of bioacoustic interactions in a comprehensive manner so as to enable a meaningful answer to be given to the perennial question posed by patients, i.e. "Is it safe?".

After a simplified non-mathematical description of the basic physics of ultrasound and its applications, the roles and effects of the two dominant physical interaction mechanisms (heat and cavitation) are discussed in detail. Special attention has been paid to the way in which the order of dominance of these and other interaction mechanisms can change as the exposure conditions are changed. This is particularly important if one is attempting to extrapolate from the results of an experiment performed on isolated cells or tissues in the laboratory, so as to predict whether or not a similar effect might occur in a patient in the clinic.

The plethora of ultrasound-induced bioeffects reported in the literature are reviewed bearing in mind the interaction mechanisms which are most likely to be dominant under that particular combination of exposure parameters. This extensive literature review has attempted to be representative, but not exhaustive, with special attention being paid to those articles which are most often referred to or which suggest potentially hazardous effects occurring at the power levels emitted by clinical equipment. Whenever possible, the potential sources of artifact which could mislead investigators are described. Sufficient detail is included with each citation so that these chapters should serve as a convenient source of reference for all investigators in the field of ultrasonic bioeffects, as well as providing the data base for the assessment of the potential safety of ultrasound.

June 1983 A. Roy Williams

ACKNOWLEDGEMENTS

I wish to thank Mrs J. L. Williams for her encouragement and active participation in the compilation of the original manuscript. I would also like to thank my friends and colleagues at the Departments of Medical Biophysics and Medical Illustration at the University of Manchester for their direct and indirect collaboration in this monumental enterprise.

CONTENTS

Preface v
Acknowledgements vi

1. MECHANICAL VIBRATIONS—BASIC PHYSICS 1
1.1 General Introduction 1
1.2 The Surface Wave Analogy 2
1.3 Physical Properties of Mechanical Waves . . 7
1.4 Compressional Waves 11
1.5 Transverse and Flexural Waves 15
1.6 Reflection and Refraction 16
1.7 Acoustic Impedance 22
1.8 Standing Waves 23
1.9 Focusing Sound 26
1.10 Resonance 30
1.11 Power Density–Intensity 32
1.12 Units of Measurement of Audible Sound . . 33
1.13 Mechanical Waves of Different Frequencies . . 36
Suggested Further Reading 38

2. ULTRASOUND 39
2.1 Generation 40
2.2 Uses and Applications of Ultrasound . . . 46
2.3 Detection and Dosimetry of Ultrasound . . 60
2.4 Acoustic Outputs of Commercial Devices . . 76
2.5 The Concept of "Dose" 77
2.6 How does Ultrasound Modify Tissues? . . . 79
References 80

3. THERMAL EFFECTS OF ULTRASOUND . . 83
3.1 Introduction 83
3.2 Attenuation, Scattering and Absorption . . . 83
3.3 Temperature Elevation Within Tissues . . . 97
3.4 Finite Amplitude Effects (Nonlinear Effects) . . 101
3.5 Other Sources of Rapidly Absorbed High Acoustic Frequencies 107

3.6 Measurements of Temperature Rise Within Tissues . . . 107
3.7 Thermal Damage to Tissues 110
3.8 Thermal Contribution to Ultrasound Bioeffects . 113
3.9 Ultrasound in Physiotherapy 118
References 121

4. MECHANICAL AND CAVITATIONAL EFFECTS GENERATED BY ULTRASONIC WAVES . . 126
4.1 Introduction 126
4.2 First Order Effects 127
4.3 Cavitation 132
References 172

5. BIOEFFECTS OF ULTRASOUND 177
5.1 Isolated Macromolecules 177
5.2 Membranes and Sub-Cellular Organelles . . 180
5.3 Cells in Suspension 183
5.4 Cells in Tissues 206
References 240

6. ASSESSMENT OF RISK FROM ULTRASONIC EXPOSURES 254
6.1 The Mechanistic Approach 256
6.2 The Bioeffects Approach 265
6.3 The Epidemiological Approach 286
6.4 Safety Levels for Diagnostic Exposures . . . 291
6.5 Recommendations 294

AUTHOR INDEX 311

SUBJECT INDEX 317

TO MY DAUGHTER KERRY AND MY PARENTS

1. MECHANICAL VIBRATIONS—BASIC PHYSICS

1.1 GENERAL INTRODUCTION

Mechanical or acoustic waves arise when the particles, of which a medium is composed, are supplied with energy and are driven to vibrate. These particles then transfer some of this energy to adjacent particles which also vibrate and pass on the energy in the form of a wave. Those mechanical waves which propagate within solid bodies give rise to what we interpret as *vibration*, whereas those which propagate in liquids and gases are known as *sound*. This relationship is illustrated in the case of someone trapped inside a sunken submarine communicating with a rescue party by hitting the metal hull with a solid object. The energy supplied by this impact sets part of the hull into vibration so that some of this energy is carried to all parts of the hull while some of it enters the outside water. This water-borne vibration reaches the rescuers where some of it may enter the air inside the diver's helmet and be interpreted as audible sound.

The human ear is a remarkably efficient device which can detect a wide range of airborne sounds, but it is unable to detect extremely low notes (which are called *infrasound*), or extremely high notes (which are called *ultrasound*). In common with other biological detection systems the average ear also has a lower limit of sensitivity for detecting any given audible sound which is known as the "threshold". Above this threshold an increase in the amount of sound power arriving at the ear is interpreted as an increase in the subjective loudness of that sound. This continues until the loudness of that sound exceeds another "threshold" (which is greatly influenced by that person's environment and psychological attitude at that time) when it becomes disagreeable or even painful. These disagreeably loud sounds may, under certain circumstances, cause irreparable damage to the hearing mechanism of that ear.

All mechanical vibrations whether they are ultrasound, audible sound or infrasound, and irrespective of whether they are propagating through solids, liquids or gases, have certain common physical properties. One or more of these properties may change gradually as you go from one type of sound to another and other properties may change dramatically as the wave travels from one medium to another. This chapter is a predominantly verbal and graphical attempt to explain these basic physical properties to those newcomers to the field who prefer a non-mathematical approach. Most of these elementary wave properties may be deduced from the observation of surface wave propagation on a pond and are readily demonstrated in a lecture theatre or classroom using a ripple tank on an overhead projector.

1.2 THE SURFACE WAVE ANALOGY

The wave properties of sound or mechanical vibrations are clearly seen in the pattern of circular ripples produced on the surface of a still pond by throwing in a small stone. These expanding ripples consist of a regular series of alternating peaks and troughs which at any one instant of time resemble the pattern shown in Fig. 1.1.

If you instantaneously freeze all motion and cut this circular pattern along a diagonal and view it along the plane of the cut surface, then the cut edge RS would look something like Fig. 1.2. The regular repeating unit which describes the wave is called the *wavelength* (λ) and is defined as the repeat distance from any measurable point to that same point after it has passed through one cycle of positive and negative displacement. The *amplitude* (A) of the wave is the maximum excursion of the wave above (or below) its original undisturbed value; thus, a big wave caused by a large stone has a larger amplitude but the same wavelength as the disturbance caused by a small stone. The amplitude therefore determines the *power* of that wave.

It should be noted that you can get the same pattern presented in Fig. 1.2 without having to freeze time. Imagine a small cork floating at any adjacent point P as shown in Fig. 1.1. This cork will stay in approximately the same position relative to the source of the disturbance (S) but will be raised up as each peak passes by and lowered below its original position as each trough passes by. The

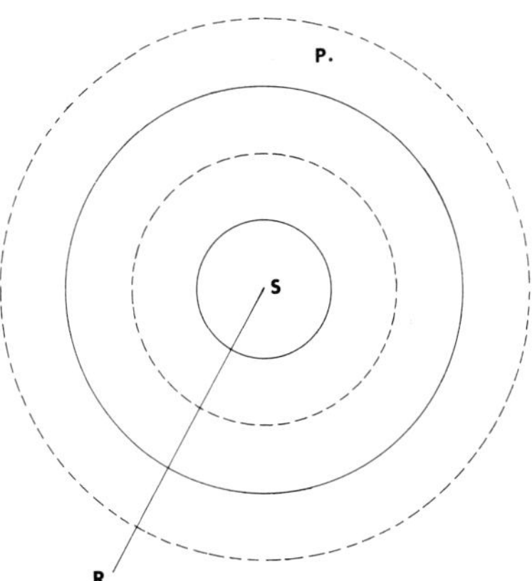

FIG. 1.1 A graphical representation of the expanding series of waves propagating away from a point source (S). The solid circles represent the wave peaks while the broken circles represent the wave troughs.

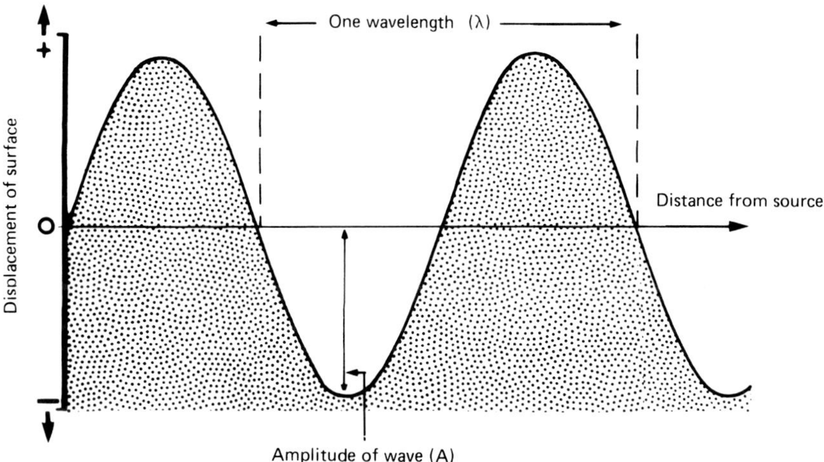

FIG. 1.2 View in the plane of the surface along the line RS in Fig. 1.1. The wavelength (λ) is defined as the length of one repeating unit consisting of a peak and a trough, while the amplitude (A) of the wave is defined as the maximum excursion of any portion of the surface from its original resting position.

displacement of the cork from its original position is displayed as a function of time in Fig. 1.3.

This regular oscillation of a body about a mean point is an example of *simple harmonic motion* (SHM). Common examples of SHM are a

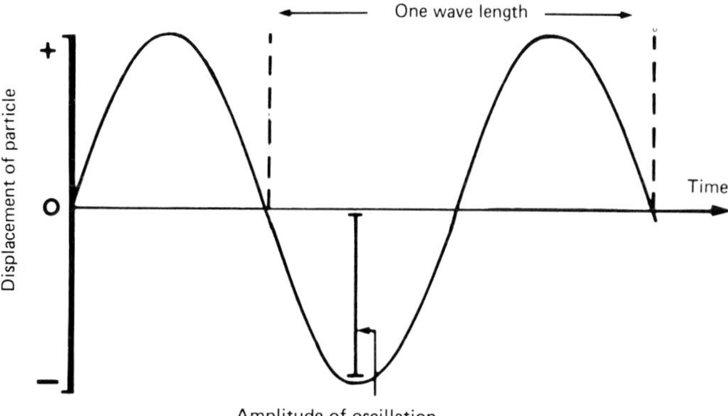

FIG. 1.3 A diagrammatic representation of the displacement of any particle in the path of an acoustic wave as a function of time.

child on a swing, a simple pendulum, a mass on a spring or a marble running down the inside of a smooth bowl (see Fig. 1.4).

Thus, particles of the medium conducting the wave (in this case the molecules at the surface of the water) oscillate with SHM, but essentially remain in the same position relative to the source of the disturbance (S) while the wave itself moves large distances. The rapidly moving wave is therefore just the *energy* put in as the original disturbance (i.e. the stone breaking the surface) causing particles of the medium to oscillate with SHM but having small amplitudes. These oscillating particles collide with or rub against neighbouring particles and so transfer a lot of their energy, which in turn causes their neighbours to be driven to oscillate. This repeating process of particle oscillation followed by energy transfer and the subsequent oscillation of neighbouring particles is commonly called *wave propagation*.

If there are no particles present to transfer the energy, then a mechanical wave cannot propagate. This is why audible sound (or any of the other members of the mechanical vibration family) cannot travel through a vacuum. It is also obvious that wave propagation must be crucially dependent upon the efficiency of energy transfer from the vibrating particle to its neighbour. This energy transfer will be influenced by both the size and number of molecules present per unit volume (i.e. the *density* of that material) as well as by the nature and magnitude of the complex inter-molecular interactions. Thus, the physical properties of the medium have a profound effect on wave propagation: this is most readily apparent if one examines the speed or *velocity* (v) of a wave in different materials (Table 1.1).

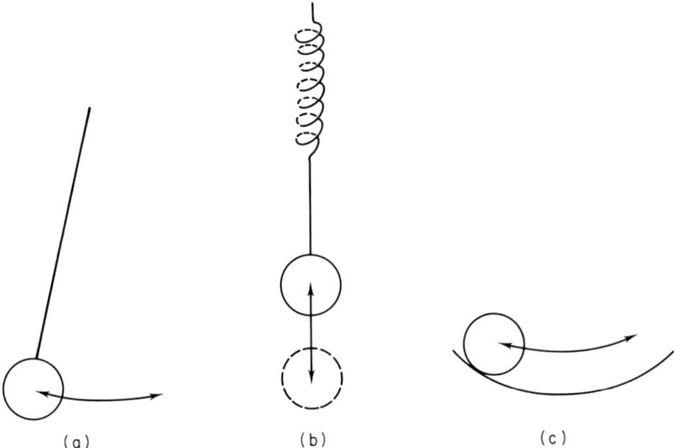

FIG. 1.4 Common examples of Simple Harmonic Motion (SHM). Diagram (a) represents a simple pendulum, (b) a mass suspended on a spring and (c) a ball rolling down the curved surface of a bowl.

TABLE 1.1

Physical properties of some common materials

Material	Density (kg/m^3)	Sound velocity (m/s)	Acoustic impedance (kg/m^2s)	
Air (dry at STP)	1·293	331·5	429	Gases
Carbon dioxide	1·977	259	512	
Water (distilled)	$0{\cdot}998 \times 10^3$	1497	$1{\cdot}494 \times 10^6$	Liquids
Water (salt)	$1{\cdot}025 \times 10^3$	1530	$1{\cdot}568 \times 10^6$	
Ethanol	$0{\cdot}79 \times 10^3$	1210	$0{\cdot}956 \times 10^6$	
Castor oil	$0{\cdot}969 \times 10^3$	1207	$1{\cdot}170 \times 10^6$	
Chloroform	$1{\cdot}49 \times 10^3$	987	$1{\cdot}471 \times 10^6$	
Carbon tetrachloride	$1{\cdot}595 \times 10^3$	926	$1{\cdot}534 \times 10^6$	
Glycerol	$1{\cdot}26 \times 10^3$	1900	$2{\cdot}394 \times 10^6$	
Mercury	$13{\cdot}6 \times 10^3$	1450	$20{\cdot}0 \times 10^6$	
Polyethylene	$0{\cdot}90 \times 10^3$	1950	$1{\cdot}755 \times 10^6$	Solids
Neoprene rubber	$1{\cdot}33 \times 10^3$	1600	$2{\cdot}128 \times 10^6$	
Polystyrene	$1{\cdot}10 \times 10^3$	2670	$2{\cdot}937 \times 10^6$	
Epoxy resin	$1{\cdot}2 \times 10^3$	2800	$3{\cdot}36 \times 10^6$	
Polymethylmethacrylate	$1{\cdot}19 \times 10^3$	2680	$3{\cdot}20 \times 10^6$	
Glass (crown)	$2{\cdot}24 \times 10^3$	5100	$11{\cdot}424 \times 10^6$	
Glass (pyrex)	$2{\cdot}32 \times 10^3$	5640	$13{\cdot}085 \times 10^6$	
Glass (flint)	$3{\cdot}88 \times 10^3$	3980	$15{\cdot}442 \times 10^6$	
Aluminium	$2{\cdot}71 \times 10^3$	6420	$17{\cdot}398 \times 10^6$	
Brass	$8{\cdot}60 \times 10^3$	4700	$40{\cdot}42 \times 10^6$	
Stainless steel (347)	$7{\cdot}9 \times 10^3$	5790	$45{\cdot}74 \times 10^6$	

For a typical solid or liquid the wave velocity (v) is related to the density (ρ) of that material by the equation:

$$v = \sqrt{\frac{K}{\rho}} \qquad [1.1]$$

Where K is the Young's modulus of a solid rod or the bulk modulus of elasticity of a liquid. Liquids expand more than solids when they are heated and so the same mass of material now occupies a larger volume and hence its density must have decreased by a significant amount. Also, since the average distance between each individual molecule of the liquid has increased, that liquid must be more compressible. Thus, the velocity of mechanical waves in a liquid usually increases with increasing temperature. For example, the velocity of sound in water rises from about 1400 m/s at 0°C to about 1550 m/s at 60°C.

For a gas, the corresponding equation is:

$$v = \sqrt{\frac{\gamma P}{\rho}} \qquad [1.2]$$

Where γ is the ratio of the specific heats at constant pressure and constant volume (Cp/Cv) and is equal to 1·40 at normal temperature and pressure (NTP, 0° Centigrade and 760 mmHg, dry, also called standard temperature and pressure, STP); and P is the ambient pressure. Gases are so compressible that one would intuitively expect the velocity of sound in a gas to increase as the pressure is increased. However, this does not occur because an increase in pressure results in a corresponding increase in density (i.e. the same mass of material now occupies a smaller volume) so that the ratio P/ρ remains constant.

Gases have the largest coefficients of thermal expansion and therefore the velocity of sound in a gas must be extremely dependent upon temperature. For example, the speed of sound in air at NTP is 331 m/s (0°C) rising to 340 m/s at room temperature (20°C). This large change in sound velocity with temperature can result in the production of acoustic mirages analogous to the optical mirages sometimes seen on very hot days. The conditions needed to produce an acoustic mirage are that the layers of air close to the ground and high up in the air are warm whereas intermediate layers are colder, i.e. there is a temperature inversion. Some of the sound waves emitted by a loud sources such as an explosion, enter the uppermost warm layer where they are refracted back to the earth's surface. A situation can then be found where the sound of the explosion can be heard at points close to and many miles away from the detonation site but not at some intermediate distances.

1.3 PHYSICAL PROPERTIES OF MECHANICAL WAVES

Figure 1.3 shows that waves flow away from their source. The number of wavelengths which pass any given point in one second is called the *frequency* (f) of that source. Once generated, this frequency remains constant irrespective of the medium that the wave is travelling through. The particles of the medium which propagate this wave also vibrate with this same frequency. It should be noted that all waves of the same type travel at the same speed in the bulk phase of any one medium, i.e. wave velocity is not dependent upon the frequency.

The velocity (v), frequency (f) and wavelength (λ) are however, related to each other by the simple equation:

$$v = f\lambda \qquad [1.3]$$

Since the frequency remains constant, but the velocity changes as the wave travels from one medium into another (Table 1.1) it follows from equation 1.3 that the wavelength must also change.

If one observes the expanding ripples on a pond, it can be seen that the amplitude of the wave progressively decreases until, in a very large still pond, it eventually "dies away". The obvious reason for this effect is that the perimeter of the circular wave is continually expanding and so the initial input of energy supplied by the falling stone has to be spread over an ever increasing area. Since the frequency does not change, this reduced amount of energy per unit area is seen as a wave of progressively decreasing amplitude.

Another contributing factor which reduces the energy content of the wave is *absorption*. The collisions or sliding contacts which transfer energy from particle to particle are never 100% efficient. Some energy must therefore be lost at each transfer; this energy appears as heat which causes a rise in the temperature of the conducting medium. For most commonly encountered situations (excepting the therapeutic applications of ultrasound), the temperature rises caused by the passage of the wave are too small to be measured.

Immediately after it has been formed, any part of the wave generated by a small source is a portion of a circle having a small radius. This source is consequently known as a *point source*. As the wave travels further and further away from its source, its radius of curvature increases until any small portion of the wave front appears to be almost flat (i.e. having an infinite radius of curvature; see Fig. 1.5). This flat front, which continues to move away from the source, is called a *plane progressive wave*.

Waves are reflected from obstacles; but if the obstacle has a hole in it, a portion of the wave front will propagate through it. Figure 1.6(a) shows what happens if a regular train of plane progressive waves encounter an obstacle having a small hole or *aperture*. This situation is analogous to a series of oceanic waves striking against the narrow entrance to an enclosed harbour. Most of each wave is reflected, but the portion that goes through is seen to form a new wave of small radius of curvature, i.e. a small aperture behaves like a new point source.

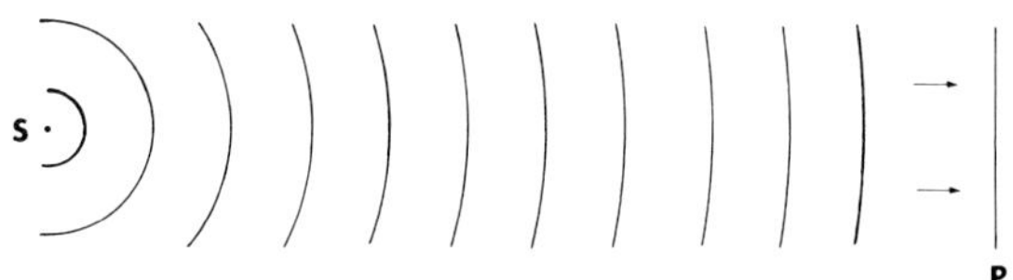

FIG. 1.5 A diagrammatic representation of the way in which the curved wave front emitted by a point source (S) becomes more planar with increasing distance from the source until it becomes a plane progressive wave (P) at an "infinite" distance.

However, Fig. 1.6(b) shows this same series of plane progressive waves striking an obstacle having a large aperture. As you can see, that portion of the plane wave entering the aperture emerges as a plane wave and it is only at the edges that the wave is part of a circular wave. Thus, an obstacle having a large aperture permits a portion of a plane progressive wave to propagate as a beam.

We are now faced with a problem of nomenclature, because, how big is a large aperture? The answer is that it depends upon the wavelength. An aperture is "large" if it is larger than the wavelength and "small" if it is equal in size or smaller than the wavelength. Thus, the same aperture can be "large" for one wave or "small" for a wave of lower frequency. For example, consider a person speaking. The velocity of sound in air is about 340 m/s and a typical frequency for human speech is about 1000 cycles per second or hertz (Hz). From equation 1.3 the wavelength (λ) is given by v/f which is 0·34 m or 34 cm. This is about ten times bigger than a person's mouth. Therefore, when a person speaks, his or her mouth must act as a small aperture and emit only circular waves (NB they will actually be portions of spherical waves since they will be radiating in three dimensions unlike the surface waves which are restricted to two dimensions). However, if we were capable of emitting an ultrasonic squeak having a frequency of say 20 000 Hz, then the wavelength in air would be decreased to about 1·7 cm which is less than half of the average diameter of a mouth. Thus, if this ultrasonic squeak could arrive at the mouth as a plane wave it would propagate as a beam and could be directed to specific locations like the beam of light from a torch.

This last point is very important, because: if ultrasound is to be used to obtain clinically useful information from patients, it is obviously better to have a directional beam that you can point at the

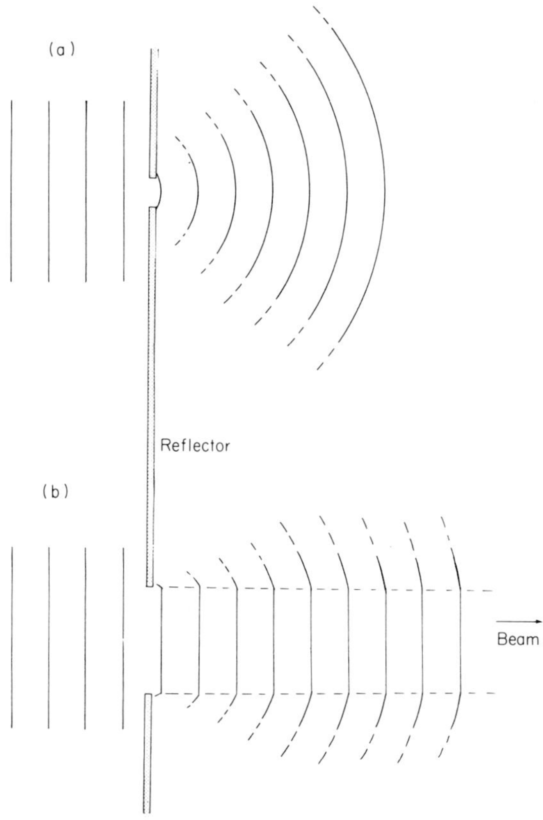

FIG. 1.6 A schematic illustration of the effect of an aperture on the progression of a plane progressive wave. Diagram (a) shows that a small aperture will only permit spherical waves to propagate, while diagram (b) shows that a large aperture will permit a plane progressive wave to propagate as a beam.

organs or structures to be investigated rather than generate a spherical wave which travels in all directions within the patient. The use of a directional beam means that you do not pick up signals from organs or structures outside your region of interest. It also means that the signals you do detect are "clearer" because there are fewer reflected waves travelling around and *interfering* with each other.

Interference is a fundamental property of all waves. If any two or more mechanical waves arrive together at the same point in space, they will merge together at that point to form a new wave which is the algebraic sum of the amplitudes of each individual wave. This is shown graphically in Fig. 1.7 where it can be seen that the two positive or two negative displacements add together to give an exceptionally large positive or negative displacement, whilst a positive and a negative displacement of equal magnitude cancel each other out. If the waves had been travelling in different directions before intersection each other at that point, then they re-form and continue their progress having the same original frequency and waveform after they have passed that point. An interference pattern

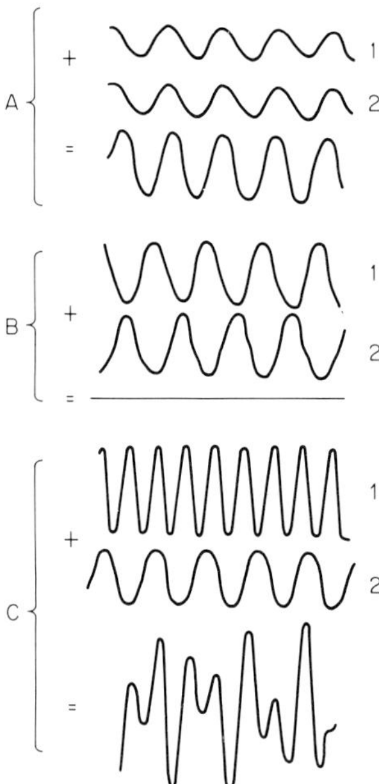

FIG. 1.7 An illustration of interference. In (A) waves 1 and 2 have the same frequency, amplitude and phase and therefore add together to form a wave having twice the amplitude. In (B) waves 1 and 2 are 180° out of phase and so their amplitudes cancel each other out. In (C) the waves are of different frequency and so their resultant has a different waveform.

is readily seen if two or more stones are thrown simultaneously into a pond.

Interference will be considered in more detail in Chapter 2, where it will be shown to be responsible for the extremely complicated patterns, called diffraction patterns, to be found in the output from many ultrasonic devices.

1.4 COMPRESSIONAL WAVES

So far we have relied heavily upon surface waves to describe the basic physics of mechanical waves. Unfortunately, this useful model has several limitations. It has been shown that the particles of the medium oscillate up and down in the vertical plane as the wave propagates along the horizontal surface. This wave is therefore called a *transverse* or *shear* wave (i.e. a wave where the particles of the medium oscillate at *right angles* to the direction of wave propagation). This means that the energy of the wave must be communicated from one vibrating particle to another by frictional or sliding contact. Solids are very efficient at conducting energy in this way. If one end of a pencil or an iron bar is twisted then the other end turns as well. However, imagine yourself holding a cylinder of air, or immerse your hands in a tank of water and imagine yourself holding a cylinder composed entirely of water; now twist one end: you should find that the other end does not turn. This experiment proves that liquids and gases are not very efficient at conducting transverse waves, in fact it is this lack of strong three-dimensional intramolecular coupling forces which enables liquids and gases to flow. It should be noted that the ability to conduct transverse waves is a property of the physical state of the material and not its chemical composition: thus, water can conduct shear waves if it is frozen into ice whereas iron can not when it is molten.

However, our surface waves are transverse waves, and as we have seen these propagate well on the surface of water. This raises the question of how the energy of this wave is transmitted. In fact, our useful liquid-borne surface waves have been a special case all along. There has in fact been a cyclical flow of water just beneath the surface from the region of the developing trough to the developing peak. It is this hidden flow which has been transferring and propagating the energy of the wave. The particles of water at the surface actually move in a circular or eliptical orbit (Fig. 1.8). Anything which interferes with this flow of water close to the surface will therefore change this method of energy propagation and consequently change the velocity (and hence the wavelength) of any given wave. Thus, liquid-borne surface waves (as distinct from surface waves borne by a solid) are unusual in that their velocity depends upon the depth of water beneath the surface. This is why ripples on a pond or waves from the sea slow down and decrease in wavelength as the water becomes very shallow (Fig. 1.9). Another consequence of this mode of energy transfer is that the shape of the wave departs from the idealized *sine wave* described in Fig. 1.2 in that the crests of each wave tend to be short while the troughs are very wide. This effect is particularly noticeable in oceanic waves.

FIG. 1.8 A diagrammatic illustration of the cyclical flow of water giving rise to the propagation of a surface wave on a liquid. At higher displacement amplitudes the circular flow becomes distorted into an ellipse which gives narrow peaks and broad troughs.

The other class of mechanical waves, which includes both audible sound and the ultrasonic beams used in medical diagnosis and therapy, differ from the transverse or shear waves discussed above in that the direction of particle vibration is *in the same direction* as the wave is propagating. These waves are called *longitudinal* or *compressional* waves. Figure 1.10 summarizes the differences between compressional and transverse waves.

When a longitudinal or compressional wave propagates as a plane progressive wave, all the particles of the medium which lie in a plane parallel to the advancing wave front are displaced together. It is therefore convenient to consider the propagating medium as being composed of an infinite number of extremely thin sheets or layers arranged parallel to the wave front (Fig. 1.11(a)). At any one time some of these layers will have been displaced away from their original position in the direction that the wave is propagating whilst other layers will have returned to their original position or overshot it to give a displacement in the opposite direction as shown in Fig. 1.11(b). From our surface wave analogy we know that each layer of liquid

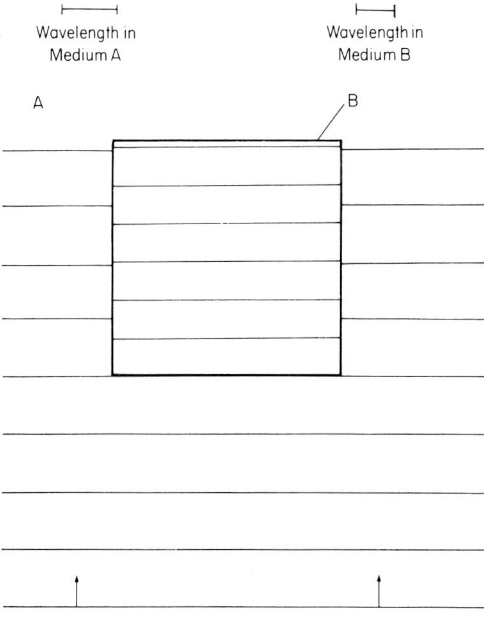

FIG. 1.9 An illustration of the decrease in the wavelength of a plane progressive liquid-borne surface wave passing over a rectangular body which is only just submerged beneath the surface of the liquid. In a shallow liquid the circular flow of fluid illustrated in Fig. 1.8 must have a smaller diameter giving a shorter wavelength.

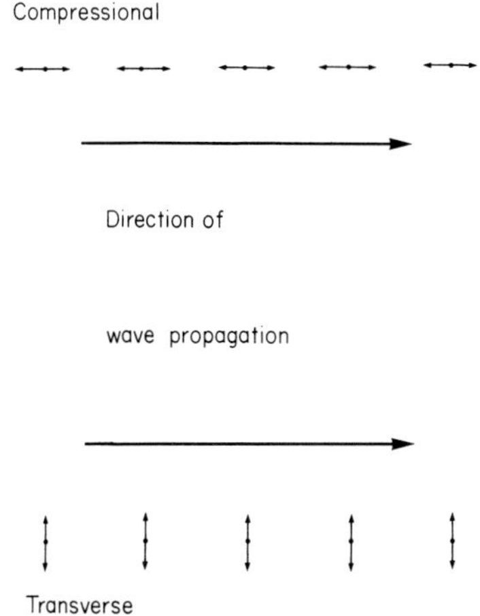

FIG. 1.10 A graphical illustration of the direction of motion of the particles of the medium relative to the direction of wave propagation for compressional (or longitudinal) and transverse (or shear) waves.

should be oscillating with simple harmonic motion. Its position or displacement at any given instant of time can therefore be found from a sine curve.

Figure 1.11(c) shows the displacement of every layer from its original position at any time, t. Certain layers (indicated by the long broken lines) are just passing through their original undisturbed positions and therefore must have zero displacement at this time. If we define a positive displacement as representing a displacement to the left and a negative displacement as a move to the right, it can be seen that the layers of the medium on either side of these points of zero displacement are either being moved towards them or away from them. If the layers of the medium are all moving towards a given

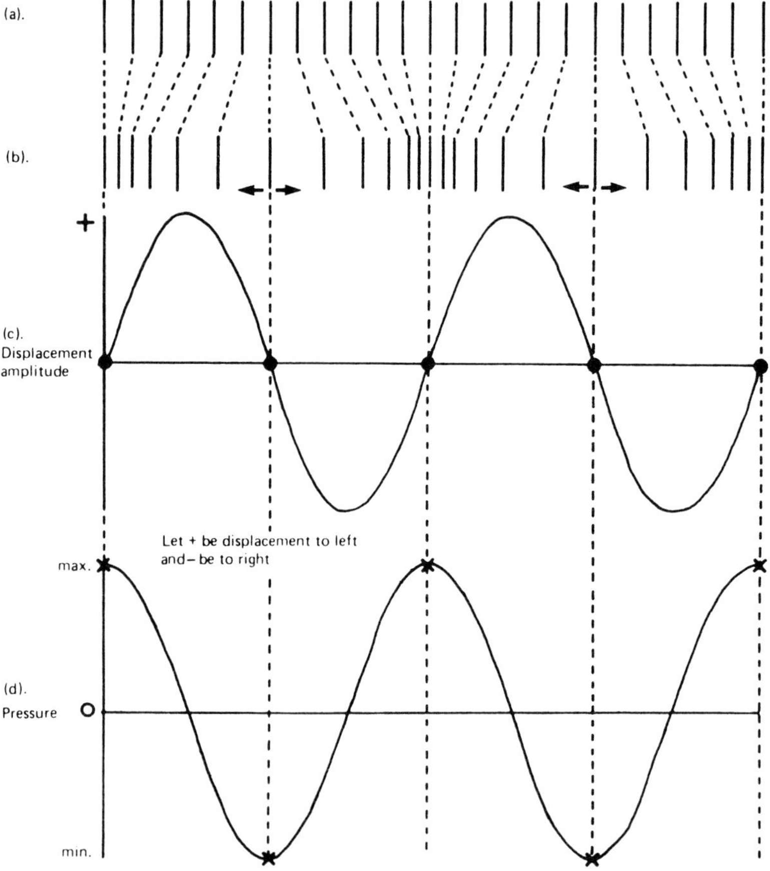

Fig. 1.11 A graphical illustration of the displacement of the layers of a medium by a compressional wave and the effect this motion has on the pressure at various points within that medium. For details see text.

point, then the medium at that point is *compressed* and therefore the *pressure* at that point must be increased. Conversely, if the adjacent layers are all moving away from that point then the pressure must be decreased. Thus, the points of zero particle displacement are the points where the pressure in the medium is at its highest or lowest. Because the particle displacements which generate these changes in pressure follow a sine wave, the pressure changes themselves must also follow the same sine wave. The point midway between the maximum and minimum pressure must therefore be a point of zero pressure (Fig. 1.11(d)). These points are also the points of maximum displacement of the particles and correspond to the point in SHM where the particle has stopped moving.

Thus, the displacement of the layers of the medium and the instantaneous pressure produced within that medium by the mechanical wave are both described by a sine curve having the same frequency, but the maximum of one curve corresponds to a zero in the other. These two curves are said to be one quarter of a wavelength *out of phase* with respect to each other. One complete cycle of a sine curve (i.e. one wavelength) also describes the motion of a particle moving in a circular orbit and corresponds to one complete revolution of 360°. It is convenient to refer to portions of a wavelength in terms of the equivalent number of degrees. Therefore, one quarter of a wavelength becomes 90°. This means that the pressure and particle displacement in a compressional wave are 90° out of phase with each other.

In general it is more convenient to measure the changes in pressure produced by a mechanical wave when it is propagating through a liquid or a gas, and to measure the displacement or acceleration of the particles of the medium when the wave is propagating through a solid.

1.5 TRANSVERSE AND FLEXURAL WAVES

As a rule transverse or shear waves travel much more slowly than compressional waves in any given solid medium. For a solid the velocity of transverse waves (Vt) is given by:

$$Vt = \sqrt{\frac{G}{\rho}} \qquad [1.4]$$

Where G is the shear modulus of elasticity and ρ the density of that material. In general, Vt is usually less than about half the velocity of longitudinal or compressional waves in that same medium.

Many musical instruments such as the piano and the violin rely upon the generation of flexural or transverse standing waves in stretched strings to generate their characteristic sounds. In this case:

$$Vt = \sqrt{\frac{T}{\varepsilon}} \qquad [1.5]$$

Where T is the tension in the string and ε is its mass per unit length. Thus, a stringed instrument is tuned by adjusting the tension on the string which changes the velocity of the wave within that string. Since the string is damped at each end giving it a finite length, different velocity waves will cause it to resonate at different frequencies and generate different standing wave patterns which in turn result in the generation of different ratios of harmonics or overtones.

Under certain circumstances, some of the energy of a compressional wave may be transformed into transverse waves. For example, a beam of therapeutic ultrasound incident at an acute angle on to a strong reflector will generate shear waves on both sides of the interface. Shear waves do not propagate well in aqueous media like the so-called "soft" tissues and are rapidly absorbed and appear as heat (see Chapter 3). It has been proposed that the sensation of pain experienced when the hand is placed in the path of a moderately high intensity beam of therapeutic ultrasound is due to the temperature rise produced within the periosteum (i.e. the nutritive and sensory outer lining of a bone) by the rapid absorption of the tissue-borne transverse waves.

1.6 REFLECTION AND REFRACTION

We are familiar with the concept of the reflection of light from a mirror and so it was not surprising that our surface wave was reflected from the material bounding the aperture as discussed earlier. Before we can investigate this phenomenon in more detail we have to be able to follow the path of a wave. The simplest approach is to apply *Huygen*'s principle which states that every point along a wave front is itself a point source and emits a circular (in two dimensions) or a spherical wave (in three dimensions). The line obtained by joining up all the wave fronts from each of these numerous point sources indicates the next shape and position of that wave front (Fig. 1.12).

Figure 1.13 depicts a plane wave front approaching a plane reflector at an angle θ. The secondary wavelets emitted from points P2 to P5 are not hindered but the wavelet from P1 strikes the surface

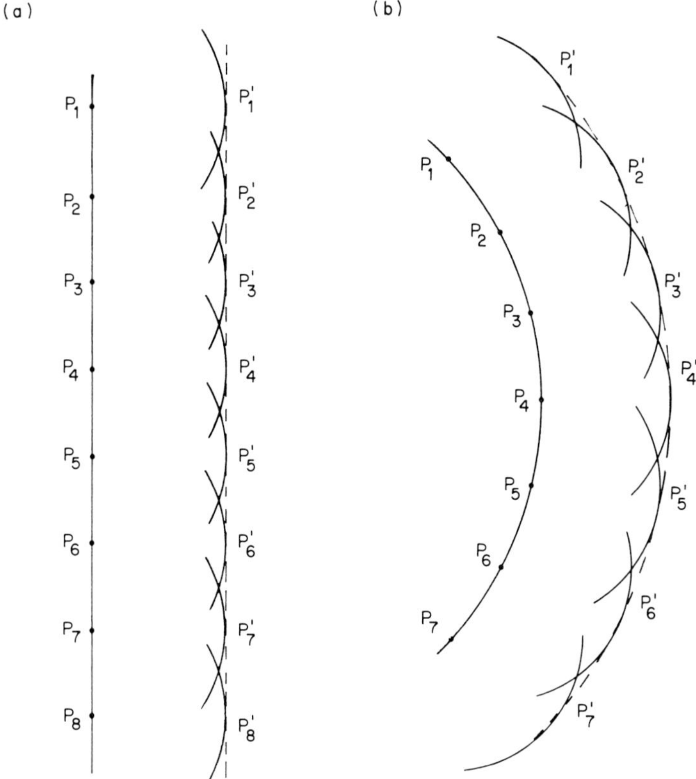

FIG. 1.12 An illustration of the application of Huygen's principle whereby any wave front is regarded as a large number of point sources each generating a spherical wave. Interference between each one of these wavelets re-constitutes a new wave front which is parallel to the source.

of the reflector so that its next position P1′ would be behind the reflector (i.e. the virtual image of P1′ which is P1′V). The effect of the reflecting surface is to reverse the direction of travel of these wavelets which strike it so that the portion of each wavelet which would have penetrated beneath the surface actually lies above it as shown in Fig. 1.13(a). It is a simple geometric proof to show that the distance from the point P1′ to the reflector is exactly the same as the distance from the reflector to its virtual image of P1′V (this is directly analogous to the reflection of light from a plane mirror).

Figure 1.13(b) shows this same wave some time later after three of the points have been reflected. It can be seen that they have reformed a plane wave which is propagating away from the reflector. As in optics, if we define the angle θ at which the wave strikes the reflector as the *angle of incidence* (i), then the reflected wave can be shown to depart at this same angle (θ) which is now called the *angle of reflection* (r).

We can use this same approach to describe the passage of a mechanical wave through the boundary between two dissimilar

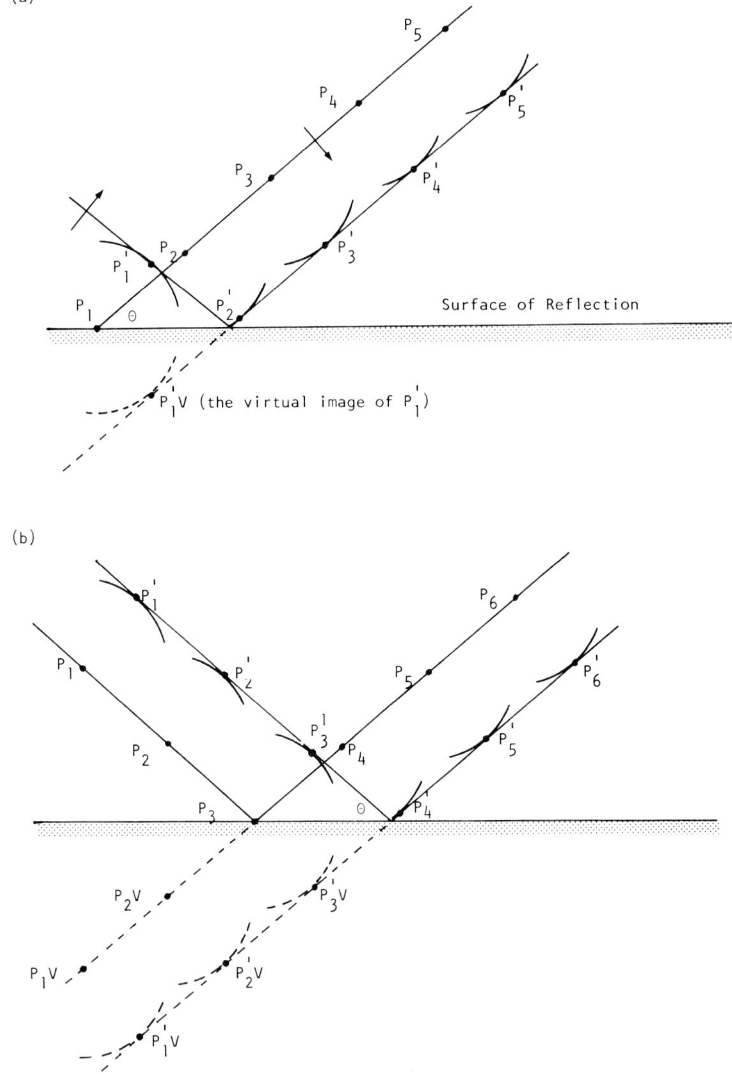

FIG. 1.13 An illustration of the application of Huygen's principle to determine the fate of a plane progressive wave front incident upon a perfect reflector. For details see text.

media: this is called *refraction*. Figure 1.14 shows the successive positions of a plane wave front approaching a plane boundary at an angle θ. Let the radius of each wavelet be the wavelength of the wave in that medium. When the wavelet from P1′ encounters the boundary it is not reflected but passes straight through it into medium 2. However, if the velocity of the wave in medium 2 is different from its velocity in medium 1, then its wavelength will also be different. In the example chosen for Fig. 1.14 the wavelength in medium 2 is less than that of medium 1 and so the radius of that portion of the wavelet in medium 2 is shorter. After most of the wave front has entered medium 2 it can be seen to have reformed a plane wave but that it propagates away from the interface at a decreased angle (θ').

If the velocity of sound in medium 2 had been greater than that of medium 1 then its wavelength would have been longer than that in medium 1 and the emerging wave would have been bent in the other direction (i.e. θ' greater than θ). This is the situation that occurs if you

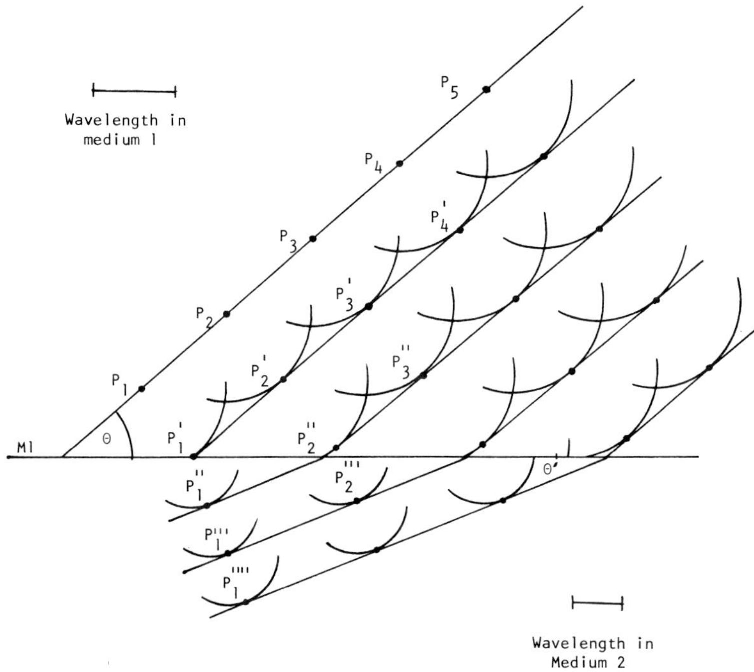

FIG. 1.14 An illustration of the application of Huygen's principle to determine the path of a plane progressive wave front incident at an angle with a material in which the wave travels at a different velocity. For details see text.

imagine the waves travelling in the opposite direction in Fig. 1.14.

In practice, both reflection and refraction occur together whenever a mechanical wave encounters a boundary between two dissimilar materials. This is shown schematically in Fig. 1.15.

This same information may be displayed more simply if we adopt the convention whereby all the individual wave fronts are replaced with a single line drawn at right angles to each front. Figure 1.16

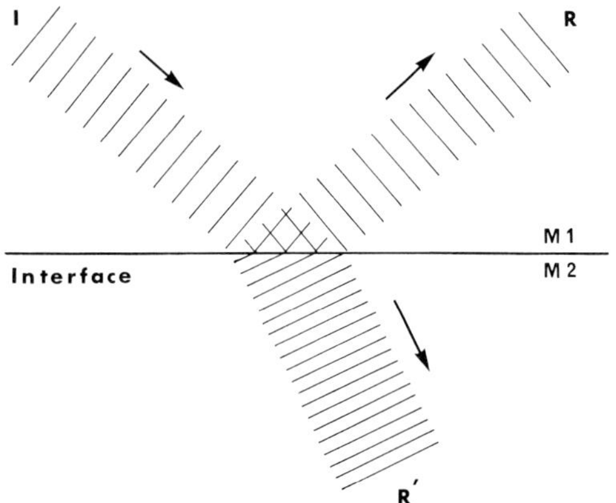

FIG. 1.15 A schematic illustration of reflection and refraction occurring at the interface between one medium (M1) and another (M2). The refracted wave (R′) is deflected because the velocity of the wave in M2 is different to that in M1. (I) is the incident wave and (R) is the reflected wave.

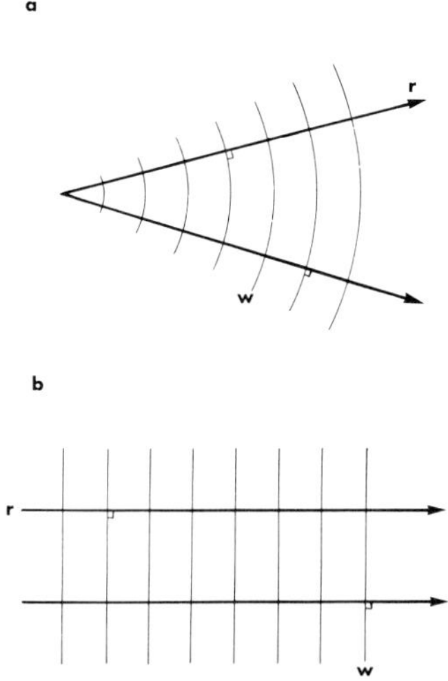

FIG. 1.16 An illustration of the convention whereby the individual wave fronts (w) are replaced by a single line (r) called a *ray*—which is drawn at right angles to each wave front.

shows how these lines (each of which is called a *ray*) can be used to describe the various types of mechanical wave fields. The expanding wave field radiating from a point source is described by a diverging set of rays (each of which is a radius of the expanding circle or sphere) while a plane progressive wave field is represented by one, two or more parallel lines. The direction of propagation of the wave is indicated by an arrow on the ray.

Thus, the information presented in Fig. 1.15 is re-drawn in Fig. 1.17. The angle (θ) between the incident wave front and the interface can be shown to be equal to the angle of incidence (i) between the incident ray and the normal drawn at the point of contact. Similarly, this angle can be shown to be equal to the angle of reflection (r) between the normal and the reflected ray.

The angle of incidence (i) and the angle of refraction (r′) are related to each other by the ratio usually credited to Willebrord Snell in 1621:

$$\frac{\sin \mathrm{i}}{\sin \mathrm{r}'} = \frac{\mathrm{v1}}{\mathrm{v2}} \qquad [1.6]$$

Where v1 is the velocity of the incident wave in medium 1 and v2 is the velocity of the refracted wave in medium 2.

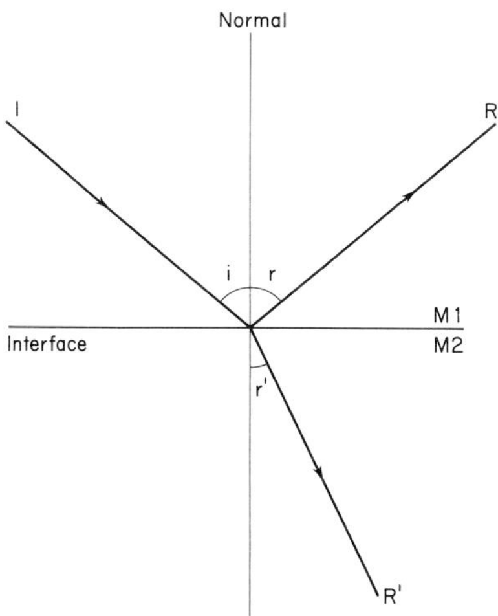

FIG. 1.17 This shows the same information presented in Fig. 1.15 except that the individual wave fronts have been replaced by rays. If the wave velocity in medium 2 had been the same as that in M1 then the refracted wave (R′) would not have been deviated, i.e. the angle of refraction (r′) would be equal to the angle of incidence (i) and the angle of reflection (r).

1.7 ACOUSTIC IMPEDANCE

But, what factors determine how much of the incident beam is reflected or refracted at any given interface? The best optical analogy is that of light illuminating a wine glass. We can clearly see the glass in air because some of the light hitting it (less than 1%) is reflected into our eye, the rest is refracted through the glass. However, if we now immerse that glass in clear water it virtually disappears. This is because the transmission properties of the glass and the water are more similar to each other than those of glass and air, consequently even less reflection occurs. This parameter which determines the transmission properties of light through a medium is the *refractive index* which is defined as the ratio of the velocity of light in a vacuum to its velocity within that medium.

The same phenomenon holds for mechanical waves, but in this case it is the *acoustic impedance* (Z) which determines the ratio of the reflected to the refracted wave. Acoustic impedance is defined as the product of the density of a material (ρ) and the velocity (v) of sound within it.

$$\text{i.e. } Z = \rho v \qquad [1.7]$$

In Figs 1.15 and 1.17, let Z1 be the acoustic impedance of material 1, and Z2 be the acoustic impedance of material 2. If Z1 equals Z2, then there will be no reflection and all of the incident wave will propagate through the interface [NB if the velocities are also the same it will not be deviated (i.e. $i = r'$)].

However, if Z1 does not equal Z2 then some of the incident wave is reflected and the remainder is refracted. The bigger the difference between them then the greater the proportion of the incident energy in the reflected wave. The mathematical relationships which describe this partitioning of the incident energy at the interface are complicated by the angle of incidence. The simplest situation (and, as we shall see later, the one which has the greatest relevance for the applications of ultrasound in medical diagnosis) is where the angle of incidence is zero, i.e. the wave approaches the interface along its normal. The reflected portion of the wave would then retrace its own path while the refracted portion of the wave would continue *in the same straight line*, i.e. it would not be deviated from the normal even though it may be travelling at a different velocity. At normal incidence, the equations which describe the partitioning of the incident energy (Ii) into the reflected (Ir) or transmitted (It) portions are:

$$\frac{Ir}{Ii} = \left[\frac{(Z_2 - Z_1}{(Z_2 + Z_1)}\right]^2 \qquad [1.8]$$

and

$$\frac{It}{Ii} = \frac{4Z_2Z_1}{(Z_2 + Z_1)^2} \qquad [1.9]$$

Where I is the *intensity* of the wave (i.e. the transmitted power per unit area). The acoustic impedances for some common materials are included in Table 1.1.

1.8 STANDING WAVES

The plane wave at normal incidence which is reflected back along its own path interferes with that portion of itself which has not yet reached the reflector to form a *standing wave field*. Figure 1.18 shows that a wave undergoes a 180° phase change on reflection—this is the same as "losing" half a wavelength. The most common example which demonstrates this phase change is seen when a transverse wave is sent along a piece of string or rubber tubing attached to a wall. If the wave travels towards the wall on the top of the string, it is reflected back underneath the string.

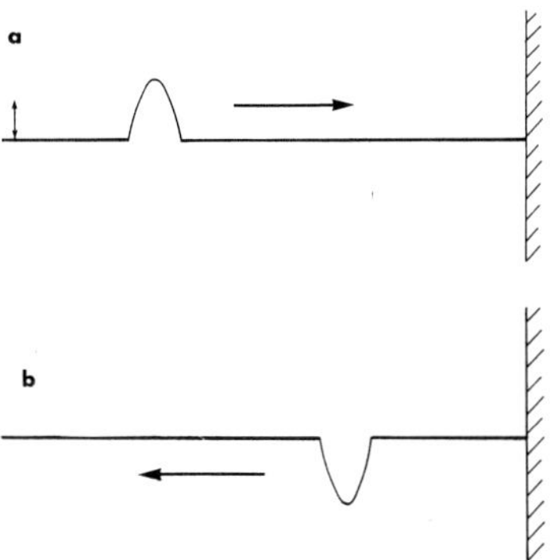

FIG. 1.18 A schematic illustration of a portion of a wave travelling along a rope or piece of rubber attached to a rigid body such as a wall. On reflection the wave is seen to have undergone a 180° phase shift and now travels underneath the rope.

Figure 1.19 shows what happens when a plane travelling wave strikes a flat perfect reflector. Diagram (a) demonstrates that the incident wave which meets the reflector at the instant of zero displacement is reflected (with 180° phase shift) so that both waves are exactly in phase. The amplitudes of both waves therefore add together to give a resultant wave whose amplitude at any given point is twice that of the incident wave. Figure 1.19(b) shows this same wave one-quarter of a wavelength (90°) later. Now, that same wave is striking the reflector at a displacement maximum. The reflected wave

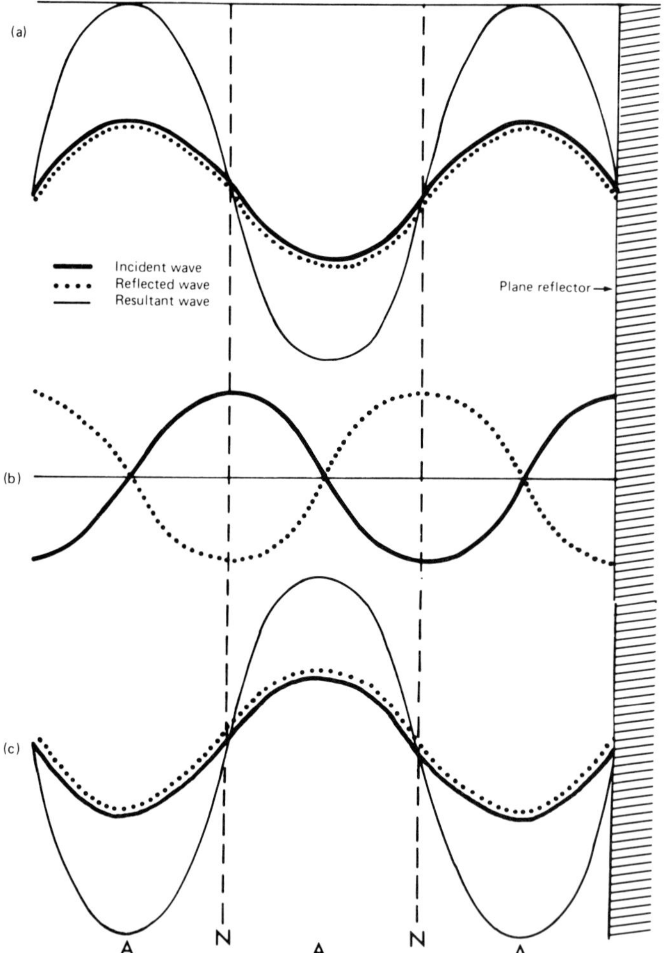

FIG. 1.19 A diagrammatic representation of the mechanism of formation of a standing wave field by reflection from a plane reflector. For details see text.

at the same point will be at a displacement minimum and therefore the reflected wave will exactly cancel out the incident wave giving a resultant of zero displacement at every point. One-quarter of a wavelength later (Fig. 1.19(c)) the incident wave again strikes at the instant of zero displacement so that the wave is reflected in such a way that the interference is constructive and both waves add together to give a resultant whose amplitude is twice that of the incident wave. However, since this wave is half a wavelength (180°) out of phase with respect to the first reflected wave, the displacement at any point is in the opposite direction to that shown in (a). A quarter of a wavelength later, the wave meets the reflector at a displacement maximum and is reflected back in such a way that complete destructive interference again occurs.

Thus, during the first quarter of any one cycle, the displacement at the points marked A will have increased to a value which is twice that of the incoming wave. During the next quarter of a cycle it will decrease to zero. During the third quarter of the cycle it again increases to a value which is twice that of the incoming wave, but it is in the opposite direction to that seen in the first quarter cycle. Finally, the displacement decreases during the last quarter cycle until it is once again zero. This pattern is repeated every cycle. These points (A) are known as *displacement antinodes*. The resultant displacement at the points marked N in Fig. 1.19 remains at zero at all times. These points are known as *displacement nodes*.

Therefore, a standing wave field consists of a regularly repeating pattern of nodes (where the displacement is always zero) and antinodes (where the displacement varies from positive to negative at the same frequency as the incoming wave, but at twice its amplitude) each separated by a distance equal to one-quarter of the wavelength. All of the energy of the incident wave (except for the little lost by absorption) is returned in the reflected wave. This means that there is virtually no net flow of energy (as defined in the section on wave power) through the medium supporting the standing wave. However, this does not mean that a standing wave field has no energy. On the contrary, a standing wave field will be seen to be a highly efficient means for concentrating the potentially hazardous effects of ultrasonic energy and applying these effects within biological tissue.

It should be noted that we would get exactly the same result as that presented in Fig. 1.19 if we took any other sinusoidally varying parameter of the wave instead of the displacement amplitude. For example, we have seen that the pressure generated by a compressional wave is 90° out of phase with the particle displacement. Therefore, the pressure measured in a standing wave field would have the same alternating pattern but would be displaced 90° so that the displacement node would be a *pressure antinode*, and the displacement antinode a *pressure node*.

It is also possible to have a *partial standing wave* if the reflector is less than 100% efficient. In this case one would have a plane progressive wave field with a superimposed standing wave component. Thus, maximum displacements or pressures measured at their respective antinodes will be less than twice that of the original incident wave. The amount of standing wave component in one of these mixed fields is usually expressed in terms of the *standing wave ratio*. This ratio is defined as the ratio of the pressure at a node to that at its adjacent antinode. This ratio is equal to one when there is no reflection and equal to infinity when the reflection is perfect.

These standing waves are also of great interest in the study of the physics of musical instruments because it is standing waves in strings which provide the music of the piano, violin, double bass etc. and airborne standing waves in tubes and pipes which provide the notes of the organ and the woodwind family.

1.9 FOCUSING SOUND

It is possible to use the properties of reflection or refraction to cause the wave energy distributed along the entire wave front to be concentrated at one small point in space. This point is known as a *focus*.

The simplest device for bringing a plane wave front to a focus is a curved reflector (this is analogous to the use of a curved mirror to focus a beam of light). Figure 1.20 shows two rays from a plane progressive wave being brought to a focus at point F after reflection from a curved surface which is a portion of a sphere. As in optics, one only obtains a point focus if the rays are close to the focus, i.e. it is a reflector of small aperture. Figure 1.21(a) show the multiple foci obtained with a spherical reflector of large aperture: the pattern formed by rays forming these multiple foci is known as a *caustic curve*. A *parabolic* mirror, on the other hand, is designed so that all rays striking the reflector are brought to the same focus (Fig. 1.21(b)).

It is this phenomenon of focusing which is usually responsible for the spectacular performance of many "whispering galleries". These are usually circular or eliptical rooms where a quiet noise emitted at a specific site is clearly heard at another specific site on the other side of the room, but not at any intervening positions. Both of these specific sites are the focal points produced by the curvature of the walls. A

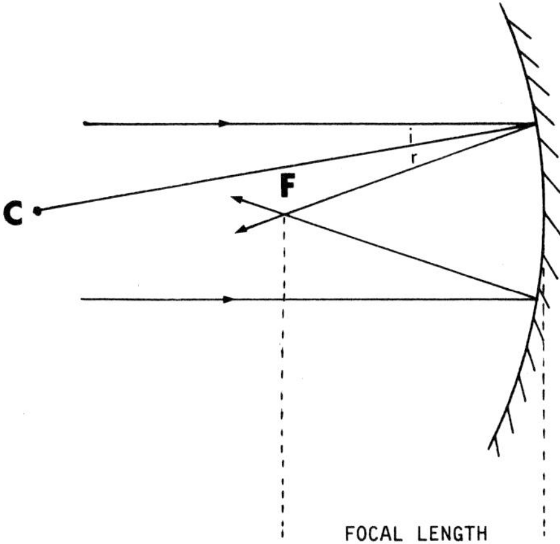

FIG. 1.20 An illustration of reflection from a curved surface to generate a point focus (F) of sound. As in optics, the point C represents the centre of the sphere from which the curved reflector has been cut. The angle of incidence (i) is equal to the angle of reflection (r) so that the focal length is half the radius of curvature.

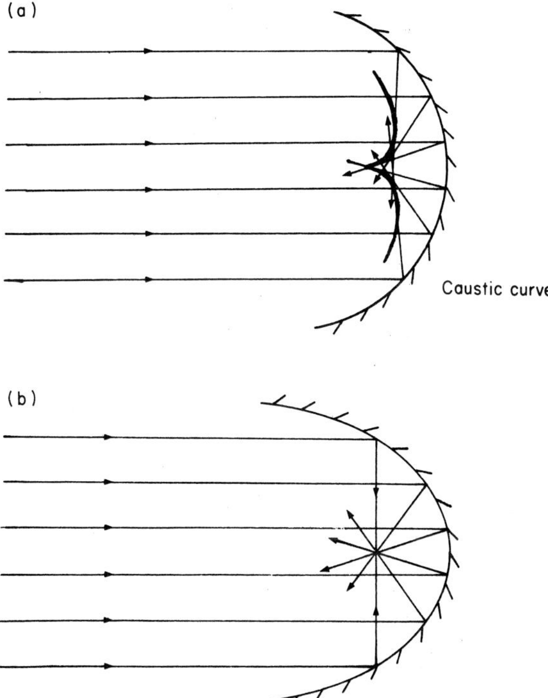

FIG. 1.21 Diagram (a) shows the multiple foci or caustic curve obtained with a curved reflector of large aperture. Diagram (b) shows that a parabolic reflector of large aperture still produces a single point focus.

sound generated at the focus of a curved reflector will be reflected as a low amplitude plane wave (i.e. the reverse of focusing). This plane wave may propagate across the room, where it is inaudible, and its amplitude will be brought back to an audible level again at the focus produced by the far wall (Fig. 1.22).

A plane wave may also be brought to a focus by refraction through a *lens*. A lens may be cut from any material whose velocity of propagation is different from that of the bulk medium. In general, solid lenses have propagation velocities greater than that of water and therefore *converging* lenses are *concave*. This is the opposite of the situation found in optics because the velocity of propagation of light in glass is slower than that in air. Figure 1.23 shows three rays from a plane progressive mechanical wave striking the flat face of a plano-concave lens. The lens material is chosen to have an acoustic

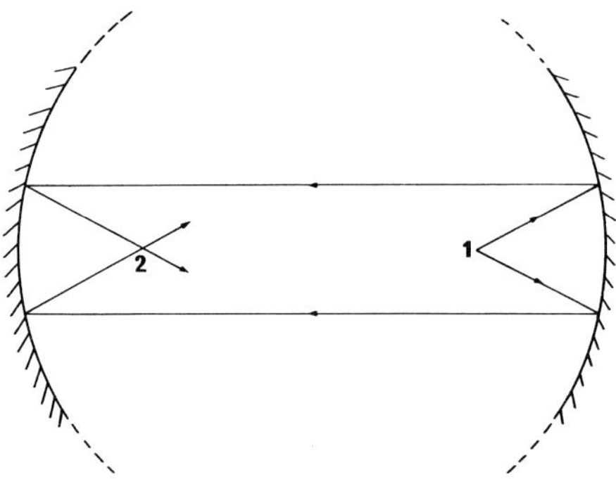

FIG. 1.22 A schematic illustration of the way in which "whispering galleries" work. The sound emitted by person number 1 is carried across the room as a broad inaudible beam and is concentrated at position 2 by the curved walls on the opposite side of the room.

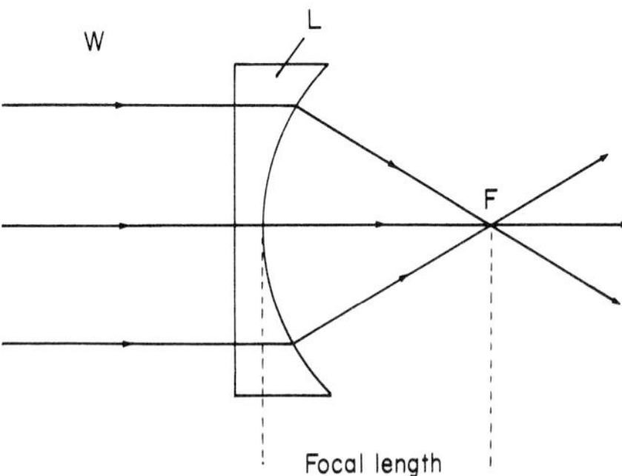

FIG. 1.23 An illustration of the path of a plane acoustic beam through a converging lens (L) suspended in water (w) which concentrates the sound at the focus (F).

impedance similar to that of water so that most of each wave enters the lens: they are not refracted because they enter at normal incidence. However, the two outer rays are refracted when they leave the lens and re-enter the water because they no longer meet the interface at normal incidence. The distance from the lens surface to its focus is known as its *focal length*.

Lenses whose curved surfaces are portions of a sphere suffer from the same defects as the curved reflector described above in that a lens having a small diameter produces a small sharp focus while a lens of larger diameter produces a larger focal volume. This latter effect is analogous to the caustic curve seen with reflector of large aperture and is called *spherical aberration*. To avoid confusion it should be noted that a lens or reflector of small aperture means that the curved surface is only a small chord of a sphere (i.e. it is a small slice from a sphere cut far away from the diameter). It should be noted that the physical dimensions of both a lens and a reflector must always be large compared with the wavelength of the beam that they are going to bring to a focus.

A curved generator can produce a focused beam without the use of a lens (Fig. 1.24). The dimensions of the generator must be large compared with the wavelength (otherwise it would act as a point source) and the wave is usually prevented from leaving the back of the generator. Curved generators are commonly used for producing focused fields of ultrasonic waves.

Mechanical waves may also be concentrated at a given point in space by the use of a waveguide. A common example of this effect is where a plane progressive series of surface ripples are trapped between two converging plane reflectors (Fig. 1.25). As the wave front decreases in length its amplitude is increased so that the total

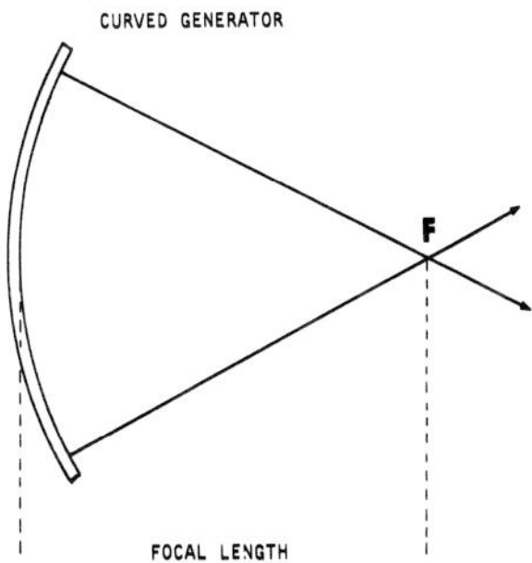

FIG. 1.24 A curved transducer generates a focus (F) whose dimensions depend upon the aperture of the transducer relative to the wavelength (i.e. a large transducer having a spherical surface must generate a large focal volume because of spherical aberration).

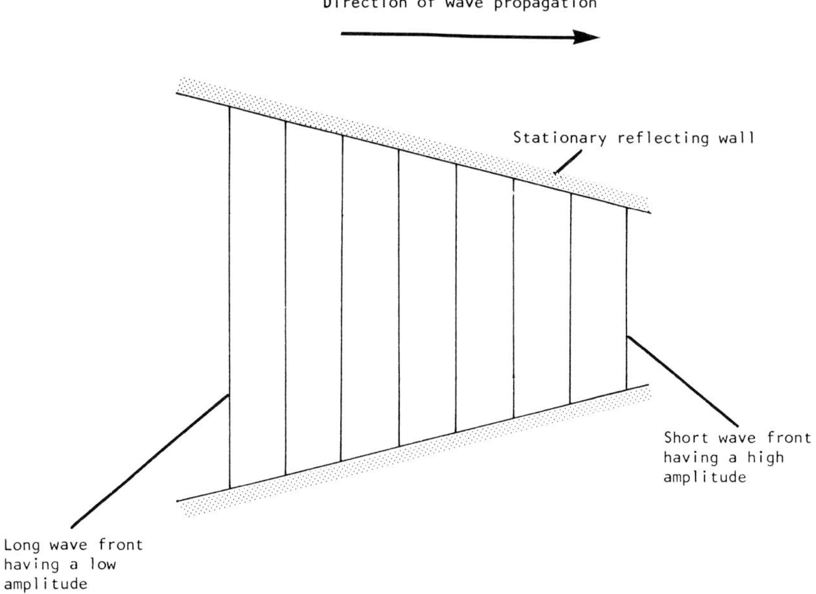

FIG. 1.25 An illustration of the mode of action of a waveguide to concentrate the spatial distribution of acoustic energy.

amount of energy contained by the wave remains constant. This effect is sometimes seen when an oceanic wave enters a tapering harbour or a river mouth.

Solid waveguides are used extensively in industry to increase the amplitude of an ultrasonic wave and to apply the vibration precisely at any desired point. These waveguides which are usually called horns or stubs may be simple tapered cylinders (Fig. 1.26(a)), two cylindrical portions of different radii cut from a solid block (b), or cones whose walls are cut with an exponential taper (c). The increase in amplitude produced at the tip of these probes is a function of the ratio of the diameter of the tip to that of the input surface. Maximum efficiency of energy transfer through these waveguides (or *velocity transformers*) is obtained when they are cut to a *resonant length*, this is usually one half wavelength or multiples thereof. Thus, if one end (e.g. its input surface) is a displacement antinode then the other end (the tip) will also be a displacement antinode.

1.10 RESONANCE

Resonance occurs when the input of energy to a system is in phase

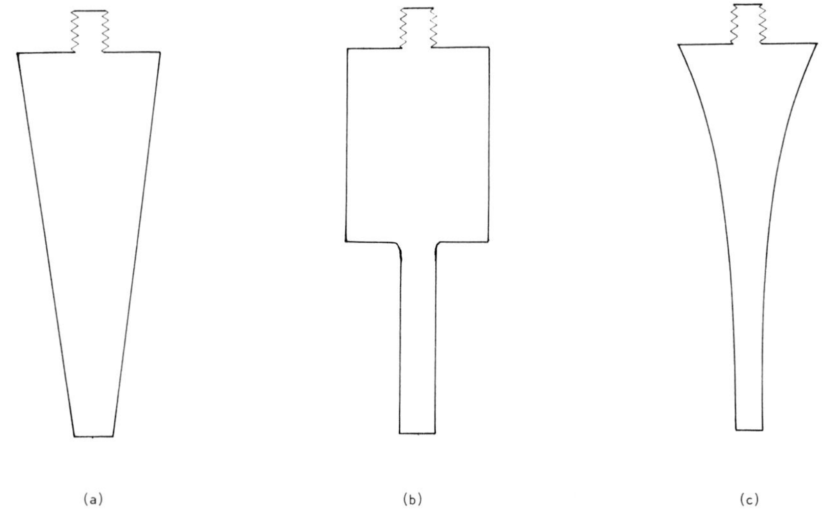

FIG. 1.26 Three representative metal waveguides commonly used in industrial applications to deliver low frequency ultrasound to the object to be sonicated. For details see text.

with the natural frequency of oscillation of that system. Common examples are a child on a swing or a ball being repeatedly bounced against the floor—if energy is supplied at just the right part of the cycle, then the amplitude of the oscillating system is maintained or increased. If the same energy is supplied in a random manner then destructive interference occurs and the amplitude decreases. An everyday example which graphically illustrates the phenomenon of resonance is the vibration at certain speeds of a car whose front wheel is not correctly balanced. As the unbalanced wheel rotates it vibrates the axle upon which it is mounted. Below or above a critical speed these vibrations are damped out by the spring and shock absorber of the suspension system. However, at a certain speed the frequency of this vibration is in resonance with the natural frequency of the spring which therefore transmits the vibration instead of damping it out and shakes the whole car.

It is important to have velocity transformers cut to a resonant length because so much of the input energy is reflected back from the tip (because of the impedance mismatch) that destructive interference would otherwise occur and waste energy. Similarly, any device which is required to transmit, generate or receive a single frequency of mechanical vibration is usually manufactured so that it is in resonance at that frequency—this greatly improves the efficiency of that device.

1.11 POWER DENSITY—INTENSITY

The amount of energy contained in the wave as it passes through any given point is commonly expressed in terms of the *intensity* at that site. Intensity is defined as the rate of flow of energy through an imaginary plane, of area one square centimetre or one square metre, orientated at right angles to the direction of wave motion. Thus, units of intensity are watts per square metre (Wm^{-2}), however, this unit is usually too large and so intensities are more conveniently expressed in units of watts per square centimetre (Wcm^{-2}) or even milliwatts per square centimetre ($mWcm^{-2}$).

$$1\ Wm^{-2} = 10^4\ Wcm^{-2} = 10^7\ mWcm^{-2} \qquad [1.10]$$

The intensity (I) of a plane progressive wave is related to the maximum pressure amplitude (Po) produced within the medium by the equation:

$$I = \frac{Po^2}{2\rho c} \qquad [1.11]$$

Where ρ is the density of that medium and c is the velocity of the wave in it (NB the product ρc is the acoustic impedance). This equation may be re-written as:

$$I = \frac{PoUo}{2} \text{ or } I = \frac{\rho c Uo^2}{2} \qquad [1.12]$$

Where Uo is the maximum velocity of the oscillating particles of the medium as the wave passes by.

Occasionally, the intensity of a wave is expressed in terms of its *root mean square* (r.m.s.) value; this is obtained if $I/\sqrt{2}$ of the maximum values of each sinusoidally varying parameter are used in the above equations

$$\text{e.g. } Po_{(r.m.s.)} = \frac{Po}{\sqrt{2}} = 0{\cdot}707\ Po$$

If there is no absorption or scattering, the intensity of a plane progressive wave remains constant whereas the intensity of a spherical wave radiating away from a point source decreases inversely with the square of the distance (d) from that source.

$$\text{i.e. } Ix = \frac{Io}{d^2} \qquad [1.13]$$

Where Ix is the intensity at any point x and Io is the original intensity emitted by that source.

1.12 UNITS OF MEASUREMENT OF AUDIBLE SOUND

In many situations, particularly when dealing with audible sound, the energy contained within a given wave is measured and compared with another (standard) source so that the result is expressed as a ratio. This ratio is therefore related to the *perceived loudness* of that sound and has the units of *bels* (B) or *decibels* (dB) and is commonly used in acoustics because of the nature of the biological detection system of the ear.

Most biological detection systems do not have a linear response curve, i.e. if the acoustic power emitted by a device is doubled, then the ear can detect the change but the new sound is not twice as loud as the original one. The response of many biological detection systems is described by the *Weber-Fechner* law which was originally based on results obtained by placing additional weights on the hand of blindfolded volunteers. If a small weight (1 g) was placed on the hand and another 1 g weight was placed on top of the first then the volunteer immediately became aware that the weight on his hand had increased. However, if the first weight had been 100 g, then the addition of a further 1 g weight would cause such a small increase in weight that the volunteer would be unable to detect it. This means that the psychological effect of adding a given stimulus (in this case a 1 g weight) is profoundly affected by the magnitude of the stimulus which was already there.

Thus, if a number of people are asked to adjust the loudness of a number of sounds (all of the same pitch) so that the differences in loudness between successive pairs of sounds are equal, we will find that equal increments in loudness correspond to equal *multiples* in intensity. If the intensities of a succession of sounds of the same pitch are in the ratio of say, 1 : 2 : 4 : 8, these will correspond to equal increases in loudness. Our ear is therefore not concerned with the difference in the intensities of two sounds, but with the difference in the logarithms of the intensities (i.e. $\log_{10} - I_1 - \log_{10} I_2$), this is commonly written as:

$$\log_{10}\frac{(I_1)}{(I_2)} \qquad [1.14]$$

Thus, one bel is the difference in perceived loudness between two sounds of the same frequency if the intensity of one is ten times that of the other, regardless of what the actual intensities may have been. The decibel (0·1 of a bel) therefore corresponds to a ratio of intensities of 1 : 1·259 or an increase in loudness of about 26%. The human ear is just able to perceive a change in loudness of about 10% which corresponds to about 0·4 dB.

The threshold intensity at which a sound is just audible varies from one person to another, and even for that same person it changes with age. An arbitrary zero of intensity (i.e. 0 dB) has therefore been fixed at 20 $\mu N/m^2$ or 10^{-12} Wm^{-2} (10^{-8} Wcm^{-2}) at 1000 cycles per second (1 kilohertz or 1 kHz). Thus, the sound intensity at the threshold of pain (roughly 120 dB at 1 kHz) corresponds to an intensity of 10^{-6} Wm^{-2} (10^{-2} Wcm^{-2}). The relative loudness of some common sounds are presented in Table 1.2.

TABLE 1.2

The relative loudness of some common sounds

Source	Sound level (dB-A)
Propeller aircraft at 5 m	130
Threshold of pain	120
Pneumatic hammer at 1 m	110
Fully laden truck at 5 m	90
Loud radio music	80
Traffic at a busy intersection	80
Normal conversation at 1 m	70
Motor car at 10 m	60
Quiet stream or river	50
Residential district without traffic	40
Quiet garden	30
Ticking of a watch	20
Quiet whisper at 2 m	10
Threshold of hearing (20 $\mu N/m^2$)	0

An additional complication is that the human ear is not equally sensitive to all frequencies in the audible range. The lower limit of audibility may therefore differ greatly from 0 dB, and so we must introduce another unit, the *phon*. A sound has a loudness of N phons if, to the ear, it seems as loud as a sound of N dB at 1000 Hz. The relationship between dB and phons is illustrated in Fig. 1.27 (these are commonly referred to as Fletcher-Munsen curves). The curved lines join points of equal loudness and hence correspond to a certain number of phons. The line marked 0 phons corresponds to the threshold of hearing and the line marked 120 phons to the threshold of pain. It should be noted that the frequency scale is logarithmic, so that equal intervals of pitch correspond to equal lengths along the x-axis.

However, the relative loudness of the sounds listed in Table 1.2 gives us little indication of the subjective loudness of this noise in the absence of information regarding its frequency distribution. For example, from Fig. 1.27 we see that a sound of 80 dB (e.g. traffic at a busy intersection) is more tolerable at frequencies less than 100 Hz and greater than about 5000 Hz and is barely audible at 25 and 15 000 Hz (which are the limits of the normal range of hearing) but is equivalent to a sound of more than 80 dB at frequencies between about 1000 and 5000 Hz. Accordingly, a sound level meter which records the total intensity (in dB) of a complex sound can give us little indication of its subjective loudness. To provide this, a sound level meter should ideally have a frequency response identical with that of the ear. It is not in fact practicable to make such a meter: but practical sound level meters do contain sets of "frequency weighting" networks known as A, B and C. Readings taken with these various networks are expressed in the units dB(A), dB(B) and dB(C) respectively. The C network has a nearly uniform response through-

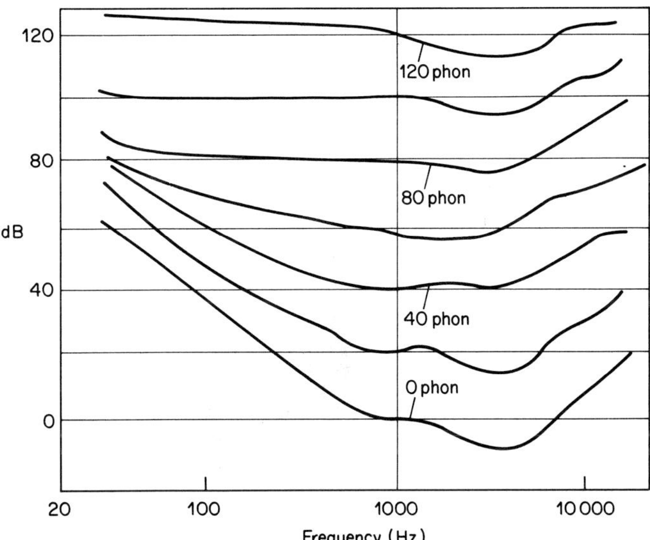

FIG. 1.27 The human ear interprets the same intensity of different frequencies of sound as being of different loudness. Thus, 60 dB of 1000 Hz sound is as loud as a motor car passing 10 m away whereas 60 dB of 30 Hz sound is barely audible. Contours of the different intensities of different frequency sounds which have the same apparent loudness are known as isophonetic curves.

out most of the audible range, so that measurements in dB(C) approximate to values in dB. The A network, on the other hand, has a frequency response curve which is similar to the inverted shape of the zero phon curve of Fig. 1.27. When this network is used the meter has a response which is similar to that of the ear so that readings in dB(A) give a reasonable indication of the subjective loudness of relatively quiet sounds. However, at high sound levels the Fletcher-Munsen curves tend to flatten out and so intense sounds should be expressed and compared in units of dB(C).

1.13 MECHANICAL WAVES OF DIFFERENT FREQUENCIES

Mechanical waves have frequencies which range from little more than zero hertz up to several hundreds of millions. For convenience, this vast range may be represented in the form of an *acoustic spectrum*, which is analogous to the more familiar spectrum of electromagnetic radiation. A typical frequency spectrum of mechanical waves is presented in Fig. 1.28 where a logarithmic scale has been chosen so that each ten-fold increase in frequency is represented by an equal distance. The spectrum can be seen to consist of three broad regions which overlap at their boundaries. The central region encompasses the audible spectrum from about 20 to 30 Hz up to about 16 000 HZ; these frequencies being the approximate lower and upper frequency limits of the average human ear. The region below about 20 Hz is designated infrasound whilst frequencies greater than about 16 kHz are called ultrasound.

Air becomes progressively less efficient at propagating sound waves as their frequency increases beyond about 20 kHz. The main reason for this is that the maximum displacement amplitudes of the vibrating gas molecules decreases with increasing frequency (for the same power input) until it becomes less than the mean path between the individual gas molecules. Thus, any high frequency ultrasound which manages to overcome the barrier due to the large acoustic impedance difference between either a solid or liquid and air will still not be able to propagate for distances of more than a few millimetres because most of the gas molecules which have been set into vibration close to the interface will not be able to collide with other gas molecules to pass on the energy of the wave.

Mechanical waves propagate by the transmission of energy from one vibrating particle of the medium to another and so the upper frequency at which mechanical waves can propagate is also determined by the inertia of the particles of that medium. The inertia of a body is a measure of its tendency to resist motion when subjected to a (transient) small force. Thus, a particle may be set into oscillation with ease at a low frequency, but more and more energy has to be supplied to maintain the same displacement amplitude as the frequency is increased (alternatively, its amplitude decreases as the frequency is increased for the same input of energy). Eventually the particle is not able to respond to the rapid changes in the direction of the applied force and so remains stationary. The wave therefore cannot propagate.

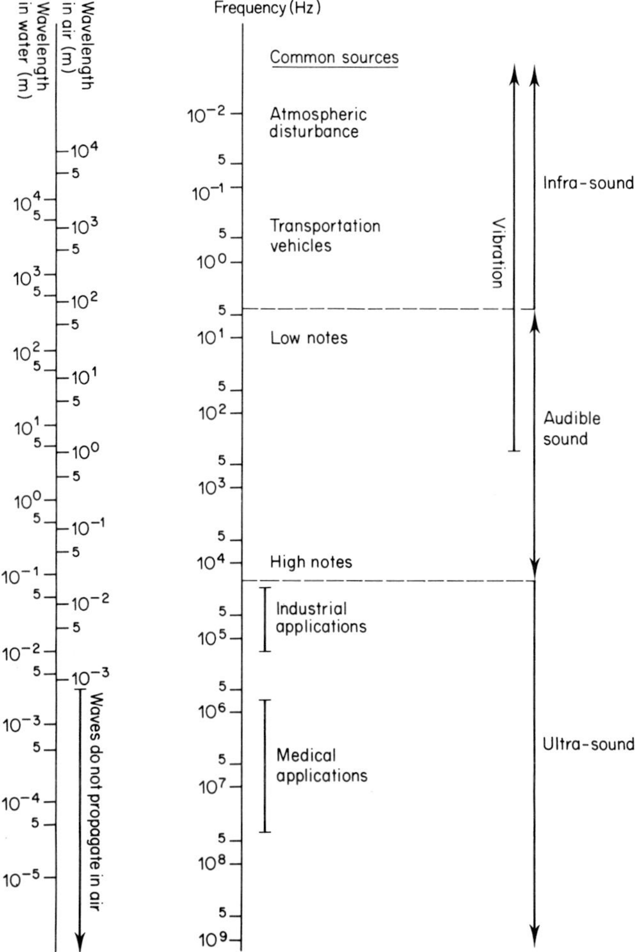

FIG. 1.28 An acoustic spectrum illustrating the range of frequencies of mechanical waves and some of their common sources of generation or application.

Electromagnetic waves (which include visible light) are propagated by sinusoidal variations of an electric and a magnetic field. These fields have no mass or inertia and so electromagnetic waves are not limited to relatively low frequencies as are the mechanical waves. For example, visible light has frequencies of the order of 10^{15} Hz while high energy X-rays and γ-rays exceed about 10^{20} Hz. The portions of the electromagnetic spectrum which have frequencies corresponding to those of mechanical waves (i.e. less than about 10^{10} Hz) are the microwaves and radio waves. Consequently, the same basic electronic arrangements are required to generate radio waves for transmission purposes or the high frequency ultrasonic beams which are used in medical diagnosis and therapy.

SUGGESTED FURTHER READING

Blitz, J. (1963). "Fundamentals of Ultrasonics". Butterworths, London.

Gooberman, G. L. (1968). "Ultrasonics, Theory and Application". English Universities Press, London.

Hueter, T. F. and Bolt, R. H. (1955). "Sonics". John Wiley, New York.

Hussey, M. (1975). "Diagnostic Ultrasound—An Introduction to the Interactions between Ultrasound and Biological Tissues". Blackie, Glasgow.

Kinsler, L. E. and Frey, P. (1962). "Fundamentals of Acoustics". John Wiley, New York.

Nyborg, W. L. (1975). "Intermediate Biophysical Mechanics". Cummings, Menlo Park, California.

Wells, P. N. T. (1969). "Physical Principles of Ultrasonic Diagnosis". Academic Press, London.

Wells, P. N. T. (1977). "Biomedical Ultrasonics". Academic Press, London.

2. GENERAL INTRODUCTION TO ULTRASOUND

Ultrasonic waves are defined as mechanical waves having frequencies above the normal limit of detection of the human ear, i.e. greater than about 16–20 kHz. However, the smaller dimensions of the vocalizing and auditory organs of small animals means that they are able to generate and detect frequencies above this arbitrary limit. In general, the smaller the animal the higher we would expect its audible range to be displaced relative to that of man. This increased frequency range is usually associated with a corresponding reduction in their low frequency range. Evans (1968) has shown that cats can hear sounds of up to 70 kHz while some rodents emit frequencies as high as 150 kHz (Sewell, 1970). These high frequency sounds are rapidly absorbed in air (Pye, 1979) and may therefore have an additional survival value in that they may allow social relationships to be established and maintained between animals which are near to each other without unduly advertising their presence to predators, e.g. a young mouse which has wandered about 20 cm from its nest and is lost in the undergrowth can alert its parents by an ultrasonic distress call without disclosing its position to a predator which may be a few metres away.

Because ultrasound is by definition inaudible, intense fields may be generated within a solid or liquid medium without discomfort to nearby humans. These intense fields can now be used for many diverse practical applications in industry, medicine and dentistry. However, the small wavelength of ultrasonic waves relative to most of the devices which generate it means that the ultrasound tends to propagate as a beam in a continuous medium (see Chapter 1). As a consequence, ultrasound has many unique physical problems associated with the measurement and display of the distribution of the radiated energy at different points within this beam. The first portion of this chapter will therefore outline the common methods employed to generate ultrasound and indicate some of its industrial and medical applications. The interference between wave fronts arising from different parts of the transducer surface to produce the complex beam geometry will be described. This will be followed by a brief description of some of the most common methods employed to quantify the amount of ultrasonic power present at various points within the ultrasonic field. After a discussion of the concepts of "dose", we will look at some of the ways in which the ultrasonic energy can be used to modify the materials through which it is propagating.

2.1 GENERATION

It is well known that bats emit ultrasound and utilize the echoes to manoeuvre around obstacles, and possibly detect their prey (Pierce and Griffin, 1938; Griffin, 1958). Under water, whales and dolphins have been shown to emit brief click-like pulses containing frequencies of up to 153 kHz which, like those of certain species of birds and bats, are only partly audible to man. Novak *et al.* (1958) and Pye (1979) have shown that these signals are generated in the larynx and may contain one or more harmonic components. Rodents also emit ultrasound when disturbed or during aggressive or mating behaviour (Sewell, 1970). It is interesting to note that one occasionally observes sudden drops of almost one octave in the ultrasonic calls of mice which are otherwise steadily increasing in frequency. Sewell proposes that this is comparable with the change of mode observed in a wind instrument, such as the recorder, when it is overblown. If a wind instrument is blown hard, a given note will change to one an octave higher, i.e. the wavelength of the standing wave field within the instrument has been halved and therefore the emitted frequency has doubled. If one now attempts to run up the scale at this higher frequency the airflow eventually becomes inadequate and the system "cracks" producing a note one octave lower. Brown (1937) has shown that similar jumps in frequency also occur with the note produced by a vibrating edge when the jet of air which has set it into oscillation is gradually withdrawn.

The simplest physical systems for generating ultrasound are the same as those employed to generate audible sound but scaled down to an appropriate size or driven at higher speeds or pressures. The Galton whistle dates from about 1883 and consists of an annular jet which impinges on the sharp edge bounding a resonant cavity whose dimensions can be changed by adjusting the position of a movable piston. This device permits a continuous change of frequency from audible to ultrasonic and is commonly employed in "silent" dog whistles. Numerous modifications and improvements have been made (e.g. the Hartmann generator of 1931 which could generate up to 50 W of acoustical energy in air even though its efficiency was only of the order of 5%) and the flowing medium can be air or a liquid. Wedge resonators are similar in that they employ a rapidly moving jet of air or liquid, but this jet impinges upon the sharp edge of a small flexible wedge which is mounted in such a way that it can be driven into flexural oscillation at 20 to 30 kHz. These wedge resonators are commonly employed in the food and cosmetic industries for the production of emulsions of immiscible liquids.

Sirens consist of a disc which can be rotated in which a number of identical holes have been cut close to the outer rim. A similar disc having the same evenly spaced pattern of holes is lined up so that the two discs are face to face. A gas or fluid is then forced through the holes in both discs while one of them is rotated. The frequency of the emitted sound is determined by the frequency of interruption of the jet flow as the holes move relative to one another and is given by the product of the number of holes in the disc times the speed of revolution. The emitted waveform is not a pure sine wave, but this is generally unimportant for the applications for which it is used, e.g. emulsification.

Most other practical applications of ultrasound employ solid generators which function by repeatedly changing their dimensions in a given plane. One of the vibrating surfaces is then coupled to the medium where the ultrasound is needed while the opposite surface is usually in contact with air. In general, solid materials are composed of symmetrical molecules which can not easily be driven to undergo cyclical shape changes by electrical or electromagnetic signals. However, two classes of materials are composed of asymmetric molecules; the application of an external force causes these molecules to change their orientation which in turn leads to a change in the shape of that solid. Magnetic materials are composed of molecules which themselves have both North and South seeking poles. When these molecules are arranged so that most of their individual North poles are aligned in the same direction, then that portion of that solid becomes the N pole, i.e. we have a magnet. When these asymmetric molecules are driven to oscillate by an applied external magnetic field the resulting shape change is called *magnetostriction*. The other class of asymmetric molecules are those which have an unequal distribution of electrical charge. When these molecules are driven to oscillate by the application of an applied external electric field the resulting shape changes are said to be due to *piezoelectricity*.

2.1.1 Magnetostriction

An alternating magnetic field is generated within a hollow coil of wire (i.e. a solenoid) when an alternating electrical current (a.c.) is passed through that wire. A bar magnet placed within the coil will change its shape at twice the frequency of the applied magnetic field because the molecular domain rearrangements which result in the shape changes are sensitive to the instantaneous value of the field strength but not its direction.

However, a magnet in an alternating magnetic field is rapidly demagnetized and the molecular domains become randomized. Therefore, a direct current (d.c.) is also passed through the same coil to provide a constant magnetic field which is superimposed onto the alternating field. This constant magnetic field means that any ferromagnetic material can be used as the core of a solenoid which then becomes an electromagnet. The core material is usually built up from strips or lamellae to reduce eddy currents, i.e. electrical currents induced within the core by the changes in the magnetic field. The two electrical supplies to the energizing coil contain a condenser (which permits a.c. to pass but blocks d.c.) and a choke (a coil of wire wrapped around an iron core which prevents the passage of a.c. but permits d.c. to flow) to isolate the different electrical generators from each other.

The changes in length of the core material are small, typically of the order of microns, and so the core is chosen to be of *resonant length* for maximum efficiency. At resonance, the ultrasonic wave generated within the core is reflected from the air-backed face at the same instant that a new wave is generated. All waves therefore interfere constructively within the transducer core so that one obtains the maximum displacement amplitude for the least consumption of power. In practice, the transducer is usually half a wavelength long; NB transducers employing cores of different materials will be of different lengths because the velocity of sound, and hence its wavelength, varies from material to material.

Velocity transformers or probes are usually employed to conduct the ultrasound to its site of application. These are also half a wavelength long and are cut to be resonant at the same frequency as the transducer. Thus, both ends of the transducer core and both ends of the velocity transformer are displacement antinodes or points of maximum vibration. The mid points of both must therefore be displacement nodes or points of zero displacement amplitude. The transducer/probe assembly may therefore be clamped or mounted at these points without interfering with the efficiency of the system (Fig. 2.1).

The Q-factor of this system is high, i.e. it operates at maximum efficiency only over a small range of frequencies. When the probe tip is applied to the load into which it is to transmit its energy, this contact changes the resonant frequency of the system. Consequently, the driving frequency has to be changed to coincide with the new

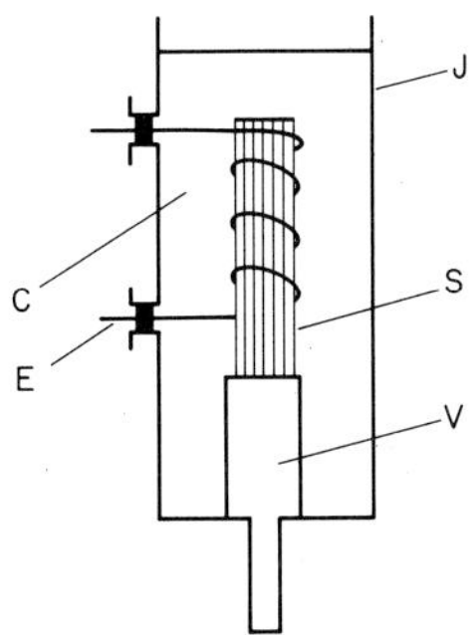

FIG. 2.1 A diagrammatic representation of a magnetostrictive transducer assembly used to generate high power, low frequency ultrasound for industrial applications (e.g. cell disruption, welding, drilling, etc.). The laminated ferromagnetic stack (S) is energized by a d.c. and an a.c. signal via the coil (E). The stack (S) is coupled to a velocity transformer (V) which is mounted at its nodal point in a tank (J) which may be filled with an oil (C) as coolant.

optimum driving frequency. Various types of sensing device are attached to the transducer or incorporated within the electrical drive circuits to ensure that the assembly is always driven at resonance.

Magnetostrictive devices were first described by J. P. Joule in 1847 and are commonly employed today for the generation of high power ultrasonic vibration for use in industry. It is difficult to manufacture the very small transducers which are necessary to generate waves having frequencies greater than about 500 kHz and so these devices are mainly used for the generation of low frequency ultrasound, especially in the range 20 to 100 kHz.

2.1.2 Piezoelectricity

The brothers Pierre and Jacques Curie discovered in 1880 that an electric charge was generated on the surfaces of certain single crystals when they were mechanically compressed. These crystals, e.g. quartz, tourmaline, lithium sulphate, cadmium sulphide and zinc oxide, all lack electrical symmetry. In the same way that all the mass of a body acts as though it were concentrated at one small point (the centre of mass of that body) so all of the individual positive and negative charges within each crystal can be represented by a single large positive or negative charge situated at specific points within that crystal. If a material is electrically symmetrical, then the centres of positive charge, negative charge and mass all occupy the same position. Materials which are electrically asymmetrical have centres of positive and negative charge which do not overlap.

If a disc or slab is cut from one of these piezoelectric materials and the two appropriate crystallographic surfaces are coated with a thin film of a metallic conductor (e.g. gold, silver or aluminium), then a voltage is obtained when the crystal is compressed or mechanically strained. The converse effect is observed when a voltage is applied across the two electroded surfaces, the crystal now undergoes a mechanical deformation.

Figure 2.2 shows that a change in the polarization of the voltage across the electrodes results in a corresponding change in the direction of distortion of the crystal. Thus, the application of an alternating voltage across the crystal would result in an oscillating change in shape. Therefore, the application of an a.c. voltage having an ultrasonic frequency would result in the crystal oscillating at that same frequency. As in the case of magnetostrictive transducers, the amplitudes of oscillation of the crystal face are extremely small and so the phenomenon of resonance is again employed to obtain maximum efficiency. As before, the crystal is cut to be half a wavelength thick and the frequency of the electrical signal is swept around the expected value of the resonant frequency until the crystal oscillates with maximum displacement amplitude for any given input power.

This resonant frequency is called the *fundamental* resonant frequency of the transducer. If the driving frequency is progressively raised above this value, the ultrasonic output decreases as you move away from resonance and approaches zero when the driving frequency corresponds to one wavelength thickness of the transducer. As the driving frequency continues to be increased the ultrasonic output also increases until another resonant condition is obtained at a frequency of three times the fundamental frequency. This frequency corresponds to 1·5 wavelengths within the transducer and is known as the *third harmonic* frequency. Yet another resonant condition occurs at the fifth harmonic when the driving frequency is five times

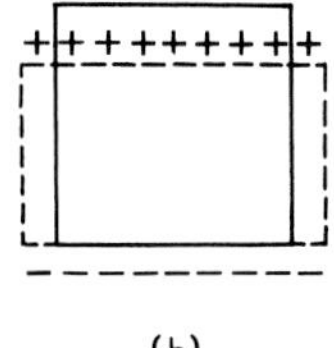

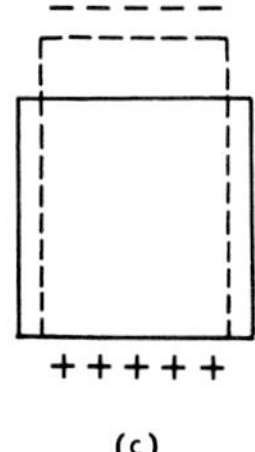

FIG. 2.2 A pictorial representation of the change in the dimensions of a piezoelectric material. The undeformed crystal (a) is flattened by one polarity of an applied voltage (b) and stretched by the reverse polarity (c). Thus, the application of a sinusoidal voltage would cause the body to oscillate between the two extreme shapes shown in (b) and (c).

the fundamental frequency and the transducer is 2·5 wavelengths thick (Fig. 2.3).

Piezoelectric materials may be fabricated from barium titanate and lead zirconate titanate (Jaffe *et al.*, 1955). These materials are mixed with various chemicals and baked at various high temperatures to yield a range of hard ceramics which are extremely stiff and capable of exerting or sustaining large forces. However, a randomly polarized mixture of these materials would not be piezoelectric and so the individual molecules have to be correctly aligned as they are within the natural piezoelectric crystals. This is achieved by first casting the ceramic to the required shape and dimensions, and then applying the conducting electrodes. A high d.c. voltage of approximately 2000 V/mm is then maintained across the electrodes while the

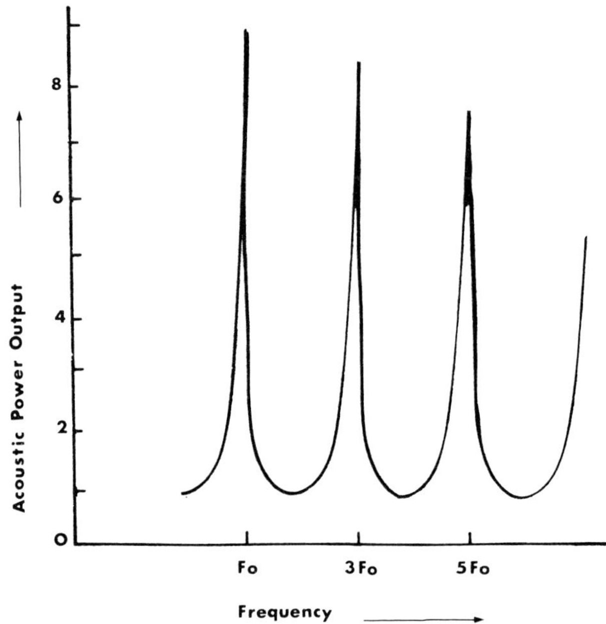

FIG. 2.3 A pictorial representation of the displacement amplitude of the transducer as a function of the driving frequency. At resonance (e.g. fo, 3fo and 5fo) the transduction process is most efficient.

ceramic is heated above its Curie temperature, and then allowed to cool. The Curie temperature is that temperature above which the molecules of the material are free to move in response to an applied force. Conversely, any piezoelectric material which is heated above its Curie point (typically 300–600°C) and allowed to cool in the absence of an applied d.c. electric field will lose its piezoelectric properties. This is analogous to the demagnetization of a metal bar after it has been heated.

2.2 USES AND APPLICATIONS OF ULTRASOUND

Ultrasound is used in so many ways for so many diverse applications in both industry and medicine that even a relatively superficial description of each would easily fill this book. Figure 2.4 outlines some of the major applications of ultrasound and lists one or more typical processes utilizing that particular physical property. This book is concerned with the biological effects of ultrasonic waves and

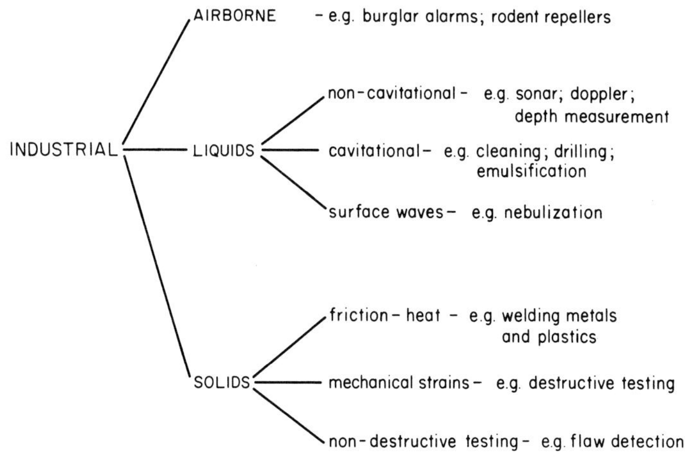

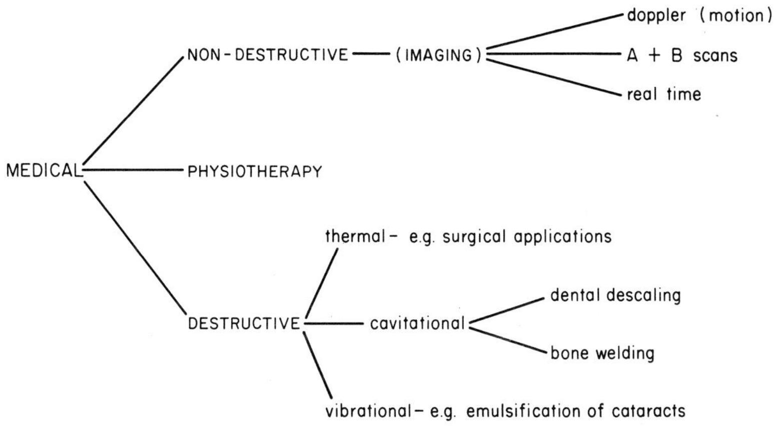

FIG. 2.4 Some commercial applications of ultrasound.

so only those applications which directly or indirectly result in human exposure to ultrasound will be described in more detail.

2.2.1 Industrial Applications

Most industrial applications utilize relatively low frequencies of ultrasound (about 20–60 kHz) where there is the possibility of human exposure via airborne conduction. For example, ultrasonic burglar alarms work by transmitting an airborne spherical wave and "tuning" the detector to ignore the echoes reflected from the room's contents. Any new source of echoes disturbs this balance and triggers the alarm—sometimes this new source of echoes may be a plant that has grown too rapidly or an insect that has flown past the emitter/detector. Those processes where ultrasound is contained within a metal velocity transformer (e.g. destructive testing of metal stubs) also emit some airborne ultrasound even though the large acoustic impedance mismatch between the metal and air ensures that most of the acoustic energy remains within the metal. However, those liquid-borne applications which utilize cavitation emit potentially dangerous levels of audible as well as ultrasonic airborne sound. The explosive shock waves and violent streaming forces generated within a transiently cavitating liquid (see Chapter 4) are used to break open cells and organelles (ultrasonic disruptors), to remove dirt and corrosion from solid surfaces (ultrasonic cleaning baths) or to chip away at hard solid surfaces by bombarding them with abrasive particles (ultrasonic drilling). These cavitation events re-radiate some of the ultrasonic energy at a wide range of frequencies which extends into the audible spectrum (white noise) and is usually heard as a loud hissing sound which sometimes resembles that of an egg being fried. In general, these cavitating devices should be operated within a closed box or, where this is not possible, the operators should wear some form of ear protection if they are to be exposed to this sound for prolonged periods.

Some operators who use these ultrasonic cleaning baths do not fully appreciate how dangerous these machines can be and ignore warnings and continue to immerse their bare hands in the bath while the transducer is still activated. One result of this negligent behaviour is that the dead and relatively impermeable outer layers of the epidermis of the skin are removed so that molecules of detergent or other solvent within the bath may come into contact with the growing cells within the skin and so greatly increase the probability of producing an allergic or irritant reaction commonly called a "contact dermatitis". Another consequence of immersing the bare hand in a cleaning bath while it is switched on is that the low frequency ultrasonic radiation is coupled into the tissues where it may generate some form of cavitational activity which will disrupt cells and interfere with many vital functions.

2.2.2 Medical Applications

Cavitation is also employed in some of the medical applications of ultrasound such as the descaling of teeth (Balamuth, 1963; Suppipat, 1974), the disruption of kidney stones within the ureter or the more hazardous procedure of removing thrombi from within blocked blood vessels (Stumpff *et al.*, 1975) and in the "welding" of bones (Wehner *et al.*, 1980). The technique of bone welding is not common but has been applied where the broken edges of the bone cannot be held together by means of a splint or pin. In this procedure bone powder is mixed into a paste with a monomer of a cyano acrylate sold commercially as Fimomed^R (which polymerizes by a free-radical mechanism) and is placed between the broken edges of the bone. The mixture is irradiated with low frequency ultrasound so that the reaction is presumably initiated by free radicals produced by cavitation in surrounding aqueous media and is accelerated by the local heating. However, the resulting mass of polymeric material will almost certainly interfere with the normal bone repair mechanisms and the long-term effects of the polymer and/or non-polymerized portions of the monomer have not been investigated.

The most common source of deliberate ultrasound exposure to humans is in the field of diagnostic imaging (Kurjak and Kratochwil, 1982). It has been estimated that in the United States of America over half of the children born in 1981 will have received at least one ultrasonic investigation before birth. This trend towards an increasing population exposure is still continuing; for example, the West German Bundesärtzekammer has recommended that participating Insurance Companies should pay for two routine ultrasonic examinations during the course of each pregnancy (one at 16–20 weeks and the other at 30–35 weeks). These examinations will certainly lead to more women having ultrasonic examinations during pregnancy, but it is not yet clear if it will result in more examinations for each individual woman.

Details of the techniques and diverse apparati used to perform these various diagnostic investigations are discussed in Wells (1977). For those who are completely unfamiliar with these techniques, there follows a brief and somewhat oversimplified description of the mode of operation of those procedures which are used most frequently.

2.2.2.1 Doppler Devices

As in the case of audible sound, an ultrasonic wave emitted by or reflected from a moving body is changed in frequency by an amount proportional to the velocity of that body at any given time (i.e. the Doppler Shift). A sample of the wave emitted by the machine and the returning echoes are fed into an electronic circuit or "black box" which compares them. If the echoes return from a stationary object then they are of the same frequency as the emitted wave and are absorbed by the black box. However, if the echoes were reflected from a moving object they would have a different frequency and this difference would come through and be fed into a loudspeaker after amplification. These devices are used to detect the foetal heart (this beats at about 120 times each minute and can therefore be distinguished from the maternal circulation which beats at 70–80 times per minute), to evaluate the functioning of the adult heart, and to detect the rates of motion of the blood cells themselves in the case of the doppler blood flowmeters.

2.2.2.2 *The "A" Scan*

The simplest device for obtaining pictorial information using ultrasound is the "A" or amplitude scan. This device is essentially similar to the industrial flaw detector in that it gives information concerning the position of boundaries having discontinuities in acoustic impedance (see Chapter 1) and its mode of operation is described schematically in Fig. 2.5. The piezoelectric transducer (T) is heavily damped by loading its rear surface with a medium such as tungsten powder embedded in an epoxy resin so that it rapidly stops vibrating after it has been driven to oscillate. Imagine this transducer coupled to the skin overlying an idealized skull by a liquid gel (c in Fig. 2.5). The transducer is now excited by a single sharp pulse of electricity which causes it to "ring" briefly just as a heavily damped bell would do after it had been hit with a hammer. This electrical pulse also starts the time base of an oscilloscope which causes a spot of light to move from left to right across the screen. The ultrasonic "pulse" emitted by the transducer passes through the coupling

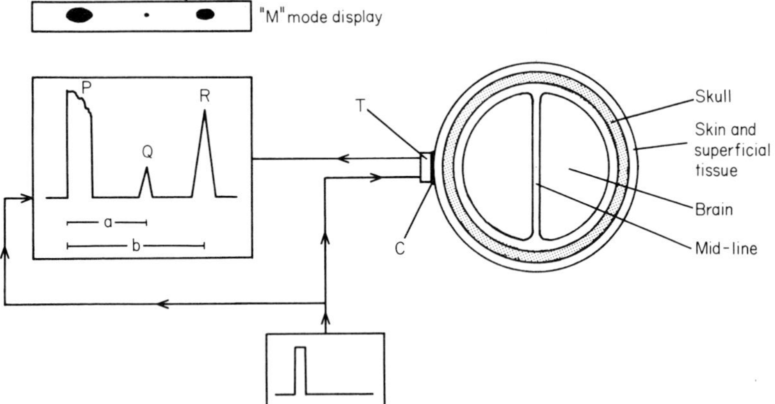

FIG. 2.5 A schematic description of the mode of operation of an "A" or amplitude mode device. For details see text.

medium and the skin and encounters the bone of the skull where, due to the moderately large mismatch in acoustic impedance, some of the energy is reflected while the remainder goes through and enters the brain. The reflected portion of the wave returns to the transducer which has almost stopped ringing and causes it to vibrate again which generates a voltage. This induced voltage deflects the spot on the storage oscilloscope giving a peak P. The remainder of the wave propagating through the brain encounters the mid-line where some of it is reflected to return to the transducer to produce peak Q where the time "a" is the time taken by that ultrasonic wave to travel from the transducer to the mid-line and back. The remainder of the acoustic wave encounters the far wall of the skull where a substantial portion of it is reflected back to the transducer to give peak R at time "b". This device tells you whether or not the mid-line of the brain is in the centre of the skull (it may be displaced by a tumour or cyst or by a pool of blood [a haematoma] leaking from a damaged blood vessel). Another pulse of ultrasound is sent out after the echoes from the first pulse have returned to the transducer. The number of ultrasonic pulses emitted per second, i.e. the pulse repetition frequency (P.R.F.), may vary from about 50 to several thousand per second. This rate is faster than the discriminating rate of the eye and so the pattern of peaks on the oscilloscope screen appears to be stationary.

2.2.2.3 The "M" Mode

If the ultrasonic beam from the "A" scan device is pointed at a moving structure such as the heart it will produce a different pattern of peaks but some of these will be seen to move from left to right and back again as the heart beats. An alternative method of display is therefore needed to present this motion in an informative manner.

The oscilloscope used in the "M" or Motion display is set up in an analogous manner to the common black and white television set except that only one line is used instead of the 405 or 625 lines used to build up the two-dimensional image. In both systems the incoming signal voltages are used to increase the brightness of the spot traversing the screen so that a high voltage gives a bright spot, a less high voltage a less bright spot and no voltage gives no visible spot. Thus, the idealized "A" scan pattern presented in Fig. 2.5 would be reduced to three bright spots along a single line with the central spot (corresponding to the voltage peak Q) somewhat less bright than the other two.

If one or more of the structures giving rise to the spots along this line was to change its position relative to the transducer then its corresponding spot would move backwards or forwards along that line. If a strip of light-sensitive paper is moved at constant speed past this pattern of spots then an oscillating pattern reflecting the change in position of the moving structures will be produced (Fig. 2.6). This figure presents one of the typical patterns which can be obtained with a normal adult heart where CW is the anterior chest wall, RV is the cavity of the right ventricle, IVS is the interventricular septum, AMVL is the anterior mitral valve leaflet, PMVL is the posterior mitral valve leaflet and P are the pericardial structures behind the heart.

2.2.2.4 The "B" Scan

This device produces a two-dimensional representation of the distribution of discontinuities in acoustic impedance within the irradiated body. It is most commonly used in obstetrics and gynaecology for confirming the presence of a foetus, identifying the number of foeti and measuring various foetal dimensions (e.g. bi-parietal diameter, chest circumference and/or crown–rump length) to assess foetal maturity and check that it is the correct size for its gestational age. A simplified description of its mode of operation is shown schematically in Fig. 2.7. The heavily damped transducer (T) emits brief pulses of ultrasound and then waits to receive the echoes as described above for the "A" and "M" modes. A series of reflectors have been placed in the path of the ultrasonic beam; S stands for a strong reflector which reflects say 20% of the incident beam while W represents a circular or cylindrical weak reflector which only reflects say 5% of the incident beam. It can be seen that the echo produced by the first strong reflector (which is at an angle of about 45° to the axis of the ultrasonic beam) does not return to the transducer and so does not appear on the "A" mode display. The ultrasonic beam strikes the other reflectors at normal incidence and so the two strong reflectors give strong echoes which produce two high-voltage peaks while the two interfaces of the weak reflector give two smaller voltage peaks on the "A" mode display. It should be noted that these peaks have not been drawn to scale and that the real "B" scan devices are usually fitted with some form of variable gain control so that the echoes coming from deep within a body may be amplified more than the echoes coming from superficial structures.

The information contained in this "A" scan picture is now transferred to a storage oscilloscope or scan converter memory tube to build up the "B" scan picture. In fact this series of voltage peaks is used to brightness modulate one line of the storage oscilloscope (as

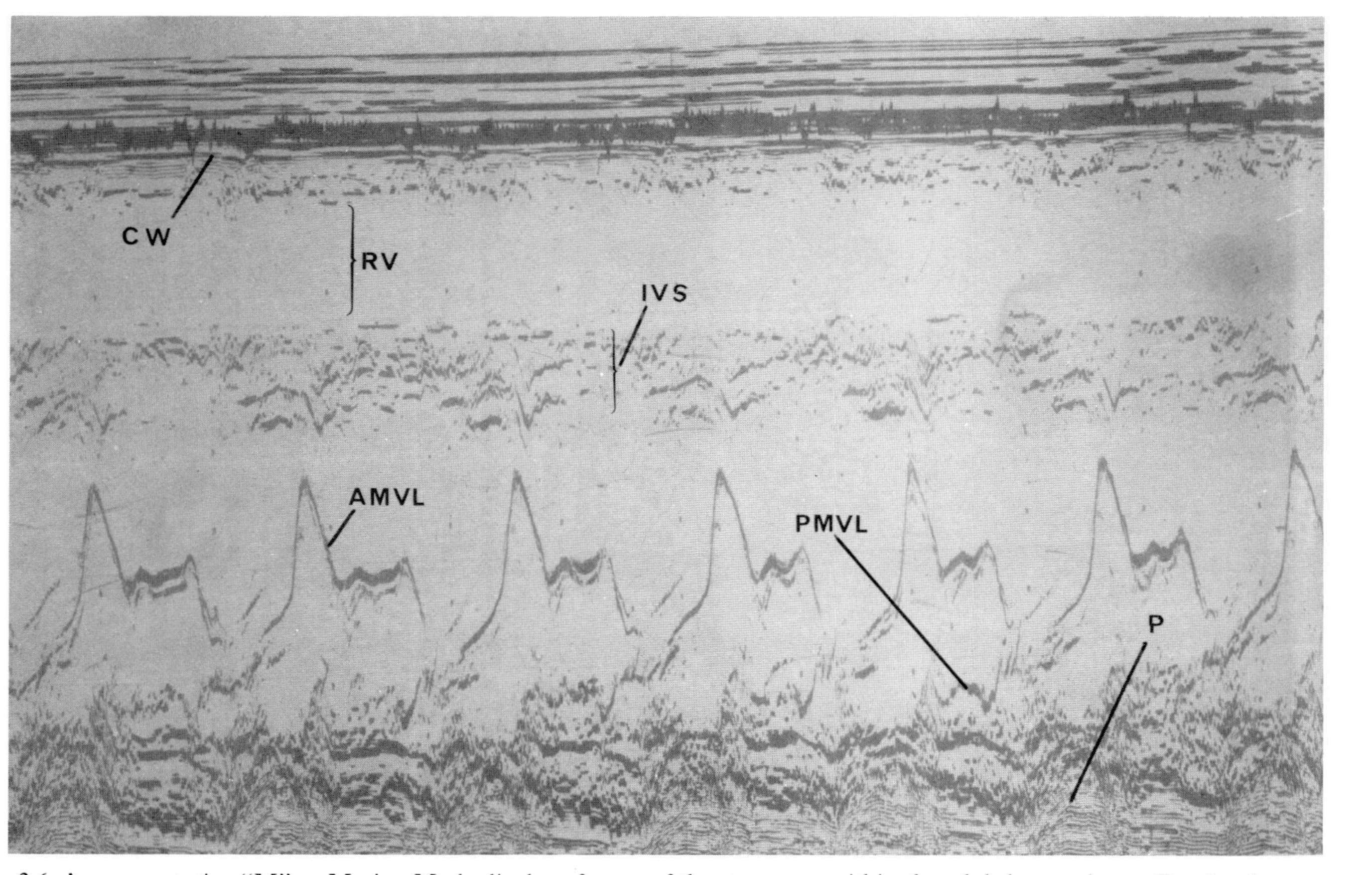

Fig. 2.6 A representative "M" or Motion Mode display of some of the structures within the adult human heart. For details see text.

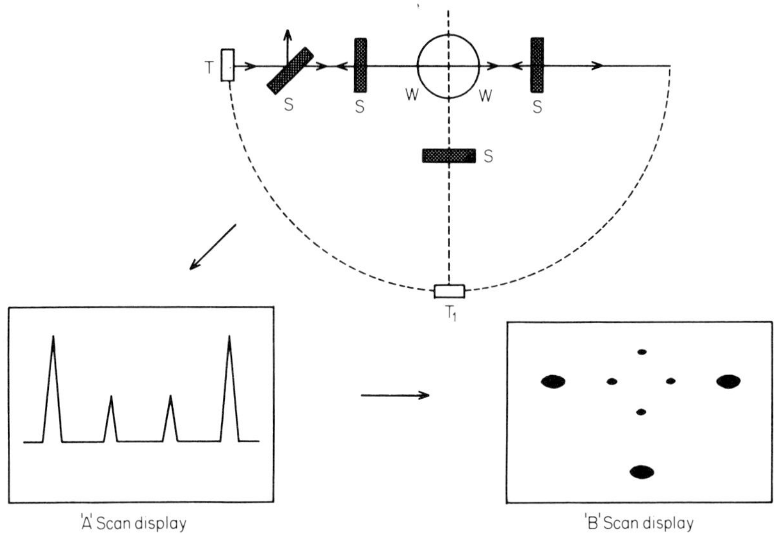

FIG. 2.7 A schematic description of the mode of operation of a "B" scan device. For details see text.

described for the "M" mode) to give the horizontal line of four bright spots indicated on the "B" mode display in Fig. 2.7.

If the transducer (T) is now moved to another position (T_1) the ultrasound beam will now pass through a new strong reflector and two different portions of the weak reflector and will therefore generate a new "A" scan picture which resembles the first three peaks of the previous scan. The problem is how to get the spots which these voltages would produce to be placed in the correct spatial position on the storage oscilloscope. This is solved electronically in a number of ingenious ways (Wells, 1977), but can be viewed conceptually as a position-sensitive detector, i.e. an electronic sensor attached to the joints in the flexible arm linking the transducer to the "B" scan machine generates an output voltage which is proportional to the angle through which that joint has been rotated. This voltage may therefore be used to direct the new spots to their correct positions on the storage oscilloscope screen. If the transducer is now swept through the arc indicated by the broken line in Fig. 2.7 it will generate a series of pairs of spots from the weak reflector which will merge together to give a complete circle.

"B" scan devices employ a bewildering variety of innovative mechanical and electronic techniques to generate and display the two-dimensional echo patterns outlined above. These are discussed at length in Wells (1977). For example, since echoes are detected with maximum amplitude only after they have been reflected at normal incidence, an extensive flat surface not quite at right angles to the propagation axis will be detected only if the transducer has a wide aperture (Kossoff *et al.*, 1968). An alternative solution is to oscillate the probe or transducer housing as it is moved around the patient (i.e. a *compound* "B" scan). This oscillatory motion increases the probability of reflections occurring at normal incidence and gives better quality pictures even though under some circumstances this may be disadvantageous (Wells, 1977).

Figure 2.8 is a typical "B" scan picture of a longitudinal section down the mid-line of a young woman's abdomen taken when she was six weeks pregnant. The uppermost pair of bright "wobbly" lines are the echoes from the tranducer/coupling medium interface (T). Below this are the two muscular and fatty layers of the body wall (W) and the large echo-free space of the bladder full of urine. It is common practice to scan women in the early stages of pregnancy when they

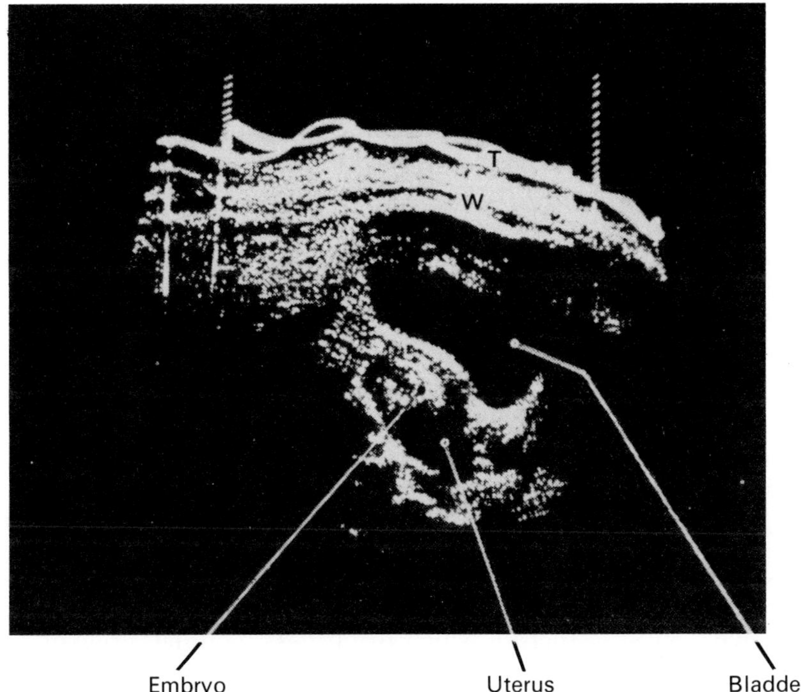

FIG. 2.8 A representative "B" mode display of the abdomen of a woman in the sixth week of her pregnancy. For details see text (from Donald and Abdulla, 1967).

have a full bladder as this displaces the uterus and its contents backwards and upwards so that the ultrasonic beam does not have to pass through the bones of the pelvic girdle (where most of the energy of the beam and its echoes would be lost by absorption and reflections).

2.2.2.5 Grey Scale Displays

The original "B" scan devices merely had a single "threshold" voltage so that a signal above this voltage gave a spot on the storage oscilloscope whereas a voltage below this "threshold" (which included most of the electronic noise inevitably associated with electronic instrumentation) did not produce a spot. These devices were satisfactory in use providing that the echoes were coming from moderately large mismatches in acoustic impedance such as skeletal structures. However, the diagnostic information contained within any given picture is greatly enhanced if instead of a single "on" or "off" threshold there are several such "thresholds" arranged in series. The interval between each pair of thresholds can be represented by a different level of grey (as implied in the above description of the "B" scan device) or by a different colour if the display is presented on a colour monitor.

The main disadvantage of the "B" scan and Grey Scale displays is that they rely on storage oscilloscopes or scan converter memory tubes. This means that the information from rapidly moving objects may be "smeared out" across the screen. Also, these devices are unable to distinguish between the useful signal and the unwanted noise and so stores both of them. If one therefore repeats the same scan many times in an endeavour to detect weak echoes from weak reflectors, the noise will eventually swamp out the signal.

2.2.2.6 Real Time Displays

These devices overcome the major disadvantages of the "B" scan and Grey Scale apparati described above and are rapidly becoming the most useful diagnostic ultrasonic display system. In its simplest form it consists of a strip of several small rectangular transducers (T) arranged side by side. Each transducer element essentially generates its own "A" scan picture which brightness modulates one or more lines on an oscilloscope display whose timebase is triggered by the electrical signal supplied to each transducer (Fig. 2.9 adapted from Bom *et al.*, 1971). As well as being able to examine moving structures such as the heart, these real time displays are able to improve their image resolution by making use of the image processing characteristics of the human eye. The reader may have noticed that the quality of the image of a television picture deteriorates as the motion is slowed down. Before the advent of the computerized image processing procedures used at present, the stop-frame picture was of such poor quality that it was barely discernible. This is because the signal information within each frame is not very much bigger than the associated noise. Yet the eye sees the moving television picture quite clearly because it averages the information carried in several frames, i.e. if a signal is in nearly the same position for several frames (each frame lasting 1/50th of a second in Britain or 1/60th of a second in the U.S.A.) then they reinforce each other and an image is produced on the retina of the eye. However, the noise is completely different in each frame so that a "peak" of noise is soon cancelled out by a "trough" of noise in another frame.

Thus real time display systems are intrinsically capable of resolving weak signals which would otherwise be obscured in a welter of noise. These devices are still at an early stage of development and yet hold the promise of being able to distinguish between cysts and tumours and the normal tissue that surrounds them even though the reflected echoes may be very weak.

2.2.2.7 Ultrasound in Physiotherapy

This is the second largest source of deliberate exposure of humans to

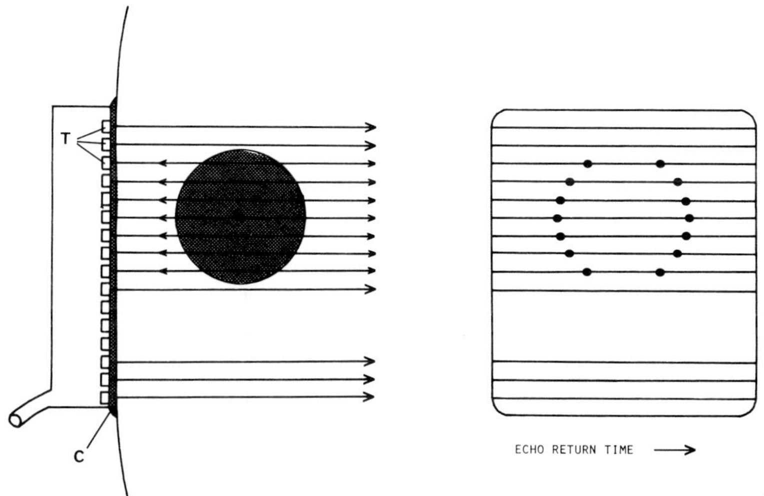

FIG. 2.9 A schematic representation of the mode of operation of a real time device (adapted from Bom *et al.*, 1971). The numerous transducers (T) are coupled to the tissues via a viscous coupling medium (C) as before. For details see text.

high frequency ultrasound but in many ways it is also the most potentially dangerous. Diagnostic devices usually operate within a limited range of low average power outputs and the main way in which the operator can increase the patient's "dose" of ultrasound is by increasing the time taken to perform the scan. The physiotherapist or chiropractor, on the other hand, has control over all of the relevant dosimetric parameters (i.e. intensity, duration and number of treatments, whether it is pulsed or continuous, and with the more modern machines even the frequency). This means that the "dose" of ultrasound which the patient receives is entirely at the discretion of the physiotherapist who usually relies heavily on his or her past experience and upon the dosage parameters recommended in certain textbooks (e.g. Summer and Patrick, 1964). In general, the total amount of ultrasonic power administered during a typical therapeutic treatment session is very much greater than that administered during a diagnostic investigation. This difference is mainly due to the longer "on" time of the therapeutic applicator since the space averaged temporal peak (S.A.T.P.) intensities used in therapy seldom exceed about 3 or 4 W/cm^2 where the S.A.T.P. intensities of some diagnostic devices may reach about 100 W/cm^2 (see Table 2.1).

Therapeutic ultrasound is usually applied directly to the skin (using a suitable viscous liquid coupling medium), if the surface to be treated is relatively flat. A liquid-filled thin-walled rubber cushion is used to conduct the ultrasound to more angular surfaces, while even less accessible anatomical sites such as the fingers may be irradiated by immersion in a water bath through which the ultrasonic beam is being propagated. The beneficial clinical effects which appear to result from ultrasonic therapy are (a) absorption of intracellular tissue fluids, (b) tissue heating, and (c) analgesia (i.e. reduction of pain). A multitude of other beneficial effects are also claimed for ultrasound therapy, some of which will be discussed in Chapters 5 and 6.

One point which is generally overlooked by physiotherapists is that the scientific rationale underlying many of the treatment regimes in current practice is at best tenuous. Most procedures have evolved over the years based upon a therapist's subjective evaluation that a patient's "condition" has apparently been improved by the treatment, rather than upon the result of a well-controlled trial to evaluate the effectiveness of that treatment. Some treatments may also be criticized on the grounds that the amount of ultrasonic power administered is so low that it is unlikely that the treatment has more than a placebo effect and may not therefore justify the cost and upheaval of bringing that patient to be treated. It is therefore

TABLE 2.1

Range of Acoustic Output Parameters from Medical Equipment (Compiled from the data of Carson *et al.*, 1978; Hill, 1977; Rooney, 1973)

Equipment type	Frequency range (MHz)	Total power output (mW)	Spatial average time average (SATA) (mW/cm^2)	Spatial peak temporal peak (SPTP) (mW/cm^2)	Typical duty factor
Pulse/echo	1–20	0·3–20	0·2–6	5×10^2–2×10^5	0·001
Foetal heart (doppler)	2–4	1–40	0·2–20	1–60	1
Peripheral vascular (doppler)	5–10	1–40	40–400	1–2×10^3	0·01–1
Therapy devices	0·5–3	0–2×10^4	0–4×10^3	0–10^4	0·2–1

recommended that physiotherapists attempt to devise trials to evaluate the efficacy of their ultrasonic treatment procedures both to improve the quality of their therapy and to eliminate those procedures which are ineffective.

2.3 DETECTION AND DOSIMETRY OF ULTRASOUND

Small animals detect ultrasound in the same way that humans detect audible sound: their cochlear apparatus is smaller and can therefore resonate in response to higher frequencies than that of man. Some animals have developed their systems for the detection and interpretation of ultrasonic echoes to a remarkable degree of sophistication. The performance of bats has been studied in great detail (Griffin, 1958). Schneider and Möhres (1960) have shown that certain species of bats move their ears to aid in echolocation, and that they became disorientated if their ears were restrained. Equally remarkable is the ability of dolphins or porpoises to discriminate between solid objects of the same shape which differ slightly in thickness (Hammer and Whitlow, 1980). It is proposed that the shape of the skull and the distribution of tissues around it assist in the conduction of ultrasonic echoes to the animal's inner ear. The lateral line organ system of many fish is also presumed to respond to echoes of ultrasonic frequencies: in this case the detection system is sensitive to acoustic pressure (baroceptors) and is related to the sense of touch.

Small microphones are not damped as heavily as loudspeakers, and so can still be used to detect low frequencies of ultrasound, i.e. less than about 150 kHz. Above this frequency we have to resort to other techniques. The major detectors of high frequency ultrasound in use today are piezoelectric. The incident acoustic wave causes a mechanical deformation of the crystal or ceramic (i.e. the reverse of the method of generation of high frequency ultrasonic waves) which results in the production of a voltage across the two electroded surfaces. The magnitude of the voltage is proportional to the distortion which in turn is a function of the incident ultrasonic intensity. Providing that the dimensions of the piezoelectric element are small compared with the wavelength of that acoustic field, it can be used as a probe to map out the distribution of acoustic energy within that medium.

Generators of audible or infrasonic sound are usually small compared with the wavelength of the sound they emit, and so act as point sources (see Chapter 1). Consequently, their acoustic fields are relatively straightforward, the intensity at any point x being found from the equation

$$Ix=\frac{Io}{d^2} \qquad [2.1]$$

Where Io is the intensity emitted by the source and d is the distance of point x away from that source. Unfortunately, the wavelengths of the ultrasonic waves used in medical applications are comparable with and may even be less than the dimensions of the generator. Under these conditions wave fronts emanating from different parts of the transducer surface must have travelled different distances to arrive at the same instant at any given point in space close to the transducer. Since the waves all travel at the same speed in any one medium, those waves which have travelled different distances will have a different phase when they meet and will therefore interfere with each other (see Chapter 1) to produce the complex pattern of peaks and troughs described below.

2.3.1 Beam Geometry

Any source whose dimensions are large compared to the wavelength generates a beam which is similar to a plane progressive wave. However, the intensity of that wave is not the same at all points within that beam. This is because individual wave fronts from different parts of the source have to travel different distances before they arrive at any given point in space. Waves which arrive in phase interfere in a constructive manner to produce a point of high intensity while waves arriving 180° out of phase interfere destructively and result in a point of zero intensity. The resulting interference pattern is commonly called a diffraction pattern and is divided into two regions. The region closest to the generator is the *near field* or *Fresnel* zone whilst the other region is known as the *far field* or *Fraunhofer* zone. The boundary between these regions occurs at a distance, d from the source which may be determined numerically from the equation:

$$d=\frac{a^2}{\lambda} \qquad [2.2]$$

Where a is the radius of a circular source and λ is the wavelength in that medium. This distance, d, is the point at which a cylindrical beam begins to diverge. If the rays bounding the diverging cone of waves are extrapolated back to their source, they meet at or near the centre of the circular generator (Fig. 2.10).

The complexities of the distribution of energy within this beam have to be examined in greater detail because this pattern is generated within people whenever they are subjected to high frequency ultrasonic irradiation. The wavelength of ultrasound used in physiotherapy is typically about 1·5 mm (at 1 MHz) in water or tissue; this is small compared with the dimensions of a typical therapeutic applicator which is about 25 mm or more in diameter.

A pressure-sensitive detector whose dimensions are smaller than a wavelength can measure the pressure produced at any given point in a medium irrespective of the pressures at adjacent points. Imagine one of these detectors placed at the centre of a large circular source and slowly moved away from it along its axis (its axis is defined as that line at right angles to the flat face of the source which passes through its centre). The output of the detector will be found to give a repeating pattern of peaks and troughs which resemble that of Fig. 2.11. The peaks represent those points where most of the individual wavelets from all points across the surface of the source have arrived with approximately the same phase and have therefore interfered constructively to give a pressure maximum. Conversely, the troughs represent those points where almost half of the wavelets have arrived approximately 180° out of phase relative to the rest and have undergone destructive interference. It should be noted that the

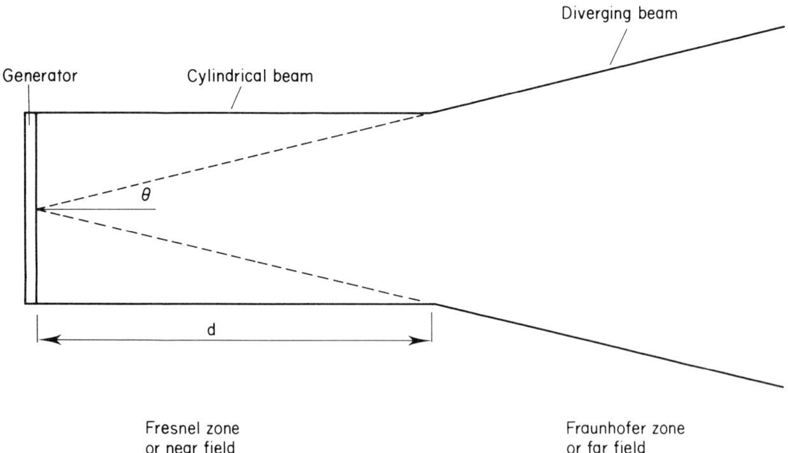

FIG. 2.10 A diagrammatic representation of the cylindrical (Fresnel) and diverging (Fraunhofer) portions of the acoustic beam produced by a large transducer similar to those used in physiotherapy.

geometry is so complex that it is virtually impossible to get complete constructive or destructive interference, hence the pressure minima never quite reach zero.

Close to the source the peaks are very sharp and close together (about one wavelength apart). The peaks become broader and further apart as you move away from the source until you get one last peak which is usually higher and wider than all the rest. This last peak (C in Fig. 2.11) is at the point, d, described above and marks the beginning of the Fraunhofer zone or the far field. It is commonly called the point of the last axial maximum. This is usually the most uniform region of

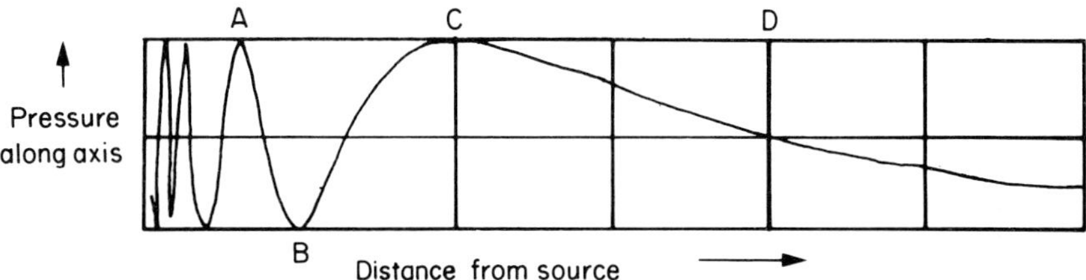

FIG. 2.11 A diagrammatic representation of the distribution of acoustic pressure along the central axis of the transducer shown in Fig. 2.10. The letters A to D refer to specific positions within the beam.

high intensity within that beam and is therefore the point frequently chosen for bioeffects work where you wish to expose a small sample of biological material under the most "ideal" physical conditions (i.e. if the sample is smaller than the dimensions of the peak, then all parts of the sample are subjected to approximately the same acoustic intensity). Similarly, no structures having dimensions greater than a wavelength can be subjected to a uniform acoustic field close to the surface of the generator, i.e. within the near field.

This distribution of acoustic intensity becomes even more complex when we consider points along lines parallel to the axis. In the near field the pattern is very similar except that troughs are usually found alongside the peaks on the axis, and peaks are found alongside the troughs on the axis. This pattern can be seen more clearly if we scan our small detector in a plane parallel with the surface of the source so that it cuts the axis at right angles (Fig. 2.12). That trace which cuts the axis at point A (in Fig. 2.11) has a central peak and two (or more) side lobes with intervening troughs. The trace which transects the axis at point B on the other hand has a central minimum and two (or more) peaks which occupy the positions of the troughs in the scan through point A. This situation is repeated at every axial maximum and minimum in the near field, but the amount of energy contained in the side lobes decreases as you get closer to the far field until only a single axial peak is usually found in the plot of the last axial maximum (scan through point C in Fig. 2.12). Beyond this point the height of the axial peak decreases as its base broadens out as the beam diverges (scan through point D in Fig. 2.12).

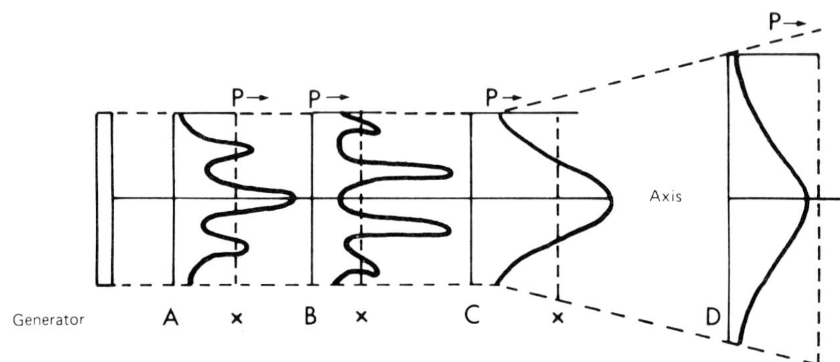

FIG. 2.12 A diagrammatic representation of the distribution of acoustic pressure across the diameter of the sound beam at the four positions (A to D) indicated in Fig. 2.11. Each individual pressure distribution profile is known as a beam plot. P represents the magnitude of the acoustic pressure and the broken line (x) indicates one specific value of that pressure.

There are several alternative ways of presenting this same information. One is to join up all points of equal intensity or pressure (this is exactly analogous to the "contour" maps of lines of equal pressure which are used to display the weather forecast). If we choose an arbitrary acoustic pressure (x) in Fig. 2.12, then we are only concerned with those points where the line of the trace crosses the broken line representing this pressure. However, Fig. 2.12 shows that each peak is three-dimensional and that each peak in Figs 2.11 and 2.12 should be regarded as the tip of a mountain, and the broken line as a plane which slices their tops off. The edges of the cut surfaces would then give circular or eliptical outlines of regions of equal power density.

Several values of x may be chosen and the resulting contours displayed together in a single diagram. Figure 2.13 presents an idealized and simplified situation where only three values of pressure have been chosen. At the lower pressure, i.e. the one which gives the largest contours, the shapes of the individual peaks have largely been lost. Nevertheless this method of presentation is most useful for

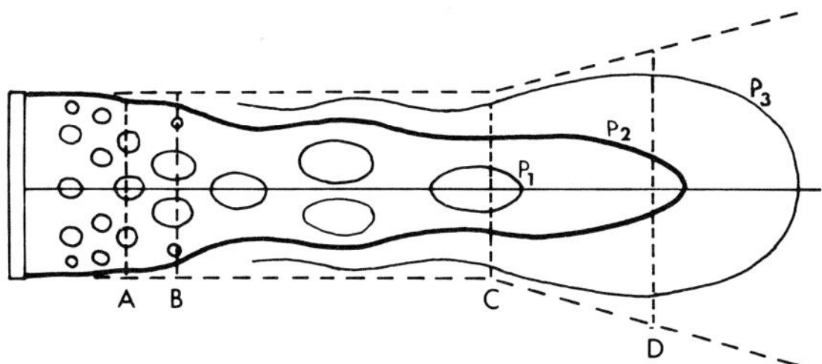

FIG. 2.13 A contour map display of the distribution of acoustic pressure across a diameter of the acoustic beam. P1 represents the highest pressure amplitude, P2 an intermediate value and P3 the lowest pressure amplitude.

showing up the "hot spots" or regions of maximum acoustic intensity within the field.

The acoustic field generated by a circular source has circular symmetry. The true three-dimensional distribution of intensity can therefore be visualized if the two-dimensional contour map of Fig. 2.13 (which is a plot along a diagonal of the beam) is spun about its long axis. Thus, the circular or eliptical "hot spots" on the axis become spheres or solid elipses which are surrounded by "shells" of decreasing and/or increasing intensity. The side lobes of Fig. 2.12 are therefore seen as toroidal bands of high intensity encircling the beam the way a chunky bracelet encircles the arm. Sections through this three-dimensional pattern are clearly seen in the so-called "ring patterns" which are obtained when a two-dimensional sheet containing some material which is sensitive to some parameter of the acoustic field is placed in a plane parallel with the transducer surface. Figure 2.14 shows the two-dimensional distribution of intensity at the same four points (A, B, C and D) from Fig. 2.11. The regions of highest intensity are indicated by the closest line spacing. The transverse plots of pressure amplitude displayed in Fig. 2.12 can be seen to be one-dimensional scans along a diagonal of each pattern.

2.3.2 Ultrasonic Dosimetry

The previous section showed that the ultrasonic intensity varied from point to point within the beam. However, if that beam is being moved evenly and repeatedly through a given organ or structure, then that portion of the body receives a time averaged "dose" which is a function of the amount of ultrasonic power being emitted by that transducer and the duration of the exposure. It is therefore important to be able to measure the quantity of ultrasonic energy emitted by a given transducer for any given driving voltage, and to have some idea

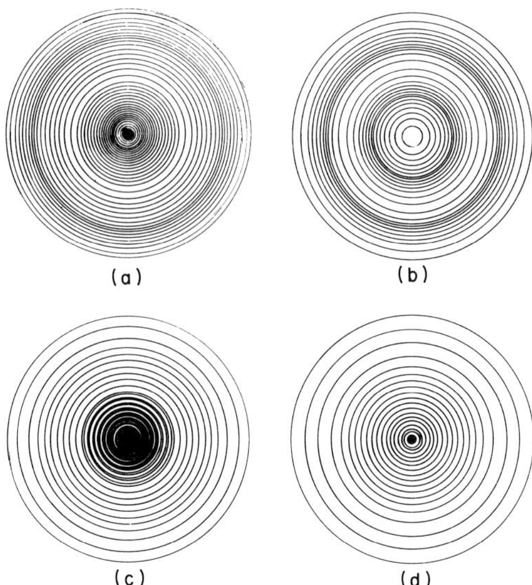

FIG. 2.14 Idealized "ring patterns" (i.e. two-dimensional "grey scale" distributions of acoustic pressure amplitude) from each of the four points A to D indicated in Fig. 2.11.

of the size and position of any "hot spots" within that beam. Thus, the numerous dosimetric techniques in common use may roughly be divided into two major groups, i.e. those techniques which attempt to quantify the total amount of ultrasonic power emitted by the transducer and those techniques which attempt to display the inherent inhomogeneities of the field.

The heterogeneous nature of the acoustic field within any one beam means that the energy distribution within that beam can be expressed in many different ways. Unfortunately, the biological information presently available does not single out any one method of presentation as being superior to another. There is general agreement that the magnitude of any biological effect is proportional to the power density or the intensity at that site. Thus, it is common practice to quote some measure of intensity (usually expressed in terms of W/cm^2 or, less commonly, W/m^2). The simplest approach is to measure the total amount of radiated energy by one technique and to divide this by the total area of the beam, obtained by some other technique. This yields the *spatial average* intensity (Ia) over the whole beam. Unfortunately, the ultrasonic beam does not have a sharp edge—its intensity gradually decreases to near zero—and so it is usually difficult to decide where the limits of the beam are. Also, the width of the beam is not constant, especially in the far field, and so the average intensity is different at different positions along the beam even when absorption can be neglected.

Many authors avoid these complications by measuring the total amount of emitted ultrasonic power and dividing it by the area of the transducer. However, under certain circumstances this can seriously underestimate the average intensity because some parts of the transducer may not be emitting ultrasound. The most common reasons for this are a fracture of the transducer or its conducting electrodes, a rigid clamping system which damps out certain parts of the transducer or an unusual arrangement of the conducting electrodes which may cause a single transducer to behave as though it was composed of a number of separate transducers.

In general, the most intense "hot spot" within the ultrasonic field generated by a plane, unfocused transducer is at the last axial maximum at the beginning of the far field (see Fig. 2.11). It is difficult to ascribe a single value to the *peak* intensity (Ip) at this site, but as a rule of thumb this value is taken to be at least three to six times the average intensity at that same site.

Additional complications occur if the transducer output is pulsed. In this case the peak and average intensities within each pulse are the same as they would be if that transducer was being driven in a continuous mode [these values are consequently called the spatial peak temporal peak intensity (S.P.T.P.) and the spatially averaged temporal peak intensity, (S.A.T.P.) respectively]. If one now allows for the "dead time" while the transducer is not emitting ultrasound, then these same values become the spatial peak time averaged intensity (S.P.T.A.) and the spatially averaged time averaged intensity (S.A.T.A.) respectively. For example, a therapeutic transducer may be driven in a continuous mode so that its spatially averaged output intensity is 1 W/cm^2. Its spatial peak intensity would then be about 4 W/cm^2. If that same driving voltage is now chopped up into pulses and delivered so that the interval between the pulses is nine times the length of each pulse, i.e. pulsed 1 : 9, then the spatial peak temporal peak intensity and the spatially averaged temporal peak intensity stay the same at 4 and 1 W/cm^2 respectively, while the time averaged spatial peak intensity and the time averaged spatially averaged intensity drop to 0·4 and 0·1 W/cm^2 respectively.

Wells (1977) presents a comprehensive account of the various physical parameters which can be measured and used to calculate the intensity distribution within an ultrasonic beam. The most commonly employed techniques are:

2.3.3 Hydrophones

These are usually piezoelectric transducers which are small compared with the dimensions of a wavelength and have resonant frequencies much greater than the source being investigated. They are sensitive to the pressure produced by the wave and are commonly used for beam plots (e.g. Figs 2.11 and 2.12). They can be used for the absolute determination of the intensity at a point but are more commonly used to determine the ratio of peak to average intensity and the size and position of "hot spots" within the beam.

2.3.4 Absorption Techniques

That portion of the energy of a wave which is absorbed by a medium is converted into heat which raises the temperature of that medium. Figure 2.15 presents a simplified version of a calorimeter which could be used to measure the total acoustic power output from a transducer. The transducer is immersed in a circulating water bath so that any heat resulting from inefficiencies in the transduction process is removed. The ultrasonic beam then enters a chamber containing a material such as castor oil or carbon tetrachloride which absorbs the sound and the resulting temperature increase is detected by means of sensitive thermocouples (Wells *et al.*, 1963).

Thermocouple junctions contained within a small volume of absorbing material were devised by Fry and Fry (1954) as point detectors to map out the distribution of ultrasonic energy within a field. They are basically small calorimeters whereby the absorbed energy warms up the medium around the thermocouple junction. They can be used for the determination of the absolute intensity at a

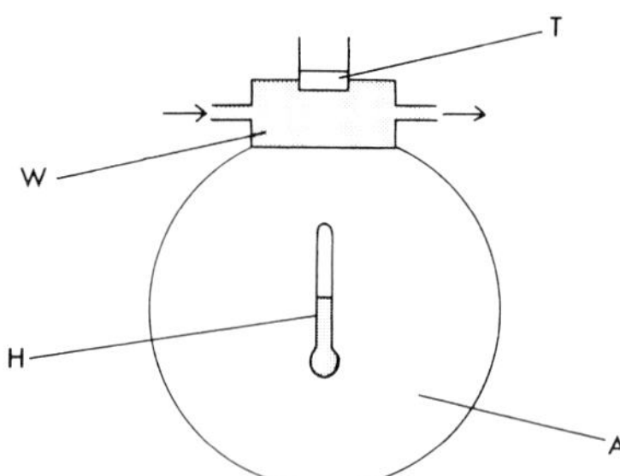

FIG. 2.15 An idealized calorimeter to measure the total amount of acoustic power emitted by a transducer (T). The ultrasound enters the thermally insulated bath of absorber (A) after passing through a non-absorbing water jacket (W) to remove any heat produced during the transduction process. The rise in temperature of the absorber is measured by means of a thermometer, thermistor or thermocouple (H).

point, but are more commonly used for beam plots. Yoshioka and Oka (1965) replaced the liquid absorbing medium by a solid slab of polyethylene. Biological tissues also absorb some of the ultrasound passing through them and so a thermocouple wire may be threaded directly through a tissue in the path of an ultrasonic beam. This gives the temperature rise produced at the site of the junction and, if the absorption coefficient of that tissue is known for that particular frequency and temperature, the ultrasonic intensity at that site may be calculated.

Thermistors are semiconductor materials whose resistance varies with temperature. The thermistor bead is therefore coated or embedded in a small volume of an absorber, e.g. polystyrene, and connected to a sensitive circuit which measures changes in its resistance. They are usually used for the same purposes as thermocouples, but the larger size of the thermistor bead relative to the thermocouple junction means that they cause a larger perturbation of the ultrasonic field.

The thermal and mechanical stresses generated within a solid block of a homogeneous transparent plastic (e.g. polymethylmethacrylate, commonly called Perspex[R] or Plexiglass[R]) cause permanent alterations which are easily seen if that block is viewed with polarized light (Lele, 1962). This technique is particularly useful for measuring the changes in the dimensions of the focal volume of a focused transducer as a function of the output power and time of exposure.

The distribution of ultrasonic intensity within the beam may also be demonstrated in a qualitative manner by means of cholesteric liquid crystals. These are liquid crystals which exhibit reversible irridescent colour effects at different temperatures. Thus, sheets of a solid material having a fairly high absorption coefficient, e.g. polyethylene, containing these crystals will produce coloured ring patterns similar to those presented in Fig. 2.14 when they are placed in front of a plane transducer (Cook and Wercham, 1971). However, these patterns do not persist for more than about 10 s before their definition is lost due to lateral heat flow.

Similar transient pictures may be obtained if the ultrasonic beam impinges normally on a slab of this same absorbing material placed in front of a thermographic camera which detects the temperature rise at various points within that slab.

2.3.5 Radiation Pressure Effects

Any medium or object in the path of an ultrasonic beam which absorbs some or all of the energy of that beam is subjected to a steady force called the radiation pressure force which tends to push that material in the direction that the wave is propagating, i.e. away from the transducer. The surface of any absorbing or reflecting target in the path of an acoustic beam must be performing a microscopic oscillation driven by the oscillations of the adjacent liquid medium. Because this surface is not stationary, the time-averaged value of the acoustic pressure at that surface is also non-zero. The resulting steady pressure on that surface (multiplied by the exposed area) is called the "radiation force". The magnitude of this force is proportional to the rate of change of intensity with distance along the beam and is therefore greatest in a strongly absorbing medium. Consequently, any liquid which is not lossless in front of a transducer tends to be pushed away from it so that a continuous circulation is established. This steady circulation is clearly seen in a tank of water if it contains some suspended particles. This circulation is commonly called "quartz wind streaming" and is illustrated in a graphical manner in Fig. 2.16.

Sjoberg *et al.* (1963) devised an apparatus whereby the magnitude of the quartz wind streaming was used as an index of the total power output emitted by a transducer. Figure 2.17 shows that the streaming was directed into a vertical conical tube (C) so that it elevated a spherical plastic bead (S). Within the conical section, the velocity of the streaming liquid is greatest in the narrow portion and least within the wide portion. Thus, the plastic sphere is pushed up until the upward force due to the motion of the liquid is equal to the downward force due to gravity. Unfortunately, the accuracy of this device is limited by a number of factors, and the output scale has to be calibrated by means of an external standard.

If a moderately intense beam of ultrasound is directed upwards towards the flat surface of a liquid, it is found that one or more portions of the surface are displaced upwards by this same radiation pressure force, i.e. the so-called "fountain effect". The pattern produced is a three-dimensional representation of the acoustic intensity distribution at that interface. The height of each peak is a

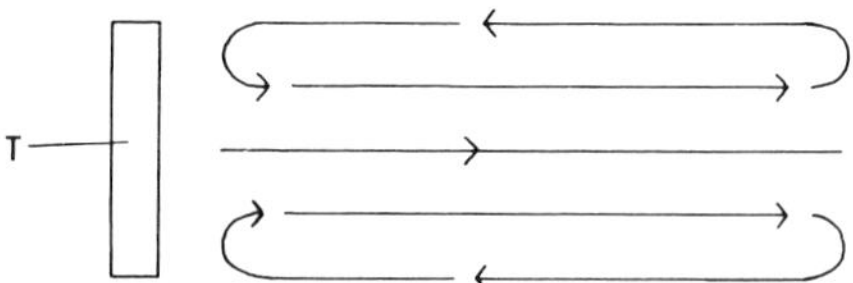

FIG. 2.16 A diagrammatic representation of the "quartz-wind" streaming pattern generated by a plane transducer directed into a large bath of an absorbing medium having a relatively low viscosity.

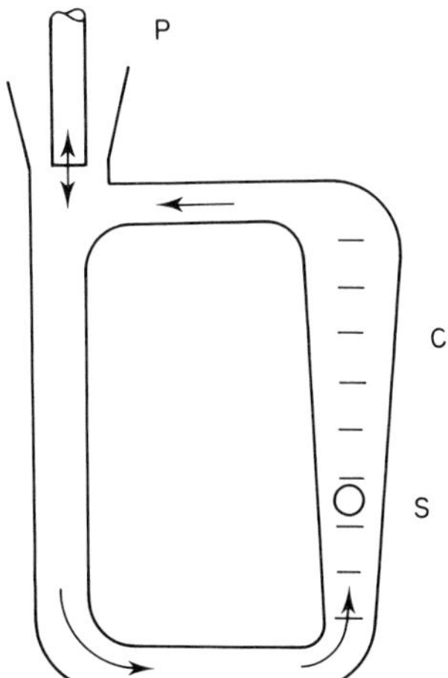

FIG. 2.17 A diagrammatic representation of a device to quantify the magnitude of the quartz-wind streaming generated by a transducer or probe (P). A plastic sphere (S) is elevated within a conical tube (C) until its upward force is just equal to the downward force due to gravity (adapted from Sjoberg *et al.*, 1963).

direct measure of intensity since the column of water rises until the upward force is exactly equal to the weight of water in that column. This phenomenon is more commonly used as a check that a transducer is emitting ultrasound or as an aid in the fine tuning of the electrical drive circuit than as a quantitative measure of intensity since it is difficult to measure the heights of the peaks with precision.

In the case of complete absorption of the ultrasonic wave, the radiation pressure force (F) is given by:

$$F \propto \frac{W}{C} \qquad [2.3]$$

where W is the ultrasonic power and C is the velocity of sound in that medium. The majority of the devices in common use for the measurement of ultrasonic power employ this principle even though the experimental arrangements appear to be quite unrelated.

Figure 2.18 is a schematic diagram describing the most straightforward approach which is eminently suitable for unfocused transducers emitting therapeutic intensities of ultrasound. A block (A) of very efficient absorbing material, e.g. SOAB rubber, is mounted on one arm of a sensitive analytical balance and immersed in a tank of degassed water. The beam of ultrasound from the transducer (T) is directed at the absorber through an acoustically transparent screen

(S) which prevents the absorber being displaced by the quartz wind streaming. Weights are added to the counterbalance pan (W) until the pan containing the absorber is restored to the position it occupied before the ultrasound was switched on. From equation 2.3 it can be calculated that a beam having a total power of 1 W in water at 20°C striking a perfect absorber at normal incidence produces a radiation pressure force of $6{\cdot}7 \times 10^{-4}$ Newtons which is equivalent to a measured weight of about 0·068 g.

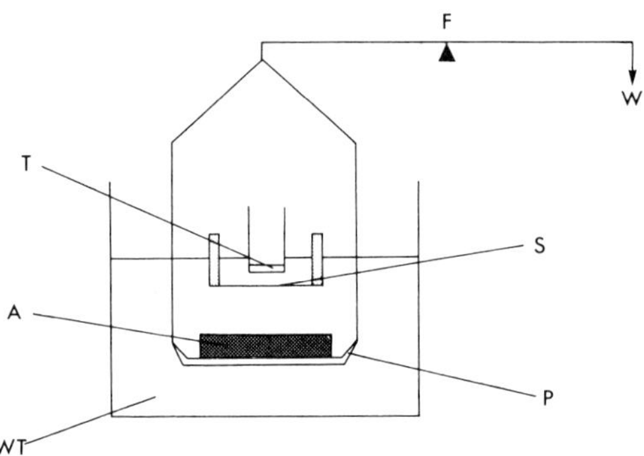

FIG. 2.18 A schematic representation of a typical arrangement to measure the acoustic power output of a transducer by means of its radiation pressure force. The ultrasound emitted by the transducer (T) passes through an acoustically transparent screen (S) to eliminate the quartz-wind streaming before being absorbed by the perfect absorber (A). The resulting force displaces the balance pan (P) within the water bath (WT) so that weights (W) have to be added to the counterbalance pan to restore the pan (P) to its original position.

If the absorber is large compared with the dimensions of the ultrasonic beam, then this device measures the total power emitted by the transducer. This value has to be divided by the cross-sectional area of the beam (obtained by some other method) to obtain the spatially averaged intensity of that beam. One procedure for estimating the area of the beam is to direct it through a series of cones of decreasing radius; these act as apertures and decrease the area of the beam which reaches the absorber. Choose a cone which allows 90% of the total amount of energy emitted by the transducer to reach the absorber. The cross-sectional area of the unshielded beam may then be taken to be 1·1 times the area of that cone.

If the absorber is not perfect, then one obtains the surprising result that the measured radiation pressures are bigger than those obtained with a perfect absorber. This is because the transport of energy by a wave is associated with a flow of momentum across the plane normal to the ultrasonic beam. At a perfect absorber there is a single momentum change but at a perfect reflector there are two momentum changes of equal magnitude because the incident wave is returned unattenuated (Wells, 1977). Thus, the radiation pressure force measured with a perfectly reflecting target is twice that obtained with a perfectly absorbing target. However, the reflecting target must be inclined or coned to deflect the ultrasonic beam into an absorber otherwise it would bounce off the transducer or other surfaces in the tank and strike the reflector again giving a falsely high value.

The radiation float assembly depicted in Fig. 2.19 is also commonly used to measure the total ultrasonic power emitted by therapeutic transducers. An air-filled thin metal cone or float (F) is attached to a metal stem (P) and weighted so that it just sinks in water. If the lower portion of the stem is immersed in a dense liquid, e.g. carbon tetrachloride (D), it is subjected to an upthrust such that the float will come to rest at a given height. If ultrasound is directed vertically downwards through a screen (S) to eliminate the quartz wind streaming, the radiation pressure force pushes the stem further into the dense liquid until it is exactly counterbalanced by the increased buoyant upthrust and a new equilibrium position is obtained. The position of the float is usually measured by means of a travelling microscope and the displacements are converted into units of weight by calibrating the device with known weights. The upper surface of the cone is usually manufactured with its apex pointing downwards so that it is self-centering in the ultrasonic beam (Kossoff, 1962).

Radiation pressure devices are also used to measure the ultrasonic power emitted by diagnostic equipment, i.e. having time and space

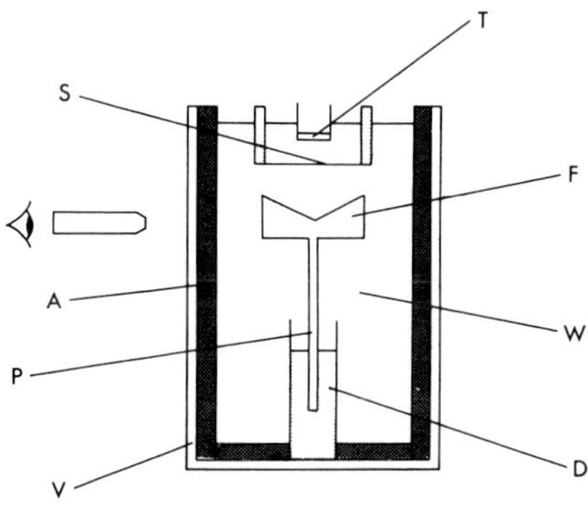

FIG. 2.19 A schematic description of a radiation float assembly which also measures the total acoustic power output of a transducer (T) by means of its radiation pressure force. For details see text.

averaged intensities of the order of 1·5 to 60 mW/cm^2 (Brendel and Ludwig, 1982). However, the displacements produced at these intensities are so small that elaborate precautions have to be taken to ensure that the measurements are not obscured by random fluctuations or damped out by too rigid suspension systems. Consequently, numerous ingenious arrangements have been devised whereby a reflector is suspended on wires and deflected horizontally (Wells *et al.*, 1964) or else the radiation force is measured by means of a modified analytical balance (Kossoff, 1965; Hill, 1970; Rooney, 1973).

The radiation pressure force exerted on a small metal sphere can also be used to estimate the acoustic intensity at any point within a beam. A stainless steel ball bearing about 2–4 mm in diameter is usually suspended from two thin nylon threads which are arranged so that the sphere may be deflected away from the transducer but cannot be deflected from side to side. The ball bearing is viewed from one side and its displacement measured with a travelling microscope. This device yields the intensity averaged over the area of the sphere (Hasegawa and Yoshioka, 1969, 1975; Dunn *et al.*, 1977).

2.3.6 Capacitance Techniques

A capacitance or a condensor essentially consists of two metal plates separated by an insulator such as air. If one of these plates is fixed and the other is set into vibration by an incident ultrasonic wave, it is periodically brought closer to and then taken further away from the other plate. Thus, the capacitance of the device is continually changing and this may be detected by a suitable electronic circuit. These are usually large devices which are used to measure the average intensity of the whole beam (Kolsky, 1956), but an alternative arrangement has been developed by Filipczyński and Lypacewicz (1972) for measuring the displacement profiles of transducer surfaces.

2.3.7 Optical Methods

The molecules of the medium through which a sound beam is propagating are alternately compressed together and then pulled apart. This results in small local variations in the density of that medium which in turn causes a time dependent variation in its refractive index. Thus, that portion of a beam of light which is passing through a sound beam is deflected by an amount which is proportional to the intensity of that acoustic field.

Schlieren systems utilize this property to produce a visual display of the shape of the ultrasonic beam passing through an optically transparent medium such as water. Figure 2.20 describes one of the simplest optical arrangements. Lenses or concave mirrors are used to project the light from a monochromatic source as a broad uniform beam through a tank of water. This beam is then brought to a sharp focus. If a small opaque object, i.e. the stop, is placed at this point then no light falls on the screen. If the beam from a therapeutic transducer is propagating through that tank, then the light passing through it is refracted and brought to a focus at an adjacent point so that it no longer falls on the stop and therefore strikes the screen forming a bright image of the beam against a dark background (Fig. 2.21). An alternative method of presenting this same information is to replace the small stop with an aperture of the same size in a large opaque card. The ultrasound beam is then seen as a dark shadow against a bright background.

2.3.8 Chemical Methods

The temperature rise caused by the absorption of an ultrasonic beam speeds up many chemical reactions. At high intensities ultrasonic waves can also initiate certain non-specific chemical reactions such as the production of free iodine from potassium iodide in the presence of small amounts of carbon tetrachloride (see Chapter 4). Thus, ring patterns may be produced on sheets of materials coated with a suitable chemical or enzymic system and arranged so that the

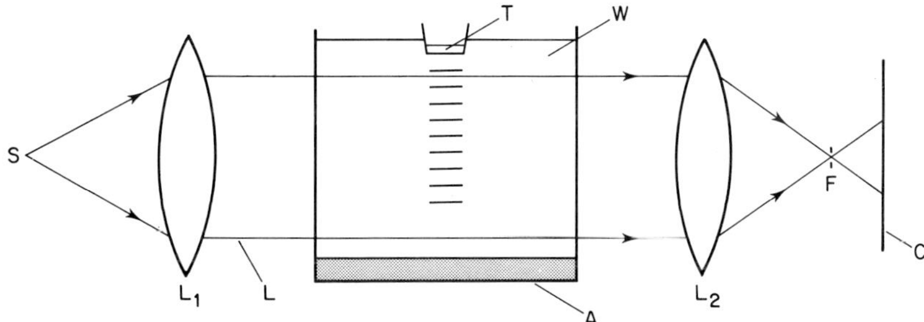

FIG. 2.20 A diagrammatic representation of the Schlieren system for the optical visualization of an acoustic beam. Coherent light from a source (S) is made into a broad parallel beam (L) by the lens (L_1) and it passes through the water bath (W) before being focused by lens (L_2). The ultrasound from the transducer (T) passes through the light beam before being reflected or absorbed by the absorber (A). An aperture or stop (F) is placed at the focus of the light beam as described in the text so that an image of the acoustic beam is formed on the screen (C).

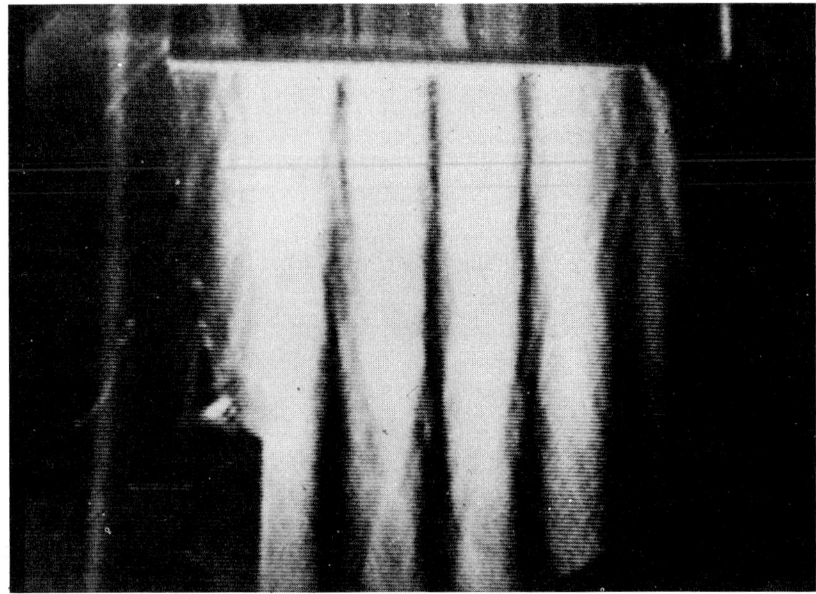

FIG. 2.21 A Schlieren photograph of one large American therapy transducer which had four separate transducer elements to provide what was claimed to be a "more uniform" distribution of intensity (photograph generously supplied by M. E. Haran, Division of Electronic Products, Bureau of Radiological Health, U.S. Food and Drug Administration).

ultrasonic beam passes through them. These techniques are not in common use.

2.3.9 Reciprocity Techniques

The reversible nature of the piezoelectric effect means that a generator may also function as a detector. Carstensen (1947) utilized this effect to enable a transducer to calibrate itself. In essence, a perfect reflector is placed in the far field of a transducer and arranged so that the echo is detected by that same transducer. No knowledge of the internal structure of that transducer is required. This principle is difficult to apply to pulsed transducers, and the newcomer to the field is well advised to stick with the inherently more simple radiation pressure force technique.

2.4 ACOUSTIC OUTPUTS OF COMMERCIAL DEVICES

Well-maintained devices for use in physiotherapy usually deliver from zero up to about 15 or 20 W of acoustic power within the frequency range 0·5–3 MHz. Their transducers are usually unfocused and have radiating areas of the order of 5 cm^2 giving maximum space averaged intensities of up to 3 or 4 W/cm^2. Their spatial peak intensities (i.e. the highest intensity obtained at any point within the beam) are usually about three times the spatial average value and can therefore reach 9–12 W/cm^2. When these devices are operated in their pulsed mode, their time averaged power is decreased, but the spatial average and spatial peak values within each pulse remain the same.

The acoustic power output emitted by diagnostic devices either cannot be altered, or is adjustable only over a small range. Dosimetric measurements have been performed by a number of investigators including Carson *et al.* (1978), Hill (1977) and Rooney (1973) and their results are summarized in Table 2.1. It can be seen that there is an enormously wide variation in acoustic power outputs, especially amongst the pulse-echo devices. This is partly a reflection of the dimensions of the radiating areas of the transducers which vary from a few square millimetres for some ophthalmic and peripheral vascular devices up to about 30 cm^2 for some real time devices. In general, pulse-echo devices generate the highest acoustic intensities (frequently of the order of 100 W/cm^2) during each pulse, but their long "off" time means that their time-averaged power outputs are extremely low. Continuous doppler devices as used to monitor the activity of the foetal heart usually emit more acoustic power and therefore have higher time averaged intensities but they do not generate high peak intensities. Peripheral vascular doppler devices emit similar amounts of acoustic power to the foetal heart monitors, but their radiating areas are usually much smaller so that their space and time averaged intensities are higher and can exceed the space and time averaged intensities recommended for some therapeutic applications of ultrasound. However, it should be noted that the peripheral vascular devices generate these intensities within a small tissue volume at high acoustic frequencies whereas therapy devices irradiate a larger tissue volume at lower frequencies.

2.5 THE CONCEPT OF "DOSE"

When a photon of ionizing radiation such as an X-ray or a γ-ray passes through a tissue, it may loose some or all of its energy. This energy which has been deposited within the tissue causes electrons to be set in motion. These moving electrons now cause the numerous biochemical alterations which manifest themselves as "radiation damage". Thus, the amount of damage produced by a beam of ionizing radiation is proportional to the number and velocity of these moving electrons, which is in turn proportional to the total amount of energy deposited within that tissue. One therefore obtains essentially the same amount of radiation damage (if repair processes can be neglected) from photons having the same energy if the same "dose" is administered in a single session at a high intensity or in numerous long sessions at a low intensity. Many photons do not loose all of their energy and therefore emerge from the tissues. The amount of energy deposited within the tissue is therefore the difference between the total amount of energy which entered the tissue and that which emerged the other side. If 1 kg of tissue has absorbed one joule of energy then it has received a radiation dose of one *gray* (Gy). Similarly, if 1 g of a given tissue has absorbed 100 erg (10^{-5} J) of energy it is said to have received a radiation dose of one rad (radiation absorbed dose) where 100 rads equals 1 gray. The total radiation dose which any tissue or organ has received is then merely the algebraic sum of all the individual exposures.

Even though a similar approach has been advocated for ultrasound by some authors, e.g. Hussey (1975), it does not appear to be suitable for the dosimetry of ultrasound. The major difference is that the energy of the ultrasonic wave which is deposited within a tissue usually does not (as far as is known at present) generate highly energetic and destructive charged species. Instead this energy appears more or less directly as heat. Thus, there does not appear to be a cumulative effect from numerous exposures to low intensities of ultrasound. Instead there is a marked correlation between the magnitude of any observed biological effect and the rate at which ultrasonic energy is being supplied to that organ or structure. In many cases this can be interpreted in terms of the rate and maximum extent of the temperature rise which occurs within that tissue; this will be discussed in more detail in Chapter 3. In addition, numerous mechanical effects occur when the energy density of an ultrasonic wave exceeds a certain "threshold" value (Chapter 4).

It therefore appears that it is the rate at which ultrasonic energy is being supplied to a tissue, i.e. the intensity of that beam, which is one of the major parameters which determines what biological effect is going to result from that exposure. The other major parameter is the time of exposure: this refers both to the overall duration of the exposure and to the way in which the ultrasonic energy is chopped up into pulses or packets. The frequency of the ultrasonic wave should also be stated as this may have a profound effect on the way in which the energy of the ultrasonic wave is converted into the form which results in the observed biological effects.

2.6 HOW DOES ULTRASOUND MODIFY TISSUES?

This question cannot be answered by one simple sweeping statement because many different potentially damaging physical situations are generated simultaneously by the ultrasonic wave. The response of a biological tissue therefore depends upon the relative magnitudes of each of these physical situations in the context of that particular combination of exposure parameters in that exposure situation and the relative susceptibility of that tissue to each individual physical agent.

Ultrasound cannot propagate through a tissue without some of its associated energy being deposited as heat. This heat will result in a rise in the temperature of that tissue if its rate of input exceeds the capacity of that tissue to dissipate it. The plethora of "purely" thermal effects of ultrasound are discussed in more detail in Chapter 3. The pressure changes associated with the propagation of a compressional wave (which is the only wave that can propagate for large distances through soft tissues) can under certain circumstances result in one or more forms of acoustic cavitation. This is a general term describing the processes whereby gaseous or vaprous bubbles or voids are created and forced into oscillation with the concomitant generation of large mechanical shear stresses which can easily disrupt biological material. This complex and in many cases unpredictable behaviour is discussed in Chapter 4. Chapter 5 describes some of the other largely non-thermal and non-cavitational forces which may also be shown to modify biological tissues, e.g. radiation pressure forces, and describes the results of those "near threshold" experiments where no single mechanism has been shown to dominate.

However, it should be noted that all of these potentially damaging physical situations co-exist together and therefore may act in concert to produce any given observed biological effect. Thus, it may be possible to demonstrate a thermal effect in the absence of cavitation, but it is virtually impossible under normal *in vivo* irradiation conditions to demonstrate a cavitational effect in the absence of a temperature rise. This temperature rise will undoubtedly "weaken" that tissue and exacerbate the destructive mechanical effects of the cavitation.

It is of some comfort to both users and recipients of ultrasound that most of the repeatable biological effects which have so far been reported indicate that ultrasonically damaged cells either recover, apparently completely, or die. A limited number of dead cells pose no serious threat to the life and well-being of a large organism such as man. The obvious exceptions to this statement are those conductive, regulatory or reproductive cells which, if they are preferentially damaged, can produce an effect out of all proportion to the number of cells which have been damaged. This situation compares favourably with that found in ionizing radiation where there appears to be an infinite range of "damage levels" affecting both the cytoplasm and the nuclear contents. Cells which are unable to repair their genetic damage before they are called upon to divide may form a tumour. This appears to be a much less probable event in cells damaged by ultrasound.

Conversely, we do not have enough information to be able to state categorically that ultrasound is "completely safe" even when used at extremely low power levels. This topic is discussed in more detail in Chapter 6 after some of the relevant bioacoustic information has been presented. The real problem is to identify which combination of physical parameters produces an environment which "damages" biological tissues so that its cells behave in an abnormal manner and so threaten the life or well-being of the entire organism.

The physical forces generated by ultrasound which damage cells are, in general, quite different to those generated by ionizing radiation. It is therefore unfair to use the same biological test systems which have been developed to assess the hazard associated with exposure to X-rays to assess the equivalent hazard of ultrasound. This automatically biases the result in favour of (and virtually guarantees) a fallacious verdict of "completely safe". The only fair method of comparison is to devise in an independent manner the most sensitive biological assay which is best suited to each modality and to use it to assess their respective potential hazards. It is generally agreed that the potential hazards arising from the use of ultrasonic irradiation in medical diagnosis are small compared with the corresponding hazards of X-rays. Nevertheless, there is sufficient cause for concern so that ultrasonic examinations should not be performed *ad libitum* (see Chapter 6).

REFERENCES

Balamuth, L. (1963). Ultrasonics and dentistry. *Sound* **2**, 15–19.

Bom, N., Lancée, C. T., Honkoop, J. and Hugenholtz, P. G. (1971). Ultrasonic viewer for cross-sectional analysis of moving cardiac structures. *Bio-med. Eng.* **6**, 500–503.

Brendel, K. and Ludwig, G. (1982). Diagnostic intensities and their measurement. *In* "Recent Advances in Ultrasound Diagnosis 3", (Eds A. Kurjak and A. Kratochwil), pp. 76–80. Excerpta Medica, Amsterdam.

Brown, G. (1937). The vortex motion causing edge tones. *Proc. Phys. Soc. Lond.* **49**, 493.

Carson, P. L., Fischella, P. R. and Oughton, T. V. (1978). Ultrasonic power and intensities produced by diagnostic ultrasound equipment. *Ultrasound Med. Biol.* **3**, 341–350.

Carstensen, E. L. (1947). Self-reciprocity calibration of electro-acoustic transducer. *J. Acoust. Soc. Amer.* **19**, 961–965.

Cook, B. D. and Wercham, R. E. (1971). Mapping ultrasonic fields with cholesteric liquid crystals. *Ultrasonics* **9**, 101–102.

Donald, I. and Abdulla, U. (1967). Further advances in ultrasound diagnosis. *Ultrasonics* **5**, 8–12.

Dunn, F., Averbuch, A. J. and O'Brien, W. D. (1977). A primary method for the determination of ultrasonic intensity with the elastic sphere radiometer. *Acustica* **38**, 58–61.

Evans, E. F. (1968). "Cortical Representation in Hearing Mechanisms in Invertebrates". Ciba Foundation Symposium. J. and A. Churchill, London.

Filipczyński, L. and Lypacewicz, G. (1972). Vibration patterns and properties of piezoelectric ceramic transducers for diagnostic application in medicine. *In* "Ultrasonics in Biology and Medicine", (Ed. L. Filipczyński), pp. 81–89. Polish Scientific Publishers, Warsaw.

Fry, W. J. and Fry, R. B. (1954). Determination of the absolute sound level and acoustic absorption coefficients by thermocouple probes-theory. *J. Acoust. Soc. Amer.* **26**, 294–310.

Galton, F. (1883). Hydrogen whistles. *Nature* **27**, 491.

Griffin, D. R. (1946). Supersonic cries of bats. *Nature (Lond.)* **158**, 46.

Griffin, D. R. (1958). "Listening in the Dark", pp. 104–111. Yale University Press, New Haven, Connecticut.

Hammer, C. E. and Whitlow, A. L. An. (1980). Porpoise echo-recognition: an analysis of controlling target characteristics. *J. Acoust. Soc. Amer.* **68**, 1285–1293.

Hasegawa, T. and Yoshioka, K. (1969). Acoustic radiation force on a solid elastic sphere. *J. Acoust. Soc. Amer.* **46**, 1139–1143.

Hasegawa, T. and Yoshioka, K. (1975). Acoustic radiation force on fused silica sphere, and intensity determination. *J. Acoust. Soc. Amer.* **58**, 581–585.

Hill, C. R. (1970). Calibration of ultrasonic beams for biomedical applications. *Phys. Med. Biol.* **15**, 241–248.

Hill, C. R. (1977). Chapter on ultrasound. *In* "Manual on Health Aspects of Exposure to Non-Ionizing Radiation", World Health Organisation (Regional Office for Europe) ICP/CEP 803.

Hussey, M. (1975). "Diagnostic Ultrasound". Blackie and Sons, Glasgow and London.

Jaffe, B., Roth, R. S. and Marzullo, S. (1955). Properties of piezoelectric ceramics in solid-solution series lead titanate–lead zirconate–lead oxide; tin oxide and lead titanate–lead hafnate. *J. Res. Natn Bur. Stand.* **55**, 239–254.

Kolsky, H. (1956). The propagation of stress pulses in viscoelastic solids. *Phil. Mag.* **1**, 693–710.

Kossoff, G. (1962). Calibration of ultrasonic therapy equipment. *Acustica* **12**, 84–90.

Kossoff, G. (1965). Balance technique for the measurement of very low ultrasonic power outputs. *J. Acoust. Soc. Amer.* **38**, 880–881.

Kossoff, G., Robinson, D. E. and Garrett, W. J. (1968). Ultrasonic two-dimensional visualisation for medical diagnosis. *J. Acoust. Soc. Amer.* **44**, 1310–1318.

Kurjak, A. and Kratochwil, A. (1982) (Eds). "Recent Advances in Ultrasound Diagnosis—3". Excerpta Medica, Amsterdam, International Congress Series 553.

Lele, P. P. (1962). Irradiation of plastics with focussed ultrasound: a simple method for evaluation of dosage factors for neurological applications. *J. Acoust. Soc. Amer.* **34**, 412–420.

Novack, A. (1958). *J. Exp. Zool.* **138**, 81–154.

Pierce, G. W. and Griffin, D. R. (1938). Experimental determination of supersonic notes emitted by bats. *J. Mammal.* **19**, 454.

Pye, J. D. (1979). Why ultrasound? *Endeavour* (New Series) **3**, 57–62.

Rooney, J. A. (1973). Determination of acoustic power in the microwatt–milliwatt range. *Ultrasound Med. Biol.* **1**, 13–16.

Schneider, H. and Möhres, F. P. (1960). Die Ohrbewegungen der Hufeisenfledermäuse (Chiroptera-Rhinolophidae), und der Mechanismus des Bildörens. *Zeit. für vergleichende Physiol.* **44**, 1.

Sewell, G. D. (1970). Ultrasonic signals from rodents. *Ultrasonics* **8**, 26–30.

Sjoberg, A., Stahle, J., Johnson, S. and Sahl, R. (1963). Treatment of Menière's disease by ultrasonic irradiation. *Acta Otolaryngol. Suppl.* 178.

Stumpff, U., Pohlman, R. and Trübestein, G. (1975). A new method to cure thrombi by ultrasonic cavitation. Proc. Ultrasonics International 1975, pp. 273–275. IPC Science and Technology Press.

Summer, W. and Patrick, M. K. (1964). "Ultrasonic therapy—A Textbook for Physiotherapists". Elsevier, London.

Suppipat, N. (1974). Ultrasonics in periodontics. *J. Clin. Periodontol.* **1**, 206–213.

Wehner, W., Müller, Th. and Neumann, A. (1980). Results of bone welding. Paper E21 at the Ultrasound Interaction in Biology and Medicine Symposium, Reinhardsbrunn, East Germany, Nov. 10–14.

Wells, P. N. T. (1977). "Biomedical Ultrasonics". Academic Press, London, New York, San Francisco.

Wells, P. N. T., Bullen, M. A., Follett, D. H., Freundlich, H. F. and James, J. A. (1963). The dosimetry of small ultrasonic beams. *Ultrasonics* **1**, 106–110.

Wells, P. N. T., Bullen, M. A. and Freundlich, H. F. (1964). Milliwatt ultrasonic radiometry. *Ultrasonics* **2**, 124–128.

Yoshioka, K. and Oka, M. (1965). Technical developments of focussed ultrasound and its biological and surgical applications in Japan. *In* "Ultrasonic Energy" (Ed. E. Kelly). University of Illinois Press, Urbana.

3. THERMAL EFFECTS OF ULTRASOUND

3.1 INTRODUCTION

Tissues cannot transmit ultrasonic energy with 100% efficiency. Some of the transmitted energy is therefore removed from the ultrasonic beam and appears as heat. The amount of heat deposited within a given tissue depends upon the absorption characteristics of that tissue and the amount of ultrasonic energy passing through it. The averâge acoustic power emitted by diagnostic equipment is so low that one would not expect significant temperature rises within the irradiated tissues. However, one would expect temperature rises of the order of a few degrees centigrade at the power levels emitted by ultrasonic therapy devices; this is believed to be one of the major mechanisms by which ultrasound exerts its therapeutic effects. Even higher temperature rises are produced at higher power densities and these can destroy the irradiated tissues; this is one of the major mechanisms enabling the use of ultrasound as a surgical technique for the production of trackless lesions deep within the brain (Meyers *et al.*, 1960) or for the treatment of Meniére's disease (Wells, 1977).

3.2 ATTENUATION, SCATTERING AND ABSORPTION

Imagine a uniform cylindrical beam of ultrasound passing through a large homogeneous tissue. If that tissue is a perfect conductor of ultrasound, then there is no loss of energy and the average intensity at any point along the axis of that beam remains constant. In practice this is found not to be the case; the intensity along the beam axis is found to decrease in an exponential manner as a function of the depth of tissue penetration (Fig. 3.1). The fact that the curve is exponential indicates that each unit portion of that tissue is removing the same fraction or percentage of the energy incident upon it. For example, imagine the tissue being composed of many layers each one centimetre thick and each layer absorbing say 10% of the incident energy. If the transducer emitted 100 units of energy, then the first centimetre of tissue would absorb 10 units of energy which leaves 90 units of energy to enter the second layer. This layer therefore absorbs 9 units of energy leaving 81 units of energy to enter the third layer which absorbs 8·1 units of energy; and so on. Thus, the amount of energy deposited within each layer of the homogeneous tissue decreases with the depth of penetration of the beam into that tissue. Since it is the absorbed energy which is converted into heat to warm up that tissue it follows that the temperature increase produced by the uniform ultrasonic beam will progressively decrease as you proceed further into that homogeneous tissue.

The exponential curve presented in Fig. 3.1 may be expressed mathematically in the form:

$$Ix = I_0 e^{-2Ax} \qquad \text{[Eq. 3.1]}$$

Where Ix is the ultrasonic intensity at any point a distance x from the transducer, I_0 is the initial intensity at the transducer surface and A is the amplitude attenuation coefficient. This amplitude attenu-

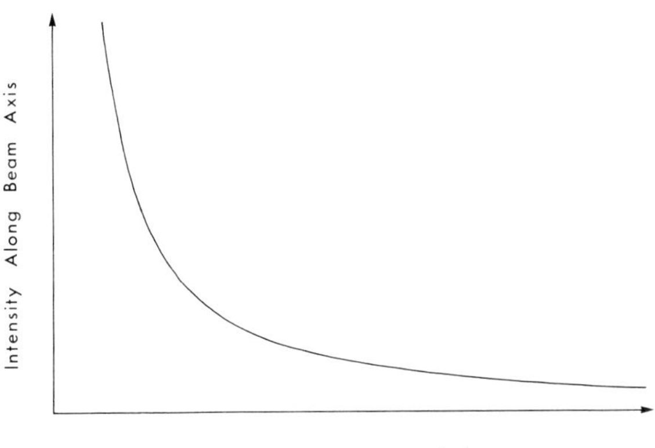

FIG. 3.1 A graphical representation of the exponential decrease in the intensity of a uniform cylindrical beam of ultrasound propagating through a homogeneous absorbing medium. The distance travelled within that medium before the intensity of the beam has been reduced to one half of its original value is known as the half intensity depth.

ation coefficient (A) is composed of two separate parts; one is the amplitude absorption coefficient (α) and the other is a measure of the scattering (s).

It has already been shown (Chapter 1) that part of the energy of an ultrasonic beam is reflected when that beam encounters an interface bounded by materials having different acoustic impedances. If the dimensions of this reflector are large compared with the wavelength of the sound and the dimension of the beam then it is called a specular reflector and a plane wave is reflected as a plane wave. However, if the dimensions of the reflector are smaller than a wavelength then the reflector acts as a point source and radiates a spherical wave (Fig. 3.2).

All biological material is inhomogeneous; different tissues have different acoustic impedances and all consist of an intricate meshwork of blood vessels, fibres, connective tissues and different types of cells. Thus, all tissues are composed of an infinite series of reflecting surfaces whose dimensions range from that of a specular reflector right through to cell nuclei whose dimensions are about one thousandth of the wavelength of 1 MHz ultrasound. These small reflectors scatter or re-radiate the ultrasonic energy in all directions and so echoes can be detected from all angles. This is the basis of the tissue scattering technique currently being evaluated by C. R. Hill and his associates (e.g. Nicholas, 1979) to see if the characteristic pattern of scattered echoes may be used as a fingerprint to aid in the identification of tissue damage (e.g. cirrhosis of the liver), or to detect the presence of abnormal tissue inclusions such as multiple metastases.

It is difficult to assess the relative contributions due to scattering (s) and absorption (α) in any one value of an attenuation coefficient (A). Apart from the obvious differences caused by different tissue composition and architecture, the result is also strongly influenced by

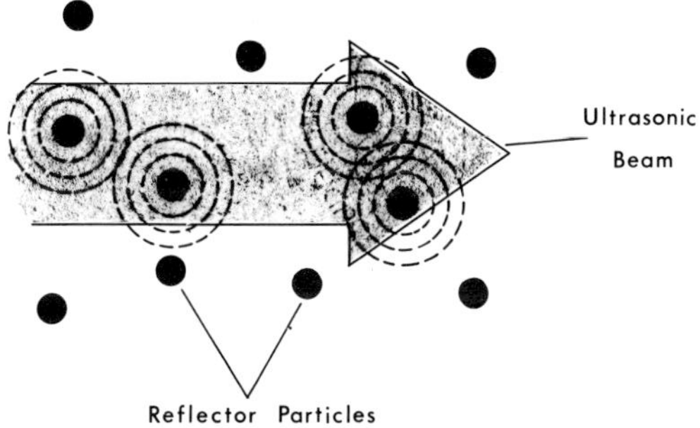

FIG. 3.2 A diagrammatic representation of the scattering patterns produced by small particles of a reflector in the path of an acoustic beam.

frequency. For example, Goss *et al.* (1979) have shown that the percentage of the attenuation coefficient due to scattering in different organs at 1 MHz was: testis (57%), brain (66%), liver (69%), kidney (72%), tendon (75%) and heart (85%). On the other hand Kremkau and Carstensen (1972) have shown that scattering has a negligible effect on the attenuation coefficient of blood at frequencies in excess of about 30 MHz.

3.2.1 Absorption in Simple Media

Homogeneous solutions of simple chemical compounds do not contain particles or structures to scatter the incident sound and so the measured attenuation coefficient (A) is the same as the absorption coefficient (α). A theory of absorption was developed by Stokes in the mid-nineteenth century and considered the effects of absorption due to shear viscosity alone (Nyborg, 1975). At any given temperature the absorption was found to be proportional to the square of the applied frequency (f). Consequently, absorption values are always quoted together with the frequency at which they were measured, or, in what should be a frequency independent form as α/f^2.

However, the mechanisms underlying acoustic absorption are so complex that this theory even underestimates the absorption characteristics of pure water. Hall (1948) proposed that this excess absorption was due to a relaxational phenomenon (i.e. the return of a system to equilibrium after it has been disturbed) resulting from the compression of the water lattice to a more dense (more closely packed) structure during the high pressure portion of the cycle and its return to a less dense (more loosely packed) structure during the low pressure cycle. Thus, at any given instant of time, the molecules are rearranging their distribution in an attempt to approach the equilibrium state applicable at that momentary value of the pressure (Nyborg, 1975).

The situation becomes even more complex when other materials are dissolved in the water. The addition of moderately low concentrations of simple molecules causes relatively little increase in the absorption coefficient at neutral pH (Hussey and Edmonds, 1971). This is not true, however, if this same concentration of simple molecules is allowed to polymerize. Figure 3.3 presents data obtained by Kessler *et al.* (1970) on the magnitude of the frequency free absorption of polyethylene glycol as a function of its molecular weight. Essentially similar results were obtained by Hawley and Dunn (1969) in their investigations of the magnitude of absorption as a function of the molecular weight of polymers of glucose (Dextrans) (Fig. 3.4). Both studies showed that the absorption of a solution

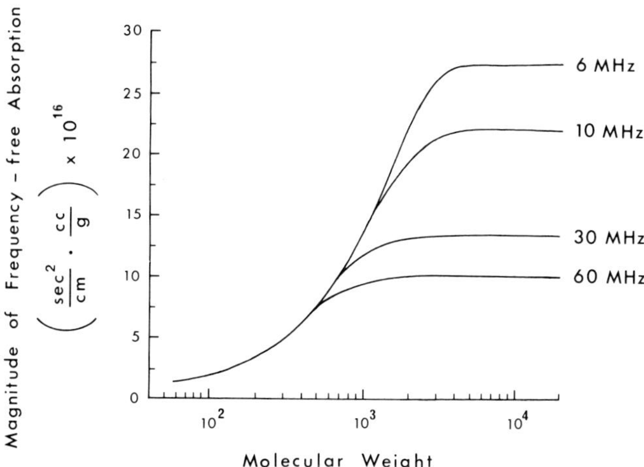

FIG. 3.3 The excess absorption of the same concentration of ethylene glycol as a function of its degree of polymerization (from Kessler *et al.*, 1970). This implies that some form of inertial effects resulting from relative motion between the macromolecules and surrounding small molecules may be contributing to absorption.

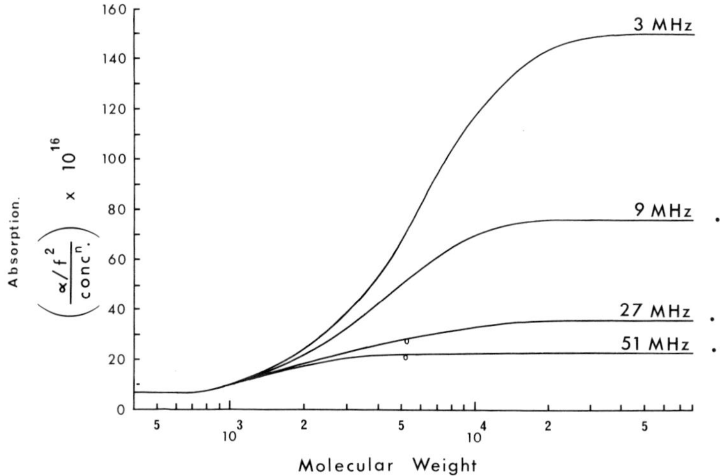

FIG. 3.4 The excess frequency-free absorption of the same concentration of glucose molecules is shown as a function of their degree of polymerization to form Dextrans (from Hawley and Dunn, 1969). The open circles represent data obtained with Inulin and the solid circles data obtained with Ficoll, a polymer of sucrose (from Kremkau and Carstensen, 1972).

containing the same quantity of material increased with the degree of polymerization. It should be noted that this effect did not continue indefinitely; above a certain limiting molecular weight the absorption became independent of the degree of polymerization. This limiting molecular weight was different for polyethylene glycol and dextran, and was also slightly less at high frequencies than for lower frequencies with the same molecule. This implies that some form of inertial effects resulting in relative motion between the suspended phase and the solvent molecules may be contributing to absorption.

Thus, absorption depends upon the molecular weight of a material and on its physical properties. It can be seen from Fig. 3.3 that the absorption of polyethylene glycol (expressed in terms of α/f^2 divided by the concentration in g/cc) at 30 MHz for molecules having a MW of 10^4 daltons is about 14×10^{-16} cm^{-1} sec^2 cc g^{-1}. However, dextran molecules of the same molecular weight have a value of about 40×10^{-16} cm^{-1} sec^2 cc g^{-1} at 27 MHz (Fig. 3.4). This difference is at least partly due to the fact that polyethylene glycol molecules form loose randomly coiled flexible chains whereas dextrans tend to form stiffer branched structures. It should be noted that all polymers of glucose appear to have approximately the same frequency dependence of absorption at constant concentration and temperature (Zaretskii *et al.*, 1972). This is also shown in Fig. 3.4 where Ficoll (a poly-sucrose which is nearly spherical and has a much lower intrinsic viscosity than dextran) and Inulin are seen to have similar absorption properties (Kremkau and Carstensen, 1972).

Proteins have even more complex structures and so it is difficult to exactly duplicate the series of experiments described above. Nevertheless, many indirect experiments indicate that the same effects occur; for example, a mixture of amino acids in approximately the same proportion and concentration as they are found in haemoglobin has been shown to have an absorption which is very much less than that of haemoglobin itself. Similarly, partial digestion of the intact haemoglobin molecule with the enzyme pronase reduces the absorption to about half of its original value (Kremkau, 1972) while complete acid hydrolysis yields a mixture of amino acids whose absorption is only slightly greater than that of water. This has also been demonstrated for other protein molecules, e.g. human serum albumin (Zorina *et al.*, 1971).

Serum albumin and haemoglobin have approximately the same molecular weight but have considerably different values of α/f^2. Gamma globulin, being another plasma protein might be expected to be similar to albumin but its absorption properties are similar to those of haemoglobin. It has been proposed that it is the three-dimensional coiling of the protein's poly-amino acid backbone (i.e. its secondary and tertiary structure) which is responsible for this difference in absorption (Wells, 1977). Thus, "coiling agents" such as sodium dodecyl sulphate which increase the α-helical content of proteins generally increase the level of absorption at any given frequency, concentration and temperature. Conversely, agents such as urea or guanidine hydrochloride which tend to weaken the forces which stabilize α-helices result in an appreciable reduction in absorption.

Another complication of absorption measurements is that absorption increases in a non-linear manner with increasing concentration. Carstensen and Schwan (1959b) and Kremkau and Carstensen (1972) showed that the specific absorption of haemoglobin increased as its concentration was increased. Part of the explanation of this effect must include competition between protein molecules for water molecules which are bound to the polar amino acid side chains to form a "hydration zone". The water molecules associated with proteins must be in a more ordered state than water molecules which are not bound. The water molecules within the hydration layer will therefore be more closely packed and therefore in a form more akin to that of an ice crystal than that of more loosely packed liquid (Frank and Evans, 1945). This will undoubtedly alter the compressibility of the water/protein mixture and hence its absorption coefficient.

3.2.2 Mechanisms of Absorption by Biological Materials

All liquids absorb acoustic energy because of frictional forces which act to oppose the periodic motion of the particles within that medium (Kinsler and Frey, 1962). In addition, water is less dense at 37°C than it is at 4°C and so it may be compressed during the high pressure portion of the acoustic cycle. Andreae and Lamb (1959) have shown that the process of lattice rearrangement occupies a finite time and so there will be a phase difference between the volume of a given liquid element and the instantaneous pressure. This phase difference results in the removal of some energy from the acoustic wave and its conversion into heat.

The dominant contributors to ultrasonic absorption in biological materials are undoubtedly relaxation processes (Wells, 1977). The energy in a system can exist in various forms such as molecular vibrational energy, lattice vibrational energy, translational energy, etc. When an ultrasonic wave propagates through this system there is an increase in the energy of one or more of these forms. If none of this energy flows into another form, then it would be returned to the vibrational energy of the wave during the decompressive part of the cycle and there would be no net absorption. However, all of the various energy forms within a system are coupled in various ways, there being a characteristic time necessary to permit the energy to flow from one form to another. Thus, some of the energy acquired during the compressive part of the cycle flows into one or more of the other available forms and is returned out-of-phase with the original propagating wave which results in absorption.

Any one single relaxation process would give a characteristic bell-shaped curve when the absorption coefficient of that solution is plotted as a function of the ultrasonic frequency. Maximum absorption occurs when the period of the wave is equal to the time constant for the relaxation process. Single relaxation frequency effects are usually only seen in simple homogeneous solutions such as aqueous solutions containing a single amino acid. Complex molecules such as proteins would be expected to have many different relaxation processes, each one being maximal at a different frequency. Figure 3.5 shows how the absorption data obtained for a solution of haemoglobin has been approximated by assuming only four discrete relaxation frequencies (from Wells, 1977; based on Schneider *et al.*, 1969).

The absorption of a solution containing amino acids or proteins varies with its pH, giving one bell-shaped peak at around pH 2–4 and another larger bell-shaped peak at around pH 11–13 (O'Brien and Dunn, 1972; Zana *et al.*, 1972). There is reasonable agreement that

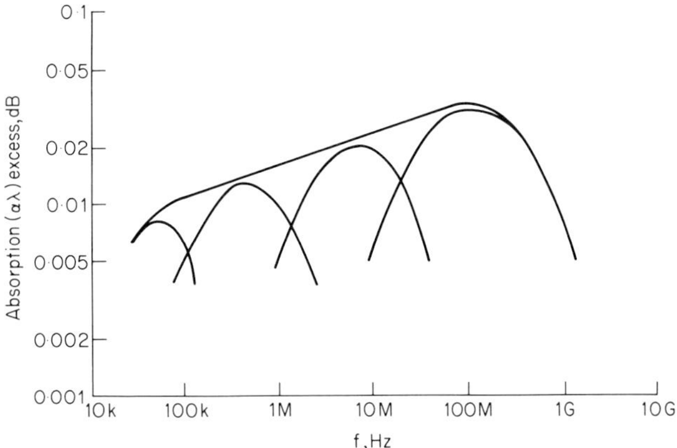

FIG. 3.5 A graphical demonstration of the way in which the experimental data of the excess absorption of a solution of haemoglobin may be closely approximated by assuming only four relaxation frequencies (from Wells, 1977; based on Schneider *et al.*, 1969).

these represent the relaxation processes resulting from the transfer of protons. This mechanism is largely confined to the pH values close to the pK values of the individual charged amino acids and so it is unlikely that these proton transfer reactions are involved in the absorption and dispersal processes in living tissues. Other relaxation processes which could occur at physiological pH are helix-coil transformations, keto-enol transformations, higher order structural transformations, and changes in the degree of water structure and solvation (O'Brien and Dunn, 1972).

3.2.3 Absorption in Cells and Tissue Extracts

With the exception of the mammalian erythrocyte which contains mainly haemoglobin, cells contain a relatively high concentration (about 15% wet weight or 60% dry weight) of a complex and variable mixture of proteins. The encapsulation of these proteins within a cell membrane means that the cell contents usually have a different acoustic impedance to the fluids bathing the outside of the cell. Thus, intact cells must scatter the incident sound field and so the measurements one obtains are usually of attenuation (i.e. absorption plus scattering) rather than absorption alone.

The simplest cellular system is that of a suspension of erythrocytes: this has been investigated extensively because it can easily be obtained in large quantities. Carstensen and Schwan (1959a) observed that the absorption coefficient of a suspension of intact erythrocytes was about 30% higher than that of the same suspension after the cells had been burst (haemolysed). This observation can be interpreted in a number of ways. First, that intact cells have a greater density and inertia than the plasma suspending medium and so fail to follow the oscillatory motion induced by the ultrasonic wave; this results in net relative motion between the cells and the plasma and the consequent loss of energy by viscous dissipation. Secondly, the concentration of haemoglobin inside the intact erythrocyte varies from about 31 to 38 g/100 ml (Morris and Williams, 1980); this is reduced to a uniform suspension containing about 15 g/100 ml after haemolysis. The reduction in the absorption due to this dilution by redistribution of the haemoglobin concentration (Carstensen and Schwan, 1959b) together with the elimination of scattering by cellular disruption is more than adequate to account for the observed decrease in absorption.

It is not known what intermolecular interactions occur between proteins inside intact cells to give rise to the observed high absorption coefficients. Intact erythrocytes are partially dehydrated when they are suspended in hypertonic saline and so the concentration of haemoglobin inside the cell may be artificially increased. If this is done, the specific absorption of a suspension of erythrocytes is increased by a factor of about three. An alternative method of increasing the intermolecular interaction is to cross-link the molecules with a fixative such as acrolein or glutaraldehyde. This technique has the advantage that the molecular interaction is increased without changing the concentration of the macromolecular species. Kremkau *et al.* (1973) showed that the absorption of fixed erythrocytes increased with increasing concentration of fixative (i.e. with increasing number of inter- and intramolecular cross-links) until it exceeded that of the unfixed cells by a factor of five.

In general, the absorption and attenuation values obtained with intact tissues are significantly larger than one would expect from extrapolation of the results obtained with purified extracts of proteins, carbohydrates and fats. In 1971 Pauly and Schwan measured the absorption of minced and homogenized liver at different frequencies. Minced liver represents an "idealized" tissue where the cellular structures are still intact but are homogeneously and randomly orientated and the whole tissue is not affected by major structural disturbances such as blood vessels and networks of collecting ducts. It has been found that over a limited range of frequencies, the absorption coefficient (α- expressed in units of dB cm^{-1}) at any given concentration may be described by an equation of the form:

$$\alpha \simeq af^{b} \qquad \text{[Eq. 3.2]}$$

where f is the frequency in MHz. It was found that the frequency dependence of the absorption of minced liver was given by $\alpha = 0{\cdot}70\ f^{1{\cdot}17}$ whereas homogenized liver was given by $\alpha = 0{\cdot}56\ f^{1{\cdot}12}$ (Pauly and Schwan, 1971). Thus, homogenized liver has an absolute value of absorption which is only about 20% less than that of the solid or minced tissue and their frequency dependencies are amost the same. This indicates that the presence of intact cells, nuclei, mitochondria and other sub-cellular structures has a minor role in determining the absorption properties of liver. This should not be confused with effects resulting from the presence of the contents of these same subcellular organelles. For example, Smith and Schwan (1971) showed that the specific absorption of the contents of liver nuclei was about three times greater than that of an equivalent concentration of haemoglobin and about five times greater than that of gelatin. In liver, a pure preparation of cell nuclei has twice the specific absorption of the original homogenate and four times the absorption of the supernatant which consists of the soluble proteins, microsomes and a small proportion of the mitochondria. This same distribution of absorption coefficients has also been shown for different cells within a tissue; for example, Lele (1977) has shown that the absorption coefficient of the sea urchin's egg is about three times that of the whole ovary in which it is contained.

3.2.4 Attenuation and Absorption in tissues

The ultrasonic literature abounds with measurements of tissue attenuation which have been made using a variety of different techniques with widely different accuracies and the results expressed in many different units. For the benefit of the newcomer to this field I wish to point out that attenuation coefficients are most commonly expressed in terms of Nepers per centimetre (Np cm^{-1}), or in decibels per centimetre (dB cm^{-1}) where 1 Np cm^{-1} is equal to 8·686 dB cm^{-1}. The techniques employed to measure attenuation include pulse transmission (Goss *et al.*, 1979), pulse reflection (Mountford and Wells, 1972), spectrum analysis (Chivers and Hill, 1975; Bamber *et al.*, 1977; Lele and Senapati, 1977), thermocouple probe measurements (Dunn and Fry, 1961) and the use of a shock-excited spherical resonator filled with a low-loss liquid (Hueter, 1958).

The most comprehensive compilation of the absorption measurements obtained with mammalian tissues is presented by Goss *et al.* (1978). This survey shows large discrepancies between the values obtained for the same tissue by different authors. At least part of this variation is caused by the different measurement techniques which were employed. The root of the problem is that tissues are not homogeneous and therefore consist of regions in which the ultrasound travels at slightly different velocities. Any two portions of the incident ultrasonic beam will travel through different cells and will therefore arrive at the detector slightly out of phase. If the detector is sensitive to the phase of the incident wave (e.g. a piezoelectric transducer) then the waves may destructively interfere with each other and so an artifactually low signal may be detected and interpreted as a falsely high attenuation coefficient. A detector which is not sensitive to phase (e.g. a radiation-force receiver or an acoustic-electric receiver) yields higher signals which are interpreted as a lower attenuation coefficient (Marcus and Carstensen, 1975; Busse *et al.*, 1977).

The dependence of the measured value of the attenuation coefficient on the method by which that value was obtained indicates that it may no longer be prudent to assign an attenuation value to a particular tissue without also specifying the method by which that value was obtained and the purpose for which it is to be used. For example the "best" attenuation coefficient for a tissue is generally thought at present to be one in which measurement artifacts due to phase cancellation, reflection, or other sources are minimized. By this criterion, the "best" value may be a factor of two or more less than that determined when such artifacts are included in the measurement method (Goss *et al.*, 1979). However, if one wished to interpret the results obtained using a clinical diagnostic ultrasound instrument then this "best" attenuation value would not be appropriate as it would not encompass all the sources of attenuation as seen by the instrument's own phase sensitive piezoelectric transducer.

Table 3.1 presents some representative ultrasonic attenuation data for five typical mammalian tissues as a function of the applied frequency. The last two columns display the best least squares linear regression power fit to the equation $A = af^b$ (Eq. 3.2) where f is the frequency in MHz. The correlation coefficient R describes the goodness of fit with $R = 1$ indicating a perfect fit (from Goss *et al.*, 1979). It can be seen that tendon which contains 35–40% of its wet weight as the densely packed protein collagen has an attenuation coefficient at low frequencies which is about ten times larger than brain tissue which contains only about 10% wet weight of protein (mainly globular proteins) and about 11% of fat.

Table 3.2 presents the absorption data for these same mammalian tissues plus testis (also from Goss *et al.*, 1979). It should be noted that these values are smaller than the attenuation values by a factor of about four, but their frequency dependencies (i.e. the exponent b) remain approximately the same (with the exception of tendon). This indicates that whatever the mechanism responsible for the differences between absorption and attenuation (i.e. scattering, reflection, measurement artifact etc.), it is also nearly linearly dependent upon frequency.

Figure 3.6 presents a brief, highly selected graphical compilation of tissue attenuation values as a function of frequency over the range commonly encountered in medical diagnostic and therapeutic applications. It can be seen that at any given frequency lung and bone have the highest attenuation followed by tendon, muscle, kidney, liver/nerve/brain, and fat with the fluid tissues such as blood having the least attenuation.

TABLE 3.1

Ultrasonic attenuation values (in Nepers/cm) for some typical biological tissues at 37°C (from Goss *et al.*, 1979).

Tissue	Frequency (MHz)						Regression analysis fit	
	0·5	0·7	1	3	4	7		
Brain	0·032	0·047	0·07	0·24	0·34	0·64	$A = 0{\cdot}07f^{1{\cdot}14}$	$R = 0{\cdot}822$
Heart	0·060	0·086	0·13	0·41	0·56	1·0	$A = 0{\cdot}13f^{1{\cdot}07}$	$R = 0{\cdot}98$
Kidney	0·049	0·070	0·10	0·34	0·47	0·87	$A = 0{\cdot}10f^{1{\cdot}09}$	$R = 0{\cdot}973$
Liver	0·038	0·055	0·08	0·29	0·40	0·75	$A = 0{\cdot}08f^{1{\cdot}13}$	$R = 0{\cdot}934$
Tendon	0·33	0·42	0·56	1·3	1·6	2·5	$A = 0{\cdot}56f^{0{\cdot}763}$	$R = 0{\cdot}998$

TABLE 3.2

Ultrasonic absorption coefficients (in Nepers/cm ±s.d.) for some typical biological tissues at 37 C (from Goss *et al.*, 1979).

Tissue	Frequency (MHz) 0·5	0·7	1	3	4	7	Regression analysis fit	
Brain		0·014 ± 0·003	0·029 ± 0·004			0·23 ± 0·09	$0{\cdot}024f^{1{\cdot}18}$	R = 0·993
Heart		0·018 ± 0·009	0·033 ± 0·006			0·21 ± 0·03	$0{\cdot}028f^{1{\cdot}04}$	R = 0·995
Kidney		0·017 ± 0·007	0·033 ± 0·004			0·20 ± 0·002	$0{\cdot}028f^{1{\cdot}02}$	R = 0·994
Liver	0·010 ± 0·006	0·020 ± 0·003	0·023 ± 0·004		0·14 ± 0·03	0·24 ± 0·02	$0{\cdot}026f^{1{\cdot}17}$	R = 0·995
Tendon	0·050 ± 0·03	0·16 ± 0·1	0·11 ± 0·04	0·53 ± 0·2	0·75 ± 0·4	1·4 ± 0·5	$0{\cdot}14f^{1{\cdot}17}$	R = 0·973
Testis	0·0078 ± 0·002	0·0085 ± 0·001	0·015 ± 0·003		0·079 ± 0·02	0·12 ± 0·02	$0{\cdot}015f^{1{\cdot}11}$	R = 0·995

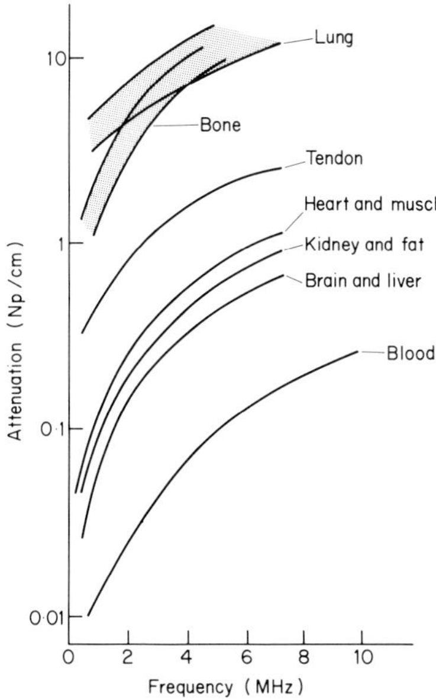

FIG. 3.6 A graphical compilation of the attenuation values of some tissues as a function of frequency over the range commonly encountered in diagnostic and therapeutic applications (based on the data presented in Goss *et al.*, 1978).

3.3 TEMPERATURE ELEVATION WITHIN TISSUES

The absorption coefficient is a measure of the amount of energy removed from an ultrasonic wave as it propagates through a given tissue. This energy appears as heat and causes an increase in the temperature of that tissue. The rate of production of heat within a unit volume of tissue ($\dot{Q}v$) is given by the equation

$$\dot{Q}v = 2\alpha I \qquad \text{[Eq. 3.3]}$$

where α is the absorption coefficient of that tissue and I is the ultrasonic intensity at that site (Nyborg, 1975). Table 3.2 shows that a typical value for the absorption coefficient of soft tissues such as liver or brain at 1 MHz is about 0·03 Np cm $^{-1}$. In these tissues an ultrasonic wave of frequency 1 MHz and an intensity of 1 W cm $^{-2}$ produces heat at the rate ($\dot{Q}v$) of 0·06 joules per millilitre per second which is equal to 0·014 cal ml $^{-1}$ s $^{-1}$ (1 calorie = 4·18 joules).

If heat is generated within a body of tissue and not transported away by conduction or convection then the temperature (T) of that body will increase at certain rate (dT/dt) given by:

$$\frac{dT}{dt} = \frac{\dot{Q}v}{\rho c_m} \qquad \text{[Eq. 3.4]}$$

where ρ is the density of that tissue and c_m is the heat capacity per unit mass of tissue. For most soft tissues the product of ρc_m is not much greater than 1 calorie per millilitre per degree centigrade rise in temperature. Thus, a 1 MHz beam at an intensity of 1 W cm^{-2} in liver or brain will cause the average temperature to rise in the absence of diffusion and convection at the rate (dT/dt) of 0·014°C/sec which is equal to 0·86°C/min.

In reality, as soon as a temperature gradient is established within a tissue, heat diffuses away in an attempt to eliminate the gradient. Nyborg (1975) has shown that a spherical source which is generating heat evenly throughout its volume at a uniform rate eventually attains a steady-state condition such that its temperature is greatest at the centre of the sphere and progressively decays with increasing distance from the surface of the sphere (S) to merge eventually with its surroundings (Fig. 3.7). The excess temperature rise at the centre of the sphere may be interpreted as a greater net diffusion of heat inward from all the heated elements of that body just beneath its surface compared with the net diffusion of heat outwards from the centre. Outside the sphere the flow of heat away from the body is much greater than the flow towards it and so the temperature decays inversely with the distance from the heated source.

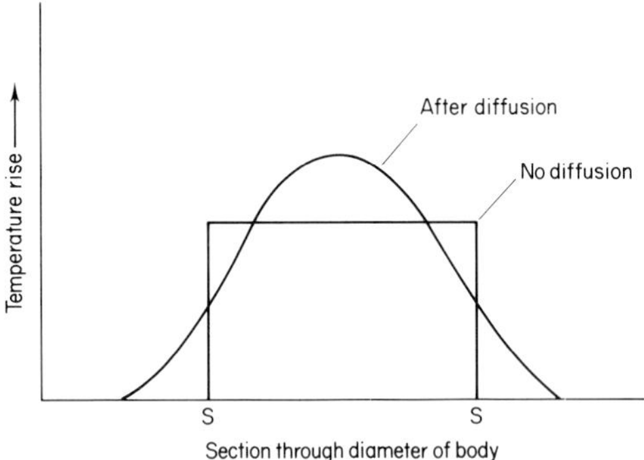

FIG. 3.7 A diagrammatic representation of the steady-state temperature distribution across a diameter of a uniformly heated spherical body suspended in a thermally conducting medium. S represents the surface of the sphere.

The excess temperature generated at the centre of a uniformly heated spherical body (To) is given by:

$$To = T_{\infty} + \frac{\dot{Q}v}{2K} R^2 \qquad \text{[Eq. 3.5]}$$

where T_{∞} is the temperature of the surrounding medium, $\dot{Q}v$ is the rate of heat production, K is the coefficient of thermal conductivity (which is equal to 0·0060 W cm^{-1} °C^{-1} for water and most biological soft tissues) and R is the radius of the sphere (Nyborg, 1975). For the example considered above the temperature rise at the centre of a sphere ($\Delta t = To - T_{\infty}$) is given by:

$$\Delta t = 5R^2{}_{(°C)} \qquad \text{[Eq. 3.6]}$$

Thus, the steady-state temperature rise at the centre of a uniformly heated sphere of radius 1 cm would be 5°C, whereas a sphere of radius 1 mm generating heat at the same rate would only develop a maximum temperature rise of 0·05°C.

Another complication is the time necessary to establish this steady-state condition. This is determined by the thermal diffusivity (D) of that tissue and is given by $K/\rho c_m$ where K is the coefficient of thermal conductivity, ρ is the density and c_m is the heat capacity per unit mass of tissue. A representative value for D in most soft tissues is 0·00144 cm^2/sec. Thus a sphere of tissue having a radius of 1 cm would take about 700 s or 12 min to reach its steady-state temperature whereas a sphere of radius 1 mm would take only about 7 s (Nyborg, 1975).

Another variable which determines the rate of temperature rise within a tissue is convection or the transport of heat away by flowing blood. It is difficult to quantify convection because different tissues have widely different degrees of vascularity and the rate of flow of blood through any one tissue can vary so much in response to the physiological needs of the animal at that time. The temperature rises predicted by the equations presented above must therefore be regarded as upper estimates unless the rate of generation is very large or in a time which is short compared with the times necessary for diffusion and convection to occur.

It should also be noted that as soon as a temperature rise is produced within most tissues *in vivo* it initiates the physiological reflexes which result in the dilation of blood vessels and an increase in the rate of blood flow through that tissue which of course increases the rate at which the excess heat is carried away.

Energy is extracted from the ultrasonic wave and converted into heat preferentially in those tissues having a high absorption coefficient. Since muscle has a higher value than fat, the ultrasound will largely pass through fatty layers and cause selective heating in underlying muscular tissues (ter Haar and Hopewell, 1982). For bone, which has an even higher absorption coefficient, the rate of heating is even higher; this fact is probably important in explaining the success of ultrasonic therapy in selectively heating joint structures.

Two additional complications must be mentioned in connection with the ultrasonic irradiation of bony structures. The acoustic impedance of bone ($3{\cdot}75$ to $7{\cdot}38 \times 10^6$ kg m^{-2} s^{-1}), is so much greater that that of the adjacent muscle tissue ($1{\cdot}65$ to $1{\cdot}74 \times 10^6$ kg m^{-2} s^{-1}), that about 30% of the incident energy is reflected (see Eqs 1.8 and 1.9). This reflected wave returns through the muscle and other tissues that it has already traversed. More energy is removed from this reflected wave during its return path and deposited as heat; thus, the temperature rise in any given soft tissue is greater if it is situated in front of a strong reflector.

The other complication is the redistribution of energy which occurs at an interface; this effect is more pronounced at large mismatches of acoustic impedance such as tissue/bone or tissue/air interfaces. Basically, four sets of boundary conditions must be equal on both sides of the interface, these are: the normal displacements, the tangential displacements, the normal stresses and the tangential stresses (Wells, 1977). Within any one medium the acoustic energy may be propagated as a longitudinal (compressive) or a transverse (shear) wave. Liquids and gases cannot propagate shear waves, with the exception of surface waves on liquids. However, soft tissues have some solid-like characteristics and so have a limited ability to conduct shear waves even though their attenuation coefficient for these waves is extremely high so that they are rapidly absorbed and converted into heat. The wave incident on a tissue/bone interface will therefore be a longitudinal wave but there will be two reflected and two refracted waves, one of each pair being a longitudinal wave and the other a transverse wave. The transverse (shear) wave generated within the bone is able to propagate but the transverse wave generated on the tissue side of the interface is rapidly absorbed and all of its energy is deposited as heat within millimetres of the bone surface. This region of soft tissue is called the periosteum and is richly endowed with blood vessels and nerve terminals (it is damage to the periosteum which causes the pain when one is kicked on the shin). The heat deposited within the periosteum raises its temperature and, if it exceeds a certain value, is interpreted as pain. If the bone is fairly close to the surface of the skin, then this is the site where the temperature rise first exceeds the "pain" threshold. Periosteal pain is therefore one of the major factors which limits the maximum ultrasonic intensity that may be tolerated during a therapeutic application. Consequently, a patient will usually be able to tolerate an ultrasonic beam of higher intensity over those portions of his or her anatomy where a thick layer of absorbent muscle tissue is interposed between the transducer and the bone.

To summarize, the temperature rise within any given tissue irradiated with ultrasound is determined by a number of factors; the most important of which are:

(1) The absorption coefficient of that tissue.

(2) The rate of input of ultrasonic energy. This in turn is a function of the ultrasonic intensity, the distribution of energy within the beam and whether the beam is continuous or pulsed.

(3) The frequency of the ultrasonic wave.

(4) The duration of the energy input (i.e. whether or not a steady state condition has been attained).

(5) The dimensions of the heated body.

(6) The presence or absence of strong reflectors either in front of or behind the tissue of interest.

(7) Whether or not nonlinear effects are occurring.

3.4 FINITE AMPLITUDE EFFECTS (NONLINEAR EFFECTS)

Any interpretation of the biological effects resulting from biomedical exposures of ultrasound may be complicated by the fact that the transmission properties of water are not always linear (Blackstock, 1970). At low power levels water appears to behave in a linear manner and a low amplitude sine wave emitted by a transducer is propagated as a sine wave. However, as the power output of the transducer is increased you reach a point where the high pressure peak of the sine wave begins to travel faster than the rest of the wave (because it compresses the medium thereby increasing its density) and the low pressure trough of the wave travels correspondingly more slowly (because it expands the medium and decreases its density). This distorts the shape of the travelling wave until it begins to resemble the reversed form of the letter N (Fig. 3.8). This wave now resembles the "sawtooth" waveform emitted by musical instruments such as the violin. These musical notes owe their distinctive shapes and sounds to the high proportion of higher frequency components or harmonics accompanying the fundamental note. This change in the waveform of the propagating ultrasonic wave therefore shows that some of its energy has been transformed into higher frequency harmonics of the original signal (i.e. 2 fo, 3 fo, etc.).

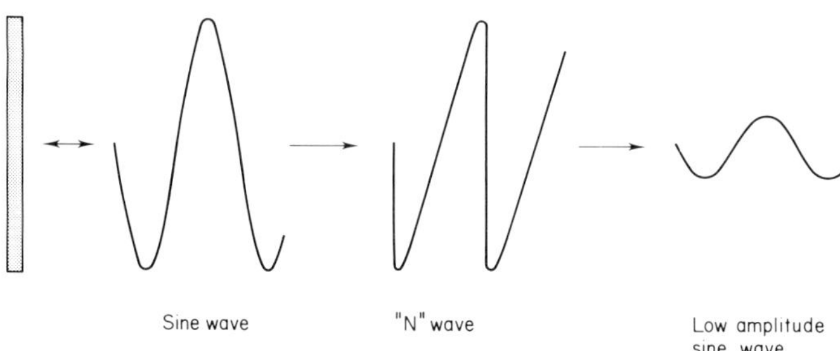

FIG. 3.8 A schematic representation of the distortion in the waveform of a high amplitude sine wave caused by finite amplitude effects. Rapid absorption of these higher harmonics (see text) results in a wave of lower amplitude which may once again become sinusoidal.

This effect is important because these higher frequency components interact more strongly with the medium through which the wave is propagating. For example, the absorption coefficient of a medium is defined as that percentage of the incident ultrasonic wave which is absorbed and converted into heat by one centimetre or one metre of that medium. This absorption coefficient increases as the square of the frequency of the incident wave even for a simple liquid like water. Thus, the rate at which energy is removed from the travelling wave and converted into heat is greatly accelerated if higher harmonics are being generated.

The "threshold" exposure conditions at which these non-linear effects begin are determined by five major parameters, each of which will be discussed separately. These are:

(1) The power (or intensity) emitted by the transducer.
(2) The frequency of the source.
(3) The transmission properties of the medium.
(4) The path length in that medium.
(5) The geometry of the acoustic beam.

It is unlikely that the intensities emitted by conventional therapeutic devices operating in their continuous wave (c.w.) mode will give rise to appreciable non-linear effects. However, if this same amount of ultrasonic energy is delivered in a pulsed mode, then the increased peak power in each pulse will increase the probability of generating non-linear behaviour (Muir and Carstensen, 1980). The rationale underlying the use of pulsed ultrasound in physiotherapy is that heating effects are minimized because the heat is "carried away" during the off-time of the transducer. In fact there is some experimental evidence that the reverse is true, i.e. that the temperature rise in a given tissue caused by pulsed ultrasound is more than the temperature rise produced by that same amount of energy given as a continuous wave. It seems probable that it is the extra absorption rate resulting from the production of higher harmonic frequencies which gives rise to this observation.

The "threshold" for the production of non-linear effects decreases with increasing frequency (Muir and Carstensen, 1980). This means that devices emitting short, intense pulses of high frequency ultrasound (e.g. pulse-echo diagnostic equipment, especially those used in ophthalmology) are also liable to result in the production of harmonics. Unfortunately, we are unable to say whether or not this makes these devices "less safe" since there is very little information available on the biological effects of waves having a high harmonic content. What is certain is that it will increase the rate of energy absorption and consequently will tend to degrade the quality of the information obtained by these devices.

The density changes occurring within a medium as a function of the applied acoustic pressure (p) are given by:

$$p = Co^2\rho + \frac{1}{2}\frac{Co^2}{\rho o}\left(\frac{B}{A}\right)\rho^2 + \ldots \qquad \text{[Eq. 3.7]}$$

where Co is the phase velocity of the wave, ρo is the undisturbed density of that material and B/A is the second order parameter of nonlinearity. If the parameter B/A was equal to zero, then only the first term ($Co^2\rho$) would remain and the propagation of sound would be linear under all conditions of frequency and intensity. However, Beyer (1974) has shown that B/A for water at 30°C has the substantial value of 5·2 and it has been shown that this gets bigger as one adds proteins (e.g. bovine serum albumin; Dunn *et al.*, 1980, 1982). There are as yet no reliable measurements of B/A for living tissues and the limiting effects of tissue absorptivity are not known or understood.

Even if the "threshold" values of intensity and frequency have been exceeded for that particular medium, it still requires a finite path length through that medium before the wave becomes distorted. Thus, close to the transducer surface the wave still propagates linearly and it is undistorted, but it becomes progressively more distorted as it progresses deeper into that medium until it reaches the limiting value where the advancing wave front is vertical (Fig. 3.8).

At this point the higher frequency harmonics are being removed by absorption as fast as they are being generated. Figure 3.9 presents a set of theoretical curves devised by Muir and Carstensen (1980) which show the relationships between the intensity of a wave and the distance it has to propagate through that medium before one should first be able to observe non-linear behaviour. It can be seen that this distance decreases with increasing frequency at any given intensity, and also decreases with increasing intensity at any given frequency. Figure 3.9 shows that a beam of 1 MHz continuous wave ultrasound at an intensity of 1 W/cm^2 in water would behave linearly until its path length exceeded about 80 cm. However, this same amount of power delivered as a pulsed beam (e.g. 1 ms on and 9 ms off so that the power density in each pulse would be 10 W/cm^2) would only be linear for about 20 cm. Each "threshold" line is displaced to the left (i.e. towards shorter path lengths) as the ratio of B/A for that medium is increased.

The other major factor which determines that point where an

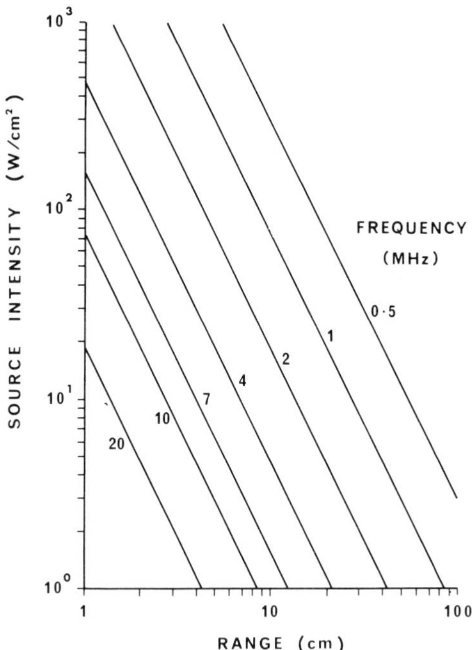

FIG. 3.9 A set of theoretical curves derived by Muir and Carstensen (1980) showing the distance a wave of a given intensity and frequency can propagate through an aqueous medium before one should be able to detect evidence of non-linear behaviour.

ultrasonic beam begins to generate significant quantities ot higher harmonic components is the geometry of the beam itself. The dimensions indicated in Fig. 3.9 refer to a plane progressive wave and become larger if that wave is diverging or become smaller if that wave is converging, i.e. being focused. It is therefore apparent that it is extremely difficult to interpret any non-thermal biological effects produced using focused ultrasonic beams because the geometric considerations plus the rapidly increasing intensity as you approach the focus means that non-linear effects are easily produced. Thus, even at a relatively low peak focal intensities up to 30 or 40% of the acoustic energy may be present as higher harmonic frequencies. This is advantageous if you are using the focused system to produce lesions within tissues as in ultrasonic surgery, since these higher harmonics are absorbed more efficiently than their fundamental. However, one would not expect a gas bubble in a liquid medium to behave in the same way when it is irradiated either by a plane wave containing no higher harmonics or by a focused system with many harmonics, even though the nominal ultrasonic intensity may be the same.

An interesting phenomenon which arises from the occurrence of these non-linear effects is acoustic "saturation" of the medium. It was mentioned above that the wave profile progressively distorts (as a function of increasing intensity or propagation distance) until its leading edge approaches the vertical (Fig. 3.8). This point cannot be exceeded and so the energy must be removed by absorption as fast as it is being converted into harmonics. If the amount of acoustic energy emitted by the transducer is now increased, then this tends to distort the wave front past the vertical; this cannot occur and so all of this extra energy goes into higher harmonics and is lost by absorption. Thus, none of the extra energy emitted by the transducer propagates past the region of extreme non-linear behaviour; this region is therefore said to be "saturated". Carstensen *et al.* (1980) report a graphic illustration of saturation in which a 5 MHz transducer placed in direct contact with an excised kidney could produce thermal lesions with ease. However, when that same organ was exposed through a 10 cm water path using the same transducer, it became impossible to produce a lesion regardless of the electrical power applied to the transducer.

The final point to be made in this section is that after a lot of the energy of the wave has been absorbed (by both the linear and non-linear mechanisms) its intensity is reduced below the "threshold" value needed to generate the non-linear effects and so the wave profile returns once again to that of a sine wave (Fig. 3.8).

Figure 3.10 is a graphical illustration of the effects of the development of non-linear effects within a homogeneous absorbing medium. Figure 3.1 showed that the intensity of the wave decreased exponentially as a function of the depth of penetration into a homogeneous medium. The energy which is removed from the wave is converted into heat and so the temperature rise within that tissue would also resemble the curve presented in Fig. 3.1. However, if non-linear effects begin to occur after a finite distance through that tissue (Dcrit.) then the absorption and hence the rate of deposition of heat will rapidly increase resulting in the formation of a "hotspot" (Fig. 3.10). So much energy will be removed at this site that very little will be left to heat the medium on the far side of the "hotspot".

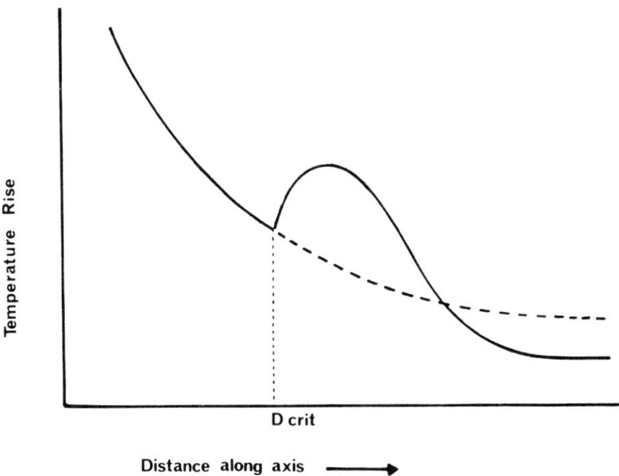

FIG. 3.10 A graphical illustration of the effect of the development of non-linear behaviour on the temperature rise produced within a homogeneous medium by a uniform acoustic beam. For details see text.

This is not likely to be too serious a problem for the physiotherapists unless they are using high frequencies to heat deep structures with high intensities of pulsed ultrasound. Nevertheless, it can be a serious problem for designers of diagnostic equipment who are trying to obtain clearer echoes from deep anatomical structures by increasing the intensity of the ultrasonic pulse. The temperature rises resulting from this non-linear absorption will still be negligibly small, but it does mean that doubling the energy of the pulse may only increase the strength of the far echoes by a few per cent.

3.5 OTHER SOURCES OF RAPIDLY ABSORBED HIGH ACOUSTIC FREQUENCIES

Physiotherapists who routinely use a number of different ultrasound machines to treat their patients sometimes find that the acoustic intensity necessary to produce the same sensation of warmth at a given site on the same individual varies widely from machine to machine. This variation is partly a reflection of inadequate dosimetry in that the output performance of each machine may have deteriorated to a different extent so that their acoustic emissions bear little resemblance to the value indicated on their output meter or setting switch. Another source of variability may lie in the waveform of the electronic signal used to excite the transducer. Some therapy devices generate square wave pulses and then "round them off" until they approximate to a sine wave. Any non-sinusoidal wave form can be constructed by the superimposition of many sine waves of different (higher) frequencies. Thus, any waveform other than a pure sine wave will also contain high frequency acoustic waves. The intensity of these high frequency waves will be extremely small compared with the intensity of the fundamental frequency, but they will be rapidly absorbed by tissue giving an artificially high skin temperature with the consequent sensation of warmth.

Similar high frequency components are produced when an electrical sine wave signal is switched off or chopped at any non-zero value of voltage. Thus, a pulsed beam of ultrasound must contain more high frequency components than the same amount of acoustic energy delivered as a continuous beam. This effect is particularly noticeable when the pulse length becomes comparable with the wavelength; in this case the waveform of the exciting voltage bears little resemblance to a sine wave.

3.6 MEASUREMENTS OF TEMPERATURE RISE WITHIN TISSUES

The object of most of the therapeutic applications of ultrasound is to raise the temperature of those parts of the body which require treatment. Lehmann and Guy (1972) state that the tissue temperatures have to exceed about 40°C to be effective. Also, that the temperature during treatment should not exceed 45°C as this would result in destructive changes. If a vigorous therapeutic response is desired then it may be necessary to elevate the temperature at the site of the pathology to be treated to temperatures which are close to the maximally tolerated levels (Lehmann and Guy, 1972).

The most obvious and easily quantifiable biological response of a mammalian tissue to an elevated temperature is dilation of the microvascular circulatory system which results in an increased blood flow through that tissue (i.e. hyperaemia). Hyperaemia of the skin may be studied in humans using a water bath as the variable heat source and a photometric device to quantify the extent of the increased redness (erythema) indicative of increased microvascular perfusion. Figure 3.11 shows this increase in microvascular perfusion as a function of the applied temperature (from Lehmann, 1953). As with most other biological responses this graph does not tell the whole story because each experimental point is itself part of a rate curve similar to that presented in Fig. 3.12 (also from Lehmann, 1953). Thus, the extent of the temperature induced hyperaemia response varies with the maximum value of temperature attained and also with the time for which that particular temperature has been maintained. It is also to be expected that the magnitude of the observed changes in response to any given temperature rise will be markedly different for tissues such as human skin which use the hyperaemic response as one of the normal mechanisms for thermoregulation as against tissues within the internal organs which do not normally encounter significant changes in temperature. Figure 3.13 presents measurements of the temperature increases produced within the muscular and bony tissues of the thigh of a pig following a 5 min irradiation with 1 MHz ultrasound at normal incidence (abstracted from Lehmann *et al.*, 1967). It can be seen that the muscle temperature is highest close to the bone and that the greatest

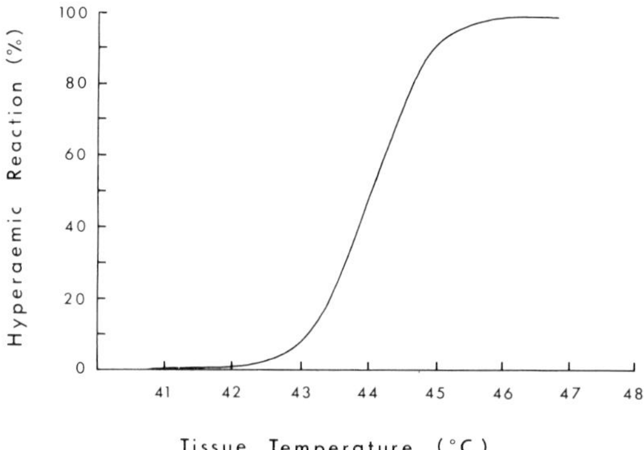

FIG. 3.11 Quantification of the magnitude of the hyperaemic response as a function of the temperature attained within that tissue (from Lehmann, 1953).

temperature rises are within the calcified bony tissue itself. The bone marrow contains the haemopoietic system and so has a lower absorption coefficient and therefore experiences a smaller rise in temperature. The upper curve (A) in Fig. 3.13 was obtained after the

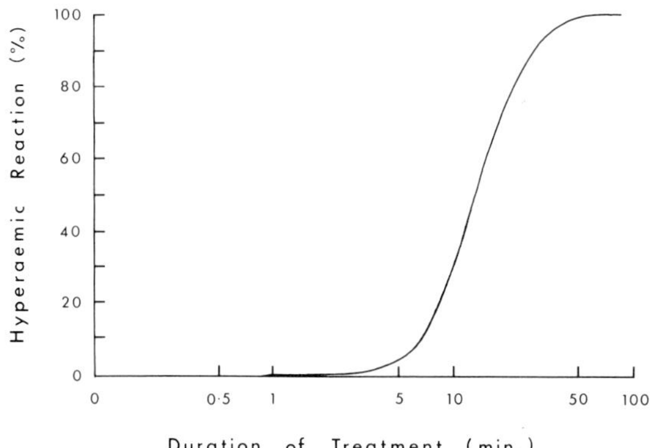

FIG. 3.12 An example of the rate of development of the hyperaemic response (from Lehmann, 1953).

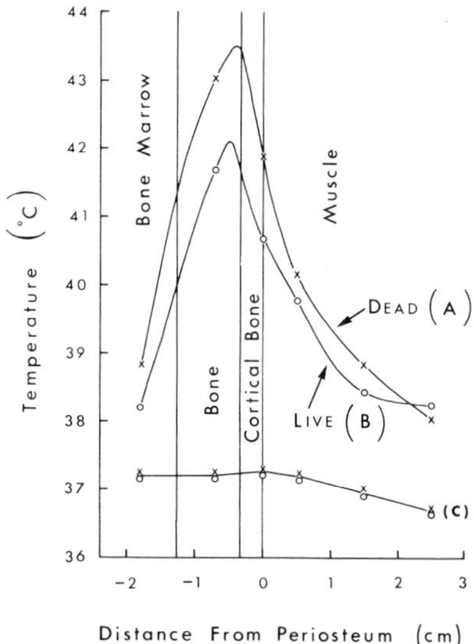

FIG. 3.13 Temperature elevations produced within the various structures of the thigh of a pig by therapeutic intensities of ultrasound (compiled from the experimental data of Lehmann *et al.*, 1967). The highest temperatures were attained with the animal which had been freshly killed but not allowed to cool (A). The blood flow within the live animal (B) carried some of the heat away. The bottom two curves (C) show the temperature distributions before the application of ultrasound.

animal had been killed but before it had had an opportunity to cool. Thus, the difference between curves A and B shows the effect of heat convection by the flowing blood (including any effects due to hyperaemia).

One of the main therapeutic applications of ultrasound is in the treatment of stiff or painful joints. Treatment is claimed to decrease pain and to increase the flexibility of the collagenous sheets which swathe and encapsulate the joint. Lehmann *et al.* (1970) have shown that rat tail tendon (and by implication all other collagenous sheets) became more extensible when it was heated to 45°C *in vitro*. If this same effect also occurs *in vivo* then it could at least partially explain the significantly greater therapeutic improvement in joint mobility seen in patients treated with ultrasound while suffering from periarthritis or fractures of the hip (Table 3.3). The control group were treated with infrared radiation which produced superficial heating and skin erythema but resulted in a negligible increase in the temperature within the joint (Lehmann and Guy, 1972).

TABLE 3.3
Gain in range of motion at the hip joint following therapeutic treatment with ultrasound or microwaves (from Lehmann and Guy, 1972).

	After treatment with	
Gain in	Ultrasound	Microwaves
Forward flexion	27·4 ±2·3°[a]	16·1° ±1·5°
Abduction	32·6 ±2·5°	21·2° ±2·1°
Rotation	45·4 ±2·8°	17·3° ±4·0°

[a] Standard error of the mean.

3.7 THERMAL DAMAGE TO TISSUES

By definition, therapy is a process whereby you deliberately change a tissue or the physical or chemical environment around and within that tissue so as to change its behaviour in a beneficial manner. This is inherently dangerous because any stimulus which induces a beneficial change can also result in or initiate an adverse or deleterious change which may damage that tissue and possibly even the whole organism. Temperature elevation is a typical example. Lehmann and his co-workers argue that if ultrasonic therapy is to be effective you should increase the tissue temperature almost up to the point where it is going to damage that tissue. Unfortunately, there is no simple answer to the question "what temperature will damage tissues?" because different tissues have different thermal susceptibilities and also the result obtained at any given temperature is critically dependent upon the time of exposure (especially at the higher temperatures).

A further complication is the criterion or function you measure to assess the "damage" caused by a given elevated temperature. For example, a rise of about two degrees centigrade in the temperature of the whole body as a consequence of an acute viral infection produces numerous unpleasant physical and psychological sensations indicating that the normal functioning of the brain and its sensory/regulatory centres has been (temporarily) perturbed. However, more crude histological or *in vitro* physiological tests would not be able to detect or discern any deleterious changes in the morphology or function of the brain at these temperatures. Thus, any single tissue would exhibit a wide range of damage "threshold temperatures" depending on the sensitivity of the assay system used to measure it. The effects produced at the lowest "threshold temperatures" would be largely reversible whereas the effects produced at higher temperatures tend to be irreversible and may result in cell death. The least sensitive assay which would therefore give the highest "threshold temperature" is probably that of histological sectioning; this is usually used to distinguish between cells which have died as a result of the thermal exposure and those which have survived. This is the assay system which has been chosen by a large number of investigators to assess the damage produced by beams of ultrasound within mammalian tissues.

Most investigators who wish to produce a discrete ultrasonic lesion within an intact tissue prefer to use a focused beam. This has several practical advantages; first, it is relatively easy to generate the high power densities (intensities) needed to produce lesions within a short time (see also the earlier section on non-linear effects); secondly, the lesion has fairly sharp lateral boundaries because of the rapid decrease in intensity with distance away from the axis in the focal region, and thirdly, the lesion does not extend all the way back to the transducer. Thus, focused transducers generate cigar-shaped (prolate ellipsoid) lesions whose length/width ratios reflect the ratio between the focal length of the system and the diameter of the transducer surface. The theory for the estimation of the temperature increase at any given point within this ellipsoid lesion is similar but more complex than that derived for a heated sphere. This has been used by Lele (1977) to estimate the temperatures produced at the boundary of a lesion, i.e. that temperature rise within the tissue which was just enough to kill the cells at the edge of the lesion. This implies that the cells just outside the lesion were heated to a slightly lower temperature and that this did not result in cell death. This interpretation should be viewed with some caution because these cells may also have been irreversibly damaged and may die at some later date, i.e. the histological lesion may grow as a function of time after irradiation.

Figure 3.14 is from Lele (1977) and shows the estimated lesion boundary temperature in both grey and white matter of cat brain as a function of the time for which that elevated temperature was maintained. The lesions were produced *in vivo* using focused ultrasonic beams of different frequencies or by directly heating the tissues by passing electrical currents through implanted wires. The animals were sacrificed after 10 days so that any cells which had been reversibly damaged would have had an opportunity to recover. It can be seen that a temperature rise of at least 60°C was necessary to kill the cells when the time of exposure was less than 10 s, but that a temperature of about 44°C killed the same cells when they were exposed to it for times in excess of 1000 s (Fig. 3.14). It should be noted that a temperature of 42·5°C did not result in morphologically

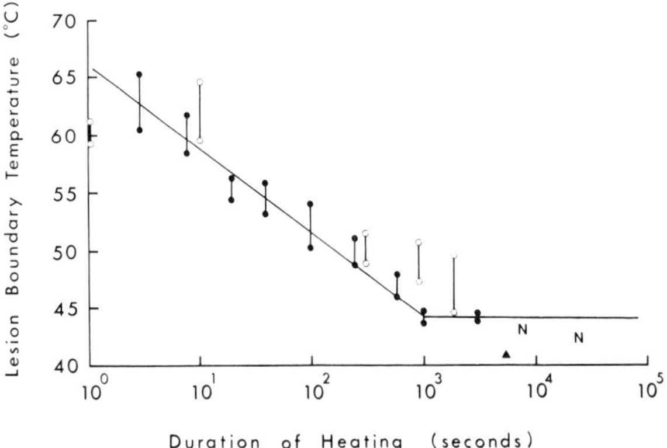

FIG. 3.14 A graphical presentation of the estimated temperature rise necessary to kill mammalian brain cells *in vivo* as a function of the time for which that temperature was maintained (from Lele, 1977). The solid circles represent data obtained by ultrasonic heating whereas the data represented by the open circles was obtained using other forms of heating. The symbol (N) represents a negative effect while the solid triangle represents a single positive data point obtained with foetal tissue.

detectable damage even when it was maintained for up to 8 h whether that temperature rise had been produced ultrasonically or by means of a heated wire. A similar temperature/time relationship was also found to be true for other tissues, e.g. lens occuli and avian egg-albumin, even though the temperature-duration thresholds in these metabolically less active tissues were higher. Conversely, active rapidly dividing tissues are more susceptible to thermal damage, e.g. a mammalian foetus subjected to a temperature elevation of 2·5°C above normal for one hour has an increased incidence of teratological changes which can be ascribed to thermal damage to the dividing and differentiating cells (Lele, 1977).

Extensive investigations by Lele and his associates on the production of tissue lesions using ultrasound, electric current, incubators, hot water baths and microwaves have shown that it is the rise in tissue temperature which causes the damage. On the basis of this data he has proposed the "Thermal Hypothesis" which states:

> For a wide class of reversible and irreversible effects caused by the action of ultrasound on tissues (and other cellular structures), the same effects can equivalently be produced by nonacoustic localised heating of the tissue, provided that the temperature history during heating and cooling duplicates the quasi-steady (averaged over a cycle) temperature history during insonation. (Lele, 1977.)

3.8 THERMAL CONTRIBUTION TO ULTRASOUND BIOEFFECTS

All biological media must absorb some of the ultrasound passing through it. This absorbed energy appears as heat causing a rise in the temperature of the irradiated medium. If the temperature is permitted to rise above a certain critical value (which is different for different tissues) and is maintained at that value for a sufficient time, then temperature alone may be responsible for any observed biological effects. However, even if the observed effect is caused by a different mechanism (such as the mechanical or hydrodynamic forces described in Chapter 4) the inevitable rise in the ambient temperature may have contributed to the effect by "weakening" the forces maintaining the integrity of those structures. For example Krizan and Williams (1977) have shown that the hydrodynamic shear stress required to disrupt human erythrocytes in laminar flow decreases with increasing temperature (Fig. 3.15). These three curves were obtained at three different exposure times (1, 5 and 10 min) and indicate that this effect is also a rate process which depends not only upon the applied temperature but also the time for which that elevated temperature was maintained.

An additional complication is that sources of non-thermal effects such as gas bubbles which act as efficient transducers converting acoustic energy into hydrodynamic shear forces (see Chapter 4) also act as local heat sources. Nyborg (1977) has shown that vibrating bubbles extract energy from the acoustic field and that only about 10% of this energy is re-radiated as an acoustic wave (i.e. scattered) at frequencies of the order of 1 MHz. The remainder is converted irreversibly into heat by several mechanisms, one of which is frictional dissipating between layers of liquid moving at different velocities (here the rate of heat production, $\dot{H}$, is given by the product of the viscosity of the flowing liquid, η, and the square of the velocity gradient, G, developed within it).

The best place to start our investigation of the role of heating in the production of ultrasonic bioeffects is by reference to Fig. 3.16. This is a compilation of the "threshold intensity" data necessary for the production of histologically detectable lesions in living tissues as a function of the duration of the applied ultrasonic exposure. The figures inside each symbol represent the frequency in MHz em-

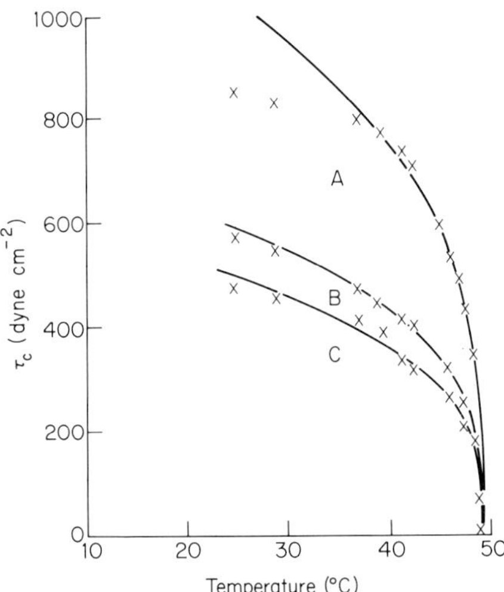

FIG. 3.15 These curves show that the hydrodynamic shear stress required to disrupt human erythrocytes in laminar flow decreased as the temperature of the suspending medium was increased. Curve A was obtained after an exposure of 1 min, curve B after 5 min and curve C after 10 min showing that this effect is also dependent upon the time for which that elevated temperature was maintained (from Krizan and Williams, 1977).

ployed by each investigator. The tissues that have been studied include cat brain (□, Fry *et al.*, 1970) and (⬡, Basauri and Lele, 1962); rat brain (○, Pond, 1968); and rabbit liver (▽), kidney (◇) and testicle (△) (all from Frizzell *et al.*, 1977). It can be seen that all the results obtained with brain tissue fall on a single straight line and that brain tissue appears to be more susceptible to ultrasonically-induced damage than the other adult tissues which have been investigated. Other researchers have repeated and extended these measurements and have investigated other tissues using different end-points (e.g. the production of visible cataracts in the lens of the rabbit eye; Lizzi *et al.*, 1978) but these have not been included in Fig. 3.16 for the sake of clarity.

It should be noted that the slope of this straight line is influenced by a number of factors, one of them being the tissue temperature which has to be attained and maintained for a given time before that tissue is damaged (Fig. 3.14). This temperature will be a function of the incident ultrasonic intensity and the duration of exposure and will also be influenced to a lesser degree by the applied frequency and the absorption coefficient of that tissue. If these results are only due to a rise of temperature then one should obtain virtually identical results

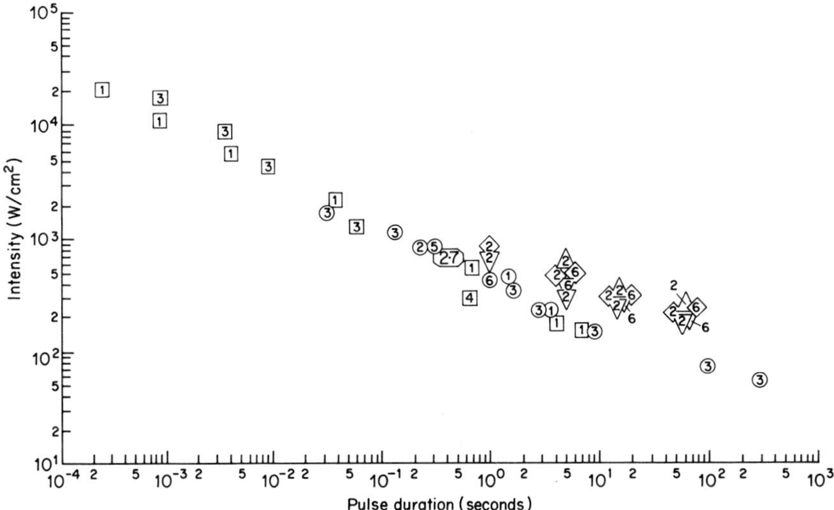

FIG. 3.16 This is a graphical compilation of the lowest acoustic intensities of single pulses of ultrasound which are capable of producing histologically detectable lesions within adult mammalian tissues *in vivo* as a function of the duration of that pulse. The number within each symbol refers to the ultrasonic frequency (in MHz) used, and the shape of each symbol refers to a particular reference which is quoted in the text.

if several short pulses of moderately high intensity are delivered in the same time taken by a single pulse of lower intensity containing the same total amount of energy (in the absence of non-linear effects).

The histological examinations used to determine the extent of tissue damage in Fig. 3.16 revealed striking differences between the morphology of the lesions produced in brain tissue at the extreme ranges of exposure conditions. Lesions produced at intensities greater than about 2000 W/cm^2 and time durations less than about 40 msec are characterized by having large empty-looking areas containing severely disrupted cells and cellular debris which are present immediately after the irradiation has been terminated. These lesions are almost certainly caused by the destructive effects of (transient) cavitation (Fry *et al.*, 1970). At the other extreme, i.e. intensities less than about 200 W/cm^2 and exposure times longer than about 10 s, the histological damage takes a minimum of 10 min to become evident and the cells remain intact. Carstensen *et al.* (1974) have concluded that tissue damage and destruction in this latter region can be explained on a purely thermal basis. If the results at one end of the straight line presented in Fig. 3.16 are dominated by cavitation whereas the results at the other end are dominated by a thermal mechanism, it can be argued that the results in the middle of the line may be affected by both mechanisms to a roughly equal extent (unless of course a third, as yet unknown, mechanism is dominating in this region).

The most important conclusion to be drawn from these experiments is that in the absence of active gas bubbles, heat production is the dominant mechanism resulting in damage to soft tissues *in vivo* during the long exposure times and low intensities used in physiotherapy. This statement needs some qualification in that the experimental evidence upon which it is based was largely obtained using focused transducers. It will be shown in Chapter 4 that focused transducers tend to suppress the effects of cavitation and it has already been mentioned that they tend to accentuate non-linear effects and so tend to magnify the role of temperature as the damaging agent.

If thermal effects are the major source of tissue damage then it would be a logical step to express the quantity of ultrasonic energy delivered to a given tissue or organ in terms of its absorbed dose or dose rate (Wells, 1973; Johnston and Dunn, 1976). However, if the tissue damage is caused, at least in part, by some non-thermal mechanism, then the concept of absorbed dose becomes less meaningful and the critical parameters at any given frequency become the peak particle displacement amplitude and the duration and interpulse interval of a pulsed waveform. In view of our present lack of knowledge concerning damaging mechanisms *in vivo* under the conditions used in medical diagnosis and therapy, we are probably best served by measurements of intensity, exposure time and pulsing parameters.

Apart from the obvious surgical applications of ultrasound in producing trackless lesions at specific points within brain tissue, ultrasound is also used in other medical and surgical applications where its heating effects appear to dominate. For example, Marmor *et al.* (1979) studied the effects of ultrasonically-induced local heating on superficial tumours in humans and observed that many irradiated nodules underwent at least partial regression. A different approach has been advocated by Lele (1979) who suggests that the whole patient should be cooled and then administered a cytotoxic drug which is only utilized by warm tissues. The tumour is then selectively heated using a movable focused transducer. Providing that the rate of clearance of the drug from the circulation is not affected by the whole body hypothermia, it is eventually removed from the circulation and the patient may be restored to normal body temperature.

Other proposed applications include the use of high intensity beams of focused ultrasound directed through the abdominal wall at the foetus and the site of implantation to induce the termination of a pregnancy (Sikov, 1973). Fahim *et al.* (1975) observed that rats whose testes had been irradiated with ultrasound at an intensity of about 1 Wcm^{-2} became sterile for 65–80 days whereas rats irradiated at an intensity of 2 Wcm^{-2} did not impregnate females at all during the 10-month duration of the study. The authors interpret their results in terms of a heat-induced sterility and suggest that this technique might be employed as a reversible and/or irreversible method of sterilization for humans. In view of other available techniques for achieving these same objectives, there does not seem to be any valid justification for subjecting humans to what could turn out to be a potentially hazardous procedure with disastrous consequences if it proved to be ineffective or dangerous.

The most well known application of ultrasound in surgery is in the treatment of Menière's disease; this is a disorder of the vestibular end organ of the inner ear which results in the spasmodic occurrence of attacks of vertigo of varying duration and severity. Arslan (1955, 1958) first suggested that ultrasound may be used to treat this condition which severely curtails the sufferer's well-being and enjoyment of life. Numerous improvements to the technique have been suggested and several specialized applications have been developed to deliver ultrasound to the vestibular apparatus without damaging the facial nerve or the cochlea (Gordon, 1962; James *et al.*, 1963; Kossoff, 1964; Johnson, 1967). Satisfactory results have been obtained in clinical trials (James, 1963; Kossoff and Khan, 1966). Small thermocouples were inserted into the temporal bones of cadavers who had been prepared for the operation and were irradiated in the normal manner. James and Halliwell (1970) showed that the temperature of the membraneous lining of the labyrinth in the region of the probe may exceed 53°C at the power levels used during a routine operation. This temperature increase is more than adequate to cause all the observed destructive changes.

3.9 ULTRASOUND IN PHYSIOTHERAPY

The surgical applications of ultrasound require the use of high ultrasonic intensities (tens to thousands of watts/cm^2) so that the elevated temperatures may kill some or all of the cells being irradiated. On the other hand, the object of physiotherapy is to assist the body in the process of repairing itself and so requires lower temperature increases to be effective. Unfortunately, it is difficult to assess the role of elevated temperature in physiotherapy because of the notable lack of well-designed and well-executed investigations into the effectiveness and mechanisms of action of any given mode of treatment (Valtonen, 1968; Clarke and Stenner, 1976). Despite the detailed comprehensive instructions on treatment regimes and methodology found in many textbooks on the subject, they are largely based upon an insubstantial framework of speculative beliefs instead of a solid data base of proven efficacy. Some noteworthy articles have appeared describing well-planned trials which have been performed as well as could be expected given the variable nature of the conditions being treated (e.g. Gieder *et al.*, 1971; Dyson *et al.*, 1976). It is regrettable that this has not encouraged the majority of the practitioners of the art of physiotherapy to emulate their example and so possibly improve the quality of their treatment even if it is only by discarding or improving those techniques which can be shown to be ineffective.

One of the major obstacles is that it is notoriously difficult to assess the beneficial (or adverse) effects of a given physiotherapeutic procedure. For example, Wells (1977) quotes the study by Roman (1960) who attempted to assess the efficacy of ultrasonic therapy for low back pain, bursitis of the shoulder and myalgia in a total of 100 patients. Some of the patients in each of these groups were treated with ultrasound in the normal manner while the remainder of the patients in that group were treated in the same way but, unknown to the patient, the ultrasound was not switched on. Sixty per cent of the patients who received the ultrasound treatment were classified as good or normal; however, 72% of the sham irradiated group were also assigned to those same categories. The result highlights the difficulties experienced by those working in the field for the most probable interpretation of these results is not that the treatment is of no use (or worse still that it actually impedes the normal healing process) but that patients derive a great deal of benefit from receiving the care and attention of highly motivated people in whom they have confidence. The improvements resulting from this "nursing" care being greater than any changes induced by the ultrasound treatment itself.

The major source of this imprecision is the reliance on the patient for his or her own subjective impressions as to whether or not they have benefited from the treatment, e.g. "does it hurt less?" This subjective personal bias is also reflected in what might at first appear as objective and even quantifiable assessments such as the range of movement of a joint "until it hurts" or to walk "as far as you can". Anyone who has indulged in strenuous exercise such as running will know that there comes a point where you feel pain and have a desire to stop. If for some reason you are motivated to continue you can overcome or at least tolerate these unpleasant symptoms and achieve higher speeds or greater distances. Competition and encouragement are two factors which increase motivation; this is why you can usually run further and/or faster when part of a group than when alone. Most patients wish to be cured and any treatment (even one which is ineffective) or encouragement will increase their motivation so that they may actually feel less pain and genuinely exhibit an improved physical performance.

Therefore, trials ought to be devised so that the endpoints may be quantified in an objective manner; e.g. if you have to treat a large scar or bruise with ultrasound, only irradiate half of it. Leave the other half as a control and compare the rates of healing of both halves using a device which objectively measures skin irregularity or colour or even by comparison with a colour chart containing a range of graded tones similar to the colour changes occurring with the bruise. An even simpler method is to produce a pair of identical bruises on a volunteer by some painless means, e.g. suction, and to treat only one of them (Williams and Clifton, 1979). This approach can be applied to many conditions; for example the placebo effects during ultrasonic treatment may be investigated by increasing the applied ultrasonic intensity until the patient can detect it and then reducing the intensity to zero. The patients then believe that they are receiving some treatment and will probably respond better than similar patients who are treated with a much larger "dose" of ultrasound which has been administered over a long period at a slightly lower intensity so that it has not caused any detectable sensations.

It is not within the scope of this book to survey and critically assess the plethora of articles reporting the applications and effectiveness of ultrasound in physiotherapy. The pioneering work performed mainly in Germany has been reviewed by van Went (1954) and the work up to the early 1960s by Lehmann (1965) and more recently by Lehmann and de Lateur (1982) who concluded that one of the main beneficial effects of ultrasound was to increase the extent of blood perfusion (hyperaemia) as a result of an increased tissue temperature (for details see Abramson *et al.*, 1960; Bickford and Duff, 1953; Buchan, 1970). Hyperaemia may be one of the major factors contributing to the improved rate of healing of varicose ulcers reported by Dyson *et al.* (1976). However, Paaske *et al.* (1973) used ^{133}Xenon washout techniques to measure the volume rate of flow of blood through cutaneous, subcutaneous and muscular tissues and concluded that ultrasound only caused inconsistent and insignificant changes in the rate of flow of blood through the micro-vasculature. It is possible that the extravasation of blood plasma due to increased capillary permeability within the heated tissues trapped some of the Xenon within the intercellular matrix which may have masked any increased rate of blood flow. This technique is notoriously difficult to use and so this negative report should be viewed with some caution.

Some of the therapeutic applications of ultrasound give cause for concern. For example, Krivitskii (1971) repeatedly used therapeutic levels of ultrasound to treat traumatic cataracts. The cornea and lens of the eye do not have an efficient blood supply to carry away any heat deposited within them. This is one of the reasons why cataracts are among the first signs of exposure to relatively high power densities of microwave irradiation. The human lens has a high ultrasonic attenuation coefficient (15 Nepers cm^{-1} at 12 MHz; Nover and Glanschneider, 1965) and so there is a real danger of inadvertently producing a thermal cataract as a result of this therapeutic procedure. Other therapeutic applications giving cause for concern are the enthusiastic administration of ultrasound to the female genital organs which seems to be especially popular in eastern Europe and Asia (e.g. Burgudzhieva, 1974; Klemenkova *et al.*, 1970; Krylova, 1971; Denk and Lewandowska, 1977; Fofanov, 1977). The question of the safety of ultrasound cannot be guaranteed and so therefore it would not be prudent to irradiate the ovaries or perhaps even an early gestation sac with therapeutic intensities of ultrasound (see Chapter 5).

REFERENCES

Abramson, D. I., Burnett, C., Bell, Y., Tuck, S., Rejal, H. and Fleischer, C. J. (1960). Changes in blood flow, oxygen uptake and tissue temperature produced by physical agents. I. Effect of Ultrasound. *Am. J. Phys. Med.* **39**, 51–62.

Andreae, J. H. and Lamb, J. (1959). Ultrasonic relaxation theory for liquids. *Proc. Royal Soc. Lond.* **B69**, 814–822.

Arslan, M. (1958). Ultrasonic surgery of the labyrinth in patients with Menière's syndrome. *Sci. Med. Ital.* **7**, 301–326.

Bamber, J. C., Fry, J. J., Hill, C. R. and Dunn, F. (1977). Ultrasonic attenuation and backscattering by mammalian organs as a function of the time after exision. *Ultrasound Med. Biol.* **3**, 15–20.

Basauri, L. and Lele, P. P. (1962). A simple method for production of trackless focal lesions with focussed ultrasound. *J. Physiol.* **160**, 513–534.

Beyer, R. T. (1974). *In* "Non-linear Acoustics", U.S. Naval Sea Systems Command, Washington D.C.

Bickford, R. H. and Duff, R. S. (1953). Influence of ultrasonic irradiation on temperature and blood flow in human skeletal muscle. *Circulation Res.* **1**, 534–538.

Blackstock, D. T. (1970). History of Nonlinear Acoustics. *In* "Nonlinear Acoustics" (Ed. T. G. Muir), Proc. 1969 ARL Symposium (2nd Intern. Symp. on Nonlinear Acoustics in Austin). Applied Research Laboratories, The Univ. of Texas, Austin.

Buchan, J. F. (1970). The use of ultrasonics in physical medicine. *Practitioner* **205**, 319–326.

Burgudzhieva, T. (1974). Effect of ultrasonics, hydrocortisone ointment and phonophoresis of hydrocortisone ointment on indices of sensory chronaximetry of patients with kraurosis vulvae, lichen scleroatrophicus vulvae and pruritus vulvae essentialis. *Vopr. Kurortol. Fizioter. Lech. Fiz. Kult.* **50**, 267–269.

Busse, L. J., Miller, J. G., Yuhas, D. E., Mimbs, J. W., Weiss, A. N. and Sobel, B. E. (1977). Phase cancellation effects: A surce of attenuation artifact eliminated by a CdS acousto-electric receiver. *In* "Ultrasound in Medicine" (Ed. D. N. White), Vol. **3**, pp. 1519–1535. Plenum Press, New York.

Carstensen, E. L. and Schwan, H. P. (1959a). Absorption of sound arising from the presence of intact cells in blood. *J. Acoust. Soc. Amer.* **31**, 185–189.

Carstensen, E. L. and Schwan, H. P. (1959b). Acoustic properties of haemoglobin solutions. *J. Acoust. Soc. Amer.* **31**, 305–311.

Carstensen, E. L., Law, W. K., McKay, N. D. and Muir, T. G. (1980). Demonstration of nonlinear acoustical effects at biomedical frequencies and intensities. *Ultrasound Med. Biol.* **6**, 359–368.

Carstensen, E. L., Miller, M. W. and Linke, C. A. (1974). Biological effects of ultrasound. *J. Biol. Phys.* **2**, 173–192.

Chivers, R. C. and Hill, C. R. (1975). Ultrasonic attenuation in human tissue. *Ultrasound Med. Biol.* **2**, 25–29.

Clarke, G. R. and Stenner, L. (1976). Use of therapeutic ultrasound. *Physiotherapy* **62**, 185–190.

Denk, J. and Lewandowska, L. (1977). Ultrasonics in the treatment of female sterility. *Ginekol. Pol.* **48**, 567–570.

Dunn, F. and Fry, W. J. (1961). Ultrasonic absorption and reflection by lung tissue. *Phys. Med. Biol.* **5**, 401–410.

Dunn, F., Law, W. K. and Frizzell, L. A. (1980). Sound propagation in liquid media. Paper presented at the "Ultrasound Interaction in Medicine and Biology Symposium", Reinhardsbrunn, East Germany, 10–14th Nov. 1980.

Dunn, F., Law, W. K. and Frizzell, L. A. (1982). Nonlinear ultrasonic propagation in biological media. *Brit. J. Cancer* **45**, 55–58.

Dyson, M., Franks, C. and Suckling, J. (1976). Stimulation of healing of varicose ulcers by ultrasound. *Ultrasonics* **14**, 232–236.

Fahim, M. S., Fahim, Z., Der, R., Hall, D. G. and Harman, J. (1975). Heat in male contraception (Hot water 60°C, infrared, microwave, and ultrasound). *Contraception* **11**, 549–562.

Fofanov, S. I. (1977). Evaluation of the various physical methods of treatment of chronic inflammation of the female genital organs. *Akush. Ginekol. (Mosk.)* **53**, 23–26.

Frank, H. S. and Evans, M. W. (1945). Free volume and entropy in condensed systems. III. Mixed liquids. *J. Chem. Phys.* **13**, 507.

Frizzell, L. A., Linke, C. A., Carstensen, E. L. and Fridd, C. W. (1977). Thresholds for focal ultrasonic lesions in rabbit kidney, liver and testicle. IEEE Trans. Biomed. Eng. *BME-24*, 393–396.

Fry, F. J., Kossoff, G., Eggleton, R. C. and Dunn, F. (1970). "Threshold" ultrasonic dosages for structural changes in the mammalian brain. *J. Acoust. Soc. Amer.* **48**, 1413–1417.

Gordon, D. (1962). An improved ultrasonic transducer for the surgery of the labyrinth. *Proc. San Diego Symp. Biomed. Engn.* 23–24.

Goss, S. A., Johnson, R. L. and Dunn, F. (1978). Comprehensive compilation of empirical ultrasonic properties of mammalian tissues. *J. Acoust. Soc. Amer.* **64**, 423–457.

Goss, S. A., Frizzell, L. A. and Dunn, F. (1979a). Ultrasonic absorption and attenuation in mammalian tissues. *Ultrasound Med. Biol.* **5**, 181–186.

Goss, S. A., Johnston, R. L., Maynard, V., Brady, J. K., Frizzell, L. A., O.Brien, W. D. Jr and Dunn, F. (1979b). Elements of tissue characterisation, part II. Ultrasonic propagation parameters. *In* "Ultrasonic Tissue characterisation II" (NBS Special Publ. 525) (Ed. M. Linzer), pp. 43–51. U.S. Govt Printing Office, Washington D.C.

Grieder, A., Vinton, P. W., Cinotti, W. R. and Kangur, T. T. (1971). An evaluation of ultrasonic therapy for temporomandibular joint dysfunction. *Oral Surg.* **31**, 25–31.

ter Haar, G. R. and Hopewell, J. W. (1982). Ultrasonic heating of mammalian tissues *in vivo*. *Brit. J. Cancer* **45**, 65–67.

Hall, L. (1948). The origin of ultrasonic absorption in water. *Phys. Rev.* **73**, 775–781.

Hawley, S. A. and Dunn, F. (1969). Ultrasonic absorption in aqueous solutions of Dextran. *J. Chem. Phys.* **50**, 3523–3526.

Hueter, T. F. (1958). Visco-elastic losses in tissues in the ultrasonic range. W.A.D.C. Technical Report No. 57–706.

Hussey, M. and Edmonds, P. D. (1971). Ultrasonic examination of proton-transfer reactions in aqueous solutions of glycine. *J. Acoust. Soc. Amer.* **49**, 1309–1316.

James, J. A. (1963). New developments in the ultrasonic therapy of Menière's disease. *Ann. Roy. Coll. Surg. Engl.* **33**, 226–244.

James, J. A., Dalton, G. A., Hadley, K. J., Freundlich, H. F., Bullen, M. A. and Wells, P. N. T. (1963). A new 3-megacycle generator for destruction of the vestibular end organ. *Acta Oto-Laryngol.* **56**, 148–153.

James, J. A. and Halliwell, M. (1970). Thermal effects of ultrasound on the temporal bone. *Med. Biol. Engn.* **8**, 477–481.

Johnston, R. L. and Dunn, F. (1976). Ultrasonic absorbed dose, dose rate and produced lesion volume. *Ultrasonics* July, 153–156.

Johnson, S. J. (1967). An ultrasonic unit for the treatment of Menière's disease. *Ultrasonics* **5**, 173–176.

Kessler, L. W., O'Brien, W. D. Jr and Dunn, F. (1970). Ultrasonic absorption in aqueous solutions of polyethylene glycol. *J. Phys. Chem.* **74**, 4096.

Kinsler, L. E. and Frey, P. (1962). "Fundamentals of Acoustics". Wiley, New York.

Klemenkova, I. G., Gisina, A. M. and Gordeeva, N. K. (1970). Ultrasonic treatment of tuberculosis in the female genitalia. *Probl. Tuberk.* **48**, 56–58.

Kossoff, G. (1964). Design of the C.A.L. Ultrasonic generator for the treatment of Menière's disease. *IEEE Trans. Sonics Ultrason.* **SU-11**, 95–101.

Kossoff, G. and Khan, A. E. (1966). Treatment of vertigo using the ultrasonic generator. *Arch. Otolaryngol. Zh.* **26**, 61–62.

Kremkau, F. W. (1972). Macromolecular interactions in the absorption of ultrasound in biological material. PH.D. Dissertation, University of Rochester, New York, U.S.A.

Kremkau, F. W. and Carstensen, E. L. (1972). Macromolecular interaction in sound absorption. *In* "Interaction of Ultrasound and Biological Tissues" (Eds J. M. Reid and M. R. Sikov), pp. 37–42. DHEW Publication (FDA) 73–8008. BRH/DBE 73–1.

Kremkau, F. W., Carstensen, E. L. and Aldridge, W. G. (1973). Macromolecular interaction in the absorption of ultrasound in fixed erythrocytes. *J. Acoust. Soc. Amer.* **53**, 1448–1451.

Krivitskii, A. K. (1971). Effectiveness of repeated ultrasonic therapy in traumatic cataracts. *Oftalmol. Zh.* **26**, 61–62.

Krizan, J. A. and Williams, A. R. (1977). Non-equilibrium co-operative model for a biomembrane under hydrodynamic shear. *Collective Phenomena* **2**, 229–234.

Krylova, T. I. (1971). Use of Ultrasound for treatment of patients with leukoplakia of the external genital organs. *Akush. Ginekol. (Mosk.)* **47**, 58–59.

Lehmann, J. F. (1953). The biophysical basis of biologic ultrasonic reactions with special reference to ultrasonic therapy. *Arch. Phys. Med.* **34**, 139–152.

Lehmann, J. F. (1965). Ultrasound therapy. *In* "Therapeutic Heat and Cold" (Ed. S. Licht), pp. 321–386. Waverley Press, Baltimore.

Lehmann, J. F. and Guy, A. W. (1972). Ultrasound therapy. *In* "Interaction of Ultrasound and Biological Tissues" (Eds J. M. Reid and M. R. Sikov), pp. 141–152. DHEW Publication (FDA) 73–8008.

Lehmann, J. F. and de Lateur, B. J. (1982). Therapeutic heat. *In* "Therapeutic Heat and Cold-Third Edition". Williams and Wilkins, Baltimore.

Lehmann, J. F., de Lateur, B. J., Warren, C. G. and Stonebridge, J. B. (1967). Heating produced by ultrasound in bone and soft tissue. *Arch. Phys. Med.* **48**, 397–401.

Lehmann, J. F., Masock, A. J., Warren, C. G. and Koblanski, N. (1970). Effects of therapeutic temperatures on tendon extensibility. *Arch. Phys. Med.* **51**, 481–487.

Lele, P. P. (1977). Thresholds and mechanisms of ultrasonic damage to "organised" animal tissues. *In* "Symposium on Biological effects and Characterizations of Ultrasound Sources" (Eds D-W. G. Hazzard and M. L. Litz), pp. 224–237. HEW Publication (FDA) 78–8048.

Lele, P. P. (1979). A strategy for localized chemotherapy of tumors using ultrasonic hyperthermia. *Ultrasound Med. Biol.* **5**, 95–97.

Lele, P. P. and Senapati, N. (1977). The frequency spectra of energy backscattered and attenuated by normal and abnormal tissue. *In* "Recent Advances in Ultrasound in Biomedicine" (Ed. D. N. White), pp. 55–58. Research Studies Press, Forest Grove.

Lizzi, F. L., Packer, A. J. and Coleman, D. J. (1978). Experimental cataract production by high frequency ultrasound. *Ann. Opthalmol.* **10**, 934–942.

Marcus, P. W. and Carstensen, E. L. (1975). Problems with absorption measurements of inhomogeneous solids. *J. Acoust. Soc. Amer.* **58**, 1334–1335.

Marmor, J. B., Pounds, D., Postic, T. B. and Hahn, G. M. (1979). Treatment of superficial human neoplasms by local hyperthermia induced by ultrasound. *Cancer* **43**, 188–197.

Meyers, R., Fry, F. J., Fry, W. J., Eggleton, R. C. and Schultz, D. F. (1960). Determination of topological human brain representations and modifications of signs and symptoms of some neurologic disorders by the use of high level ultrasound. *Neurology* **10**, 271–277.

Morris, D. R. and Williams, A. R. (1980). The internal viscosity of the human erythrocyte may determine its lifespan *in vivo*. *Scand. J. Haematol.* **24**, 57–62.

Mountford, R. A. and Wells, P. N. T. (1972). Ultrasonic liver scanning; the quantitative analysis of the normal A-scan. *Phys. Med. Biol.* **17**, 14–25.

Muir, T. G. and Carstensen, E. L. (1980). Prediction of nonlinear acoustic effects at biomedical frequencies and intensities. *Ultrasound Med. Biol.* **6**, 345–357.

Nicholas, D. (1979). Ultrasonic diffraction analysis in the investigation of liver disease. *Brit. J. Radiol.* **52**, 949–961.

Nover, A. and Glanschneider, D. (1965). "Untersuchungen über die Fortpflanzungsgeschwindigkeit und Absorption des Ultraschalls im Gewebe. Experimentelle Beitrage zur Ultraschall diagnostik intraocularer Tumoren". *Graefes. Arch. Klin. Exp. Ophthal.* **168**, 304–321.

Nyborg, W. L. (1975). "Intermediate Biophysical Mechanics". Cummings, Menlo Park, California.

Nyborg, W. L. (1977). "Physical Mechanisms for Biological Effects of Ultrasound", pp. 1–59. HEW Publication (FDA) 78–8062.

O'Brien, W. D. Jr and Dunn, F. (1972). Ultrasonic absorption by biomacromolecules. *In* "Interaction of Ultrasound and Biological Tissues" (Eds J. M. Reid and M. R. Sikov), pp. 13–19. DHEW Publication (FDA) 73–8008.

Paaske, W. P., Hovind, H. and Sejrsen, P. (1973). Influence of therapeutic ultrasonic irradiation on blood flow in human cutaneous, subcutaneous and muscular tissues. *Scand. J. Clin. Lab. Invest.* **31**, 389–394.

Pauly, H. and Schwan, H. P. (1971). Mechanism of absorption of ultrasound in liver tissue. *J. Acoust. Soc. Amer.* **50**, 692–699.

Pond, J. (1968). A Study of the biological action of focussed mechanical waves (focussed ultrasound). Ph.D. Thesis, University of London.

Roman, M. P. (1960). A clinical evaluation of ultrasound by use of a placebo technic. *Phys. Ther. Rev.* **40**, 649–652.

Schneider, F., Muller-Landau, F. and Mayer, A. (1969). Acoustical properties of aqueous solutions of oxygenated and deoxygenated haemoglobin. *Biopolymers* **8**, 537–544.

Sikov, M. R. (1973). Ultrasound; its potential use for the termination of pregnancy. *Contraception* **8**, 429–438.

Smith, A. and Schwan, H. P. (1971). Acoustic properties of cell nuclei. *J. Acoust. Soc. Amer.* **49**, 1329–1330.

Valtonen, E. J. (1968). A historical method for measuring the influence of ultrasonic energy on living tissue under experimental conditions. *Acta Rheum. Scand.* **14**, 35–42.

Wells, P. N. T. (1973). Letter to the Editor. *Ultrasound Med. Biol.* **1**, 53.

Wells, P. N. T. (1977). "Biomedical Ultrasonics". Academic Press, London and New York.

Went, J. M. van (1954). "Ultrasonic and Ultrashort Waves in Medicine". Elsevier, Amsterdam.

Williams, A. R. and Clifton, H. (1979). Effects of therapeutic ultrasound on the rate of fading of human bruises. Paper presented at the London meeting of the British Bioacoustical Discussion Group at the British Institute of Radiology. July 6th.

Zana, R., Lang, J., Tondre, C. and Sturm, J. (1972). Interactions of ultrasound with proteins and nucleic acids in solution. *In* "Interaction of Ultrasound and Biological Tissues" (Eds J. M. Reid and M. R. Sikov), pp. 21–26. DHEW Publication (FDA) 73–8008.

Zaretskii, A. A., Zorina, O. M., Fursov, K. P. and El'Piner, I. E. (1972). Characteristics of the absorption of ultrasound by protein solutions. *Soviet Phys. Acoust.* **17**, 397–399.

Zorina, O. M., Fursov, K. P. and El'Piner, I. E. (1971). Attenuation of ultrasound in solutions of serum albumin and its hydrolyzates. *Soviet Phys. Acoust.* **17**, 129–130.

4. MECHANICAL AND CAVITATIONAL EFFECTS GENERATED BY ULTRASONIC WAVES

4.1 INTRODUCTION

It is perhaps not too surprising that mechanical waves such as ultrasound should be able to generate mechanical forces, and that these forces may, under certain circumstances, damage the tissues through which the wave is propagating (Dunn and Pond, 1978). Each molecule within the medium in the path of the wave is driven to oscillate about a fixed point (i.e. a first order effect) with an amplitude proportional to the square root of the intensity at that site. Thus, adjacent portions of the same macromolecular or membraneous structure may be vibrating with different displacement amplitudes and so may be "fatigued" and rupture.

However, the largest and most damaging mechanical forces which are generated by an ultrasonic wave are the fluid shear forces and/or shock waves produced by the oscillatory behaviour of minute bubbles containing gas or vapour (i.e. *cavitation*). The general term cavitation encompasses a wide range of bubble behaviour from the relatively gentle linear pulsations of gas-filled bodies to the violent and highly destructive formation and collapse of vapour-filled voids and cavities. Stabilized gas bubbles driven to oscillate by ultrasonic intensities similar to or even lower than those emitted by the diagnostic machines in current use generate steady-state (i.e. not oscillatory) hydrodynamic shear fields as a second order effect which can produce a variety of lethal and sub-lethal damage in adjacent cellular structures.

The third major class of mechanical forces generated by ultrasonic waves are the radiation pressure forces which effectively push or pull cells and gas bodies to different portions of an inhomogeneous acoustic field. These effects are most evident in situations where a wave is reflected back along its own path (i.e. in a standing wave field) and they may greatly enhance the development of cavitation-like activity, and hence the destructive effects resulting from its occurrence. In addition, these radiation pressure forces may, under certain circumstances, result in the temporary or permanent blockage of small capillaries with "plugs" of erythrocytes (Dyson *et al.*, 1971).

4.2 FIRST ORDER EFFECTS

These are a direct result of the alternating (a.c.) motion of the particles of the medium induced by the propagation of the wave through it. Each particle is initially displaced from its resting position and overshoots this position on its return journey so that it oscillates back and forth about this point with gradually decreasing amplitude until it is eventually stationary in its original position after the wave has departed. The maximum displacement amplitude of each particle is determined by the amount of energy contained within the wave at that site and is proportional to the square root of the incident intensity at any given frequency. The frequency is important because the particles are driven to vibrate faster at high frequencies and will therefore require the input of more energy to be driven at the same maximum displacement amplitude as the frequency is raised. Thus at any given intensity, the maximum displacement amplitude (Ao) of the particles of the medium decreases with increasing frequency (i.e. $Ao \propto 1/f$). For a wave of frequency 1 MHz and an intensity of 1 Wcm^{-2}, Ao is equal to about 0·018 μm.

The velocity or speed at which this particle is moving varies from zero, when the particle has reached its maximum displacement amplitude, to its maximum value when the particle is passing through its point of origin, i.e. through the point of zero displacement amplitude. The maximum value of this velocity is also a function of the energy content of the wave and is again proportional to the square root of the applied intensity. For the same 1 MHz wave at an intensity of 1 W/cm^{-2} the maximum value of the particle velocity is 12 cm s^{-1}.

The oscillatory motion of each particle means that it must be accelerated from its stationary position (i.e. its position of maximum displacement amplitude) until it attains its maximum velocity.

Thereafter, it must be decelerated at the same rate until its velocity is again zero when it reaches its other maximum displacement amplitude. The maximum value of this acceleration is also proportional to the square root of the intensity at that site. As the frequency of the wave increases each particle has to undergo more cycles of acceleration and deceleration per second and so the maximum values developed have to be increased. These instantaneous acceleration forces are very large compared with the forces we normally encounter; for example, a 1 MHz wave at an intensity of 1 $W cm^{-2}$ develops a maximum particle acceleration of 71 000 times the acceleration due to gravity here on earth (i.e. 71 000 g).

Thus, a "typical" ultrasonic beam which might be used in ultrasonic therapy produces minute particle displacements with small particle velocities but is associated with enormously large values of instantaneous particle accelerations.

4.2.1 Can These Forces Damage Biological Tissues?

The crucial question is not only what are the magnitude of the forces generated by a given ultrasonic beam but how are those forces distributed within the irradiated tissues. Damage usually results when different portions of the same structure are subjected to different forces so that it is twisted or torn. A simple analogy is that of a well-balanced see-saw: this remains level whether the weights at either end are both one gram or both one ton. It is only when each end bears a different weight that a torque is developed which results in motion.

Unlike the particle velocity, the velocity of the parent ultrasonic wave is constant in any one medium and is independent of the applied frequency (this is not strictly true but effects due to dispersion etc. are typically small; Carstensen and Schwan, 1959). For water and most soft tissues this velocity is about 1500 $m s^{-1}$. The wavelength is given by the velocity divided by the frequency and so a 1 MHz wave has a wavelength of about 1·5 mm *in vivo*. Within this wavelength at any one instant of time some of the particles will be in the process of being displaced in the forward direction whilst others are being displaced in the opposite direction. The distance between the regions of maximum opposite motion is one half wavelength or about 750 μm at 1 MHz. Thus, structures having dimensions of this order or larger will be subjected to cyclical strains equal to twice the maximum particle displacement amplitude. At an intensity of 1 $W cm^{-2}$ a 1 MHz wave would therefore stretch and compress a structure 750 μm long by about 0·04 μm or 0·005% of its length. Structures which are much smaller than a wavelength, e.g. a "typical" cell which may be only about 40 μm in diameter, are subjected to an even smaller strain because the whole cell and its immediate neighbours are all being displaced together in the same direction at essentially the same time.

Although these strains are small, they are being applied a million times every second in a continuous 1 MHz beam. It is therefore possible that biological structures, particularly membranes, may be "fatigued" over a long period of time. It is difficult to demonstrate fatigue in biological systems, partly because the organism may detect the applied force and compensate for it by strengthening the structures you are attempting to fatigue. For example, bone is a solid material which is designed to withstand loads and bear strains. If it is not strained, as may occur in astronauts or in people who are confined to bed for long periods, it tends to lose calcium salts and become weaker. Conversely, if it is repeatedly strained it is made stronger by the deposition of more calcium salts.

4.2.2 "Fatigue" in Biological Systems

The only cell membrane for which there is convincing evidence of the occurrence of a fatigue-type process is that of the human erythrocyte, or red blood cell. This cell is unique in that it consists of only a plasma membrane enclosing a homogeneous fluid in which the protein haemoglobin is dissolved. The cell is extremely flexible because the surface area of the membrane is "too large" for the volume it encloses and so the cell may be deformed to an infinite variety of shapes without stretching the membrane. If these cells are suspended in an isotonic saline medium containing an inert thickening agent such as Dextran (a polyglucose) and subjected to hydrodynamic shear stress in laminar flow, they will distort to give a cigar-shaped prolate elipsoid which will be aligned parallel to the direction of flow (Schmid-Schönbein and Wells, 1969). The cell maintains its stable orientation because its membrane effectively rotates about the centre of the cell in a manner analogous to the flexible tread of a military tank rotating around its set of drive and load-bearing wheels. Morris and Williams (1979) demonstrated that the membrane of the human erythrocyte underwent a time and stress dependent series of progressive structural rearrangements during its rotation which eventually resulted in the failure of the membrane and the release of the haemoglobin contents (i.e. it was fatigued). Similarly, Williams (1971) has shown that those erythrocytes which survived repeated exposure to ultrasonically-induced hydrodynamic shear stresses *in vitro* exhibited an enhanced rate of autolysis during sterile incubation at 37°C (Fig. 4.1). This result may also be interpreted in terms of progressive sub-lethal damage to the erythrocyte membrane which, if it was continued, would result in irreversible damage to the cell.

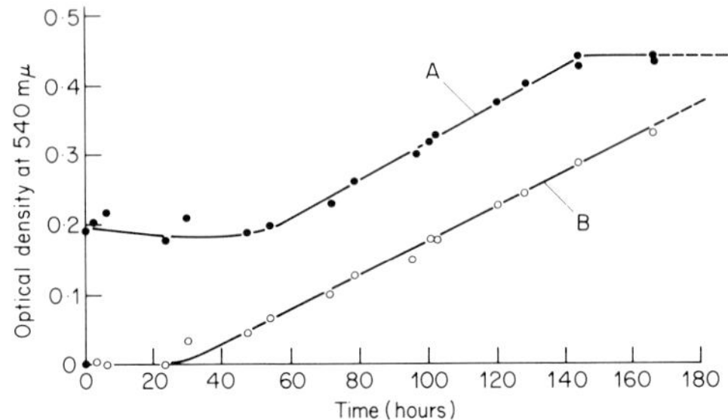

FIG. 4.1 The time-dependent autolysis of human erythrocytes during sterile incubation at 37°C. Cells subjected to "sub-lethal" acoustic microstreaming forces (A) lyse at a faster rate than control cells (B) suggesting that their membranes may have been "fatigued" (from Williams, 1971).

4.3.3 Effects on Thixotropic Structures

Another physical property of biological tissue and its constituents which may be altered by these first order forces is that of thixotropy. In simple terms, a thixotropic material is one which has a three-dimensional (solid-like) structure in the absence of an applied shear field but becomes a liquid when it is subjected to shear. The most common example of a thixotropic liquid is that of a non-drip paint which exists in the form of a gel until it is stirred or squeezed between the brush and the surface to be painted. After it has been applied in the normal manner it is no longer subjected to a shear force and slowly returns to its gel-like state so that it should not "run" as much as non-thixotropic paints.

Concentrated suspensions of many biological materials develop some thixotropic behaviour because the molecules form weak intermolecular linkages with each other and/or physically intertwine or "tangle up" if the molecules are loose random coils. These intermolecular forces are usually long-range electrostatic or hydrogen bonds and are therefore easily broken by an applied velocity gradient. For example, Williams (1969a) has shown that dilute solutions of native mammalian DNA are thixotropic at velocity gradients less than about 0·001 sec^{-1}. Similarly, the haemoglobin solution inside the intact human erythrocyte may be shown to exhibit some thixotropic behaviour at the concentrations present inside the oldest and most senile cells within the normal population (Williams, A. R., unpublished observations).

It has been proposed that structures such as the mitotic spindle formed during cell division are thixotropic. This is supported by the early visual observations of mitosis in plant cells where it was noted that the mitotic spindle broke down when the microscope stage was accidently or deliberately struck. The shock wave propagating as an acoustic wave through the sample as a result of the sharp blow to the microscope stage contains a wide range of frequencies but is otherwise essentially similar to the ultrasonic waves generated by the equipment used in medical diagnosis and therapy. It is therefore possible, but has not been demonstrated, that an ultrasonic wave may liquefy the mitotic spindle during cell division. If this was to occur *in vivo* it would almost certainly lead to the death of the cells concerned. An additional potential hazard is that if this was to occur *in vivo* during the terminal stages of meiosis it might result in the production of viable gametes which could be deficient in one or more chromosomes or conversely contain one or more extra chromosomes. The results of the genetic experiments performed to date suggest that this is not a likely occurrence, or even if it does occur the gametes are either non-viable or are unsuccessful at or incapable of causing fertilization. Nevertheless, this is a potential hazard which ought to be investigated in more detail.

To summarize, the first order forces produced within tissues by ultrasound at the frequencies and intensities commonly employed in diagnosis and therapy are characterized by extremely small particle displacements, moderate particle velocities and extremely high particle accelerations. It is fortunate that the dimensions of the cellular structures are usually much smaller than the wavelength of the travelling wave so that the entire cell tends to be displaced as a single entity which reduces the possibility of cell damage due to a fatigue mechanism. The strains produced in structures which are larger than a wavelength are extremely small compared with the strains that those same structures normally bear during strenuous exercise. Finally, it was suggested that ultrasound may liquefy thixotropic structures including the mitotic and/or meiotic spindles. There is no firm evidence to substantiate this proposal, but it merits further investigation.

4.3 CAVITATION

This is an extremely complex topic which easily merits an entire book of its own. This section is therefore grossly oversimplified in an attempt to present a graphic portrayal of the various types of cavitation and their possible roles *in vivo*. Cavitation is best regarded as the oscillatory activity of highly compressible bodies such as gas and/or vapour-filled "bubbles" or "cavities" which are powered by and therefore extract energy from the incident acoustic field. Some of this energy (about 10%) is re-radiated as an acoustic wave but the remainder is transformed into other forms and may appear as heat (resulting in the production of highly energetic short-lived chemical species), or as shock waves or hydrodynamic shear fields which easily disrupt biological tissues. There are so many diverse physical and chemical mechanisms associated with the occurrence of cavitation that it should not be regarded as a single entity but rather as a broad spectrum of related phenomena. This continuous spectrum ranges from the linear pulsation of gas-filled bodies in low amplitude sound fields (i.e. "stable" acoustic cavitation) through to the violent and highly destructive non-linear "explosive" behaviour of vapour-filled cavities known as "transient" or "collapse-type" acoustic cavitation. The continuous nature of this phenomenon is illustrated by the fact that a bubble which is exhibiting all the characteristics associated with stable cavitation changes its behaviour as the intensity of the incident ultrasonic field is increased until it eventually exhibits most, if not all, of the characteristics associated with transient cavitation at high acoustic intensities.

Cavitation is a difficult phenomenon to study; it tends to be unpredictable in its occurrence because it requires far more energy to cause a bubble or cavity to be formed than it takes to drive a pre-existing bubble or cavity into oscillation. These bubbles and cavities apparently grow from minute pockets of stabilized gas known as micronuclei. Thus, the occurrence of any cavitational activity in a medium is crucially dependent upon the number and availability of these micronuclei, the intensity and its temporal characteristics, and the frequency of the ultrasonic source, the availability and composition of dissolved gas as well as many other variables. This has led many investigators to examine simpler systems such as stable cavitation which is apparently subject to fewer uncontrollable variables. However, even stable cavitation requires the presence of stable gas bubbles of the right size in the right place at the right time. Some investigators have gone one step further and have dispensed with gas bubbles altogether and generate physical phenomena similar to those developed around oscillating bubbles by causing small metal wires or probes to oscillate while submerged in a liquid.

This section will therefore begin with a description of the physical forces generated by these simple metal models of oscillatory bubbles and will indicate some of the biological effects they can produce. These results will then be compared with those obtained using single bubbles or arrays of stable gas bubbles. Finally, we will look at the characteristics of "free-bubble" cavitation and at some of the factors affecting its occurrence *in vitro* and *in vivo*.

4.3.1 Models of Stable Acoustic Cavitation

Holtzmark *et al.* (1954) have shown that a complex steady-state or time-independent streaming pattern is established around a stationary solid cylinder immersed in a fluid through which an acoustic wave is propagating. Exactly the same solution is obtained if the fluid is kept stationary and the cylinder is driven to oscillate at the same acoustic frequency. This steady-state streaming pattern results from an unequal distribution of time-independent forces around the surface of the cylinder (Nyborg, 1965) and is illustrated in Fig. 4.2. The dimensions of the patterns are determined by the boundary layer thickness (δ) which is given by the equation:

$$\delta = \left(\frac{2\eta}{\omega\rho}\right)^{\frac{1}{2}} \qquad \text{[Eq. 4.1]}$$

where η is the viscosity of the surrounding liquid, ρ is its density and ω is the angular frequency of the vibrating cylinder which is equal to 2π times its frequency of vibration. This acoustic boundary layer is shown in Fig. 4.2 as the circular velocity contour separating the inner and outer streaming vortices in each quadrant.

The dimensions of this boundary layer thickness are small; for example, it is about 4 μm at 20 kHz in water or isotonic saline, and only 0·56 μm at 1 MHz in the same media. Thus, the streaming pattern is not readily visible at ultrasonic frequencies and is commonly referred to as acoustic microstreaming. These small dimensions also mean that the velocity gradients, i.e. the rate of change of velocity with distance, and hence the hydrodynamic shear stresses (i.e. the product of the velocity gradient and the kinematic viscosity), are large even though the actual streaming velocities which produce them are relatively low (of the order of a few cm sec^{-1}).

Figure 4.3 presents the steady-state streaming velocity as a function of the distance from the surface of the vibrating body, i.e.

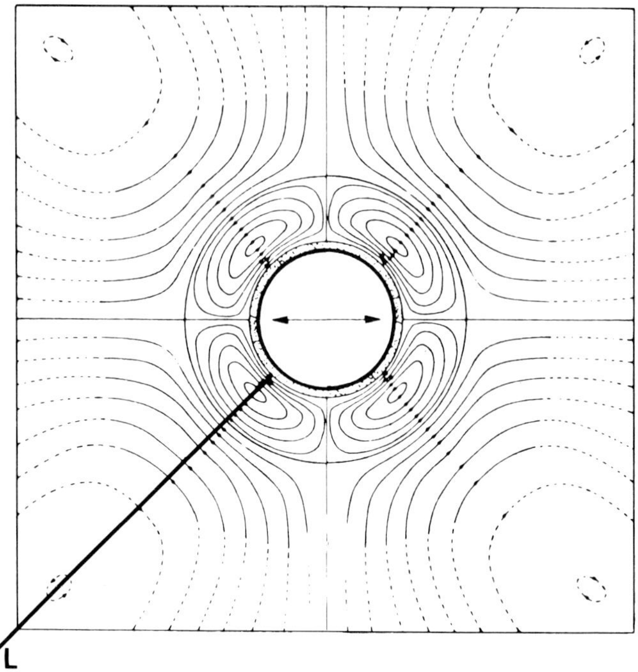

FIG. 4.2 A theoretical prediction of the acoustic microstreaming field generated around a solid cylinder oscillating within a stationary fluid (modified from Holtzmark *et al.*, 1954).

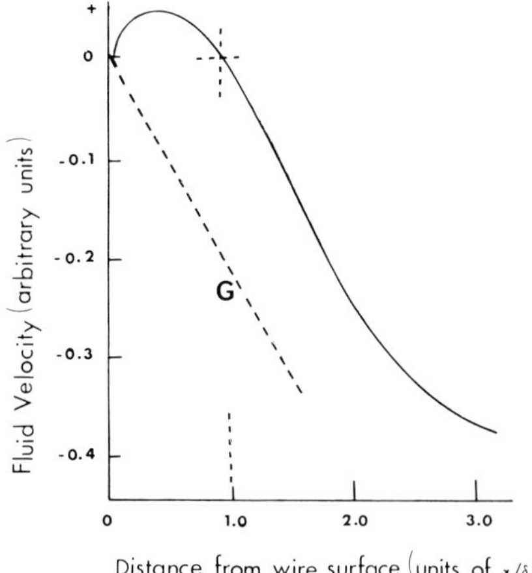

FIG. 4.3 A graphical representation of the distribution of streaming velocities along the line L shown in Fig. 4.2 (modified from Holtzmark *et al.*, 1954). The symbol δ refers to the boundary layer thickness and the broken line N is the slope of the linear approximation of the maximum velocity gradient given by equation 4.2.

the fluid velocity at every point along the line L indicated in Fig. 4.2. It can be seen that the streaming velocity increases with increasing distance from the solid surface, reaches a maximum value at a distance equal to half the boundary later thickness and subsequently decreases again to be equal to zero at a distance equal to δ. Beyond this point it increases again, but in the opposite direction to that of the fluid close to the solid surface. It can be seen that the velocity gradient, i.e. the slope of the curve, is not linear but its central portion may be approximated by a straight line (Nyborg, 1965). The slope of this straight line (G in Fig. 4.3), which is therefore a linear approximation of the maximum velocity gradient (G) generated by the acoustic microstreaming field is given by:

$$G = \frac{\omega \varepsilon o^2}{a\delta} \qquad \text{[Eq. 4.2]}$$

where ω is again the angular frequency, εo is the displacement amplitude of the vibrating body, a is its radius and δ is the boundary layer thickness.

Thus, a solid or gaseous body of radius 2 μm oscillating with a displacement amplitude of 1 μm in water or saline at a frequency of 1 MHz has a boundary layer thickness of 0·56 μm and generates a velocity gradient of the order of $5{\cdot}6 \times 10^6$ s^{-1} and a hydrodynamic shear stress of the order of $5{\cdot}6 \times 10^4$ dyn cm^{-2} or $5{\cdot}6 \times 10^3$ Nm^{-2}. These stresses are more than large enough to disrupt most biological cells and tissues.

Figure 4.2 may also be used to show the acoustic microstreaming patterns generated by the tips of vibrating wires or probes. For example, cover the lower half of the figure so that remaining hemispherical portion of the vibrating body may be imagined to be the hemispherical tip of a wire or probe (the shank of which continues down beneath the covered portion) vibrating in its transverse mode. Similarly, if either the left or right halves of Fig. 4.2 are covered then the remaining portion of the figure describes the streaming pattern generated around the hemispherical tip of a wire or probe oscillating in its longitudinal mode. It should be noted that the microstreaming pattern is the same for both modes of vibration and that the same acoustic theory as outlined above may be applied in each case.

It is not practicable to drive an unsupported length of this wire to vibrate in a uniform manner at high frequencies while immersed in a liquid. Nevertheless, it is relatively easy to produce a transverse or flexural standing wave pattern along the length of a wire as shown in Fig. 4.4 (Williams *et al.*, 1970). In this arrangement a steel or tungsten wire 250 μm in diameter was clamped by one end to the tip of a steel or titanium velocity transformer driven in its longitudinal mode at 20 kHz. The wire was mounted at right angles to the long axis of the velocity transformer and adjusted to a resonant length, i.e. an integral number of half wavelengths plus a quarter of a wavelength. The free end of the wire was ground to a hemispherical tip and this was the portion of the wire which oscillated with the greatest displacement amplitude at any given driving voltage. It should be noted that the limits of transverse displacement indicated in Fig. 4.4

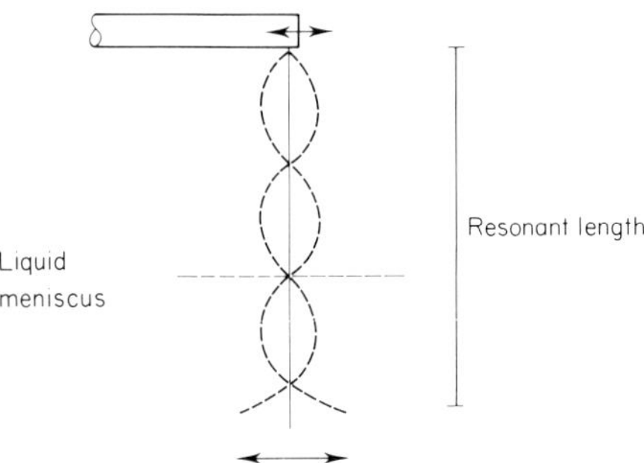

FIG. 4.4 A diagrammatic representation of the flexural standing wave pattern developed along a resonant length of a metal wire when one end is driven to vibrate in a transverse mode.

have been greatly exaggerated for the sake of clarity. In practice, the maximum displacement amplitude of the wire tip seldom exceeded about 30 μm which was small compared with the dimensions of the wire which was usually about 250 μm diameter. The free terminal portion of the wire was immersed in the suspension to be irradiated so that the liquid meniscus met the wire at a displacement node, this prevented the formation of surface waves and the possible loss of sample in the form of an aerosol.

A reasonable length of thin wire cannot be driven in its longitudinal mode at an ultrasonic frequency without generating other additional more complex modes such as circular and transverse motion. These additional modes may be inhibited if an exponentially tapered velocity transformer or probe is used (Fig. 4.5) so that it is only the terminal portion which has been ground down to the dimensions of a wire. This device is commonly called the Mason Horn after its originator Dr Warren P. Mason.

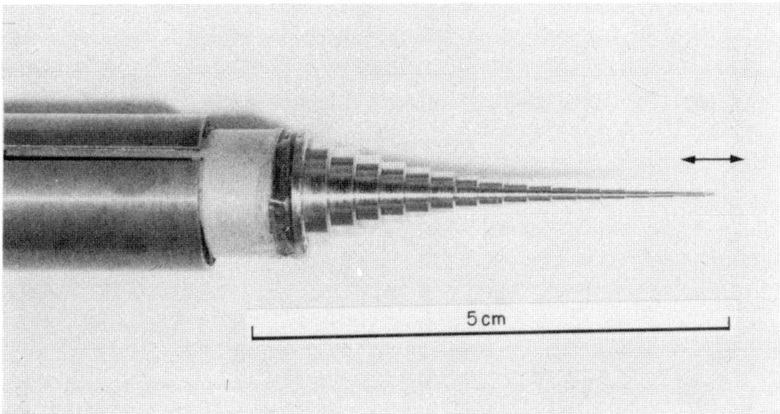

FIG. 4.5 An exponentially tapered metal velocity transformer or probe (Mason horn) whose tip has been ground to approximate the dimensions of a small gas bubble (52 μm diameter) and which vibrates in its longitudinal mode at 85 kHz.

4.3.2 Biological Effects of Acoustic Microstreaming

4.3.2.1 Cells in Suspension

The displacement amplitude of the vibrating tips, and hence the magnitude of the velocity gradients and shear stresses generated by these models of stable acoustic cavitation, may be increased by increasing the voltage supplied to the driving transducers. At low power levels one would therefore expect relatively gentle shear forces which could disaggregate clumps of cells without damaging them whereas higher power levels will damage and disrupt the cells and possibly even their intracellular organelles. Parry *et al.* (1971) have shown that sheets of exfoliated stratified squamous epithelial cells from the human cervix may be disaggregated without causing detectable damage to the cells themselves. Williams and Slade (1971) showed that aggregates of the common soil bacterium *Sarcina lutea* could also be disaggregated by the transversely oscillating wire at 20 kHz and that the dispersive force was proportional to the square of the displacement amplitude of the wire tip as predicted by equation 4.2. Similarly, Williams (1974a) has shown that human platelets in their own plasma were induced to liberate their serotonin at a rate which was again proportional to the square of the displacement amplitude of the wire tip (Fig. 4.6). It should be noted that the extrapolated "threshold" for consistent release is about 225 μm^2

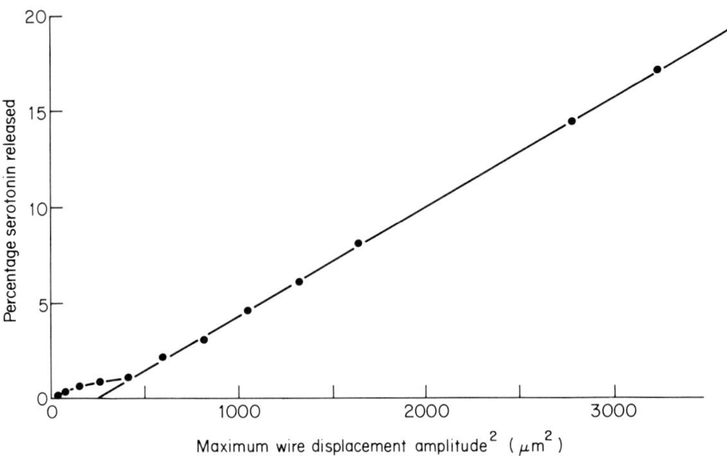

FIG. 4.6 The release of serotonin from human platelets as a function of the square of the displacement amplitude of a transversely oscillating wire (from Williams, 1974a). For explanation see text.

which corresponds to a displacement amplitude of 15 μm and a hydrodynamic shear stress of about 900 dyn cm^{-2}. This stress is smaller than that required to disrupt human and canine erythrocytes using a similar apparatus (about 5600 dyn cm^{-2}; Fig. 4.7 from Williams *et al.*, 1970) which indicates that human platelets are exceptionally fragile cells which are easily disrupted by acoustic microstreaming forces.

It should be borne in mind that the velocity gradients generated at a given displacement amplitude are crucially dependent upon the geometry of the tip of the wire or probe (Sanders *et al.*, 1972). Also, the function used to estimate the maximum velocity gradients (equation 4.2) may not be as valid at high displacement amplitudes. A further complication is that the dimensions of that portion of the velocity gradient which is approximately linear decreases as the oscillatory frequency increases until it becomes smaller than the cells being sheared. This means that the numerical values of shear stress obtained using equation 4.2 are best regarded as order-of-magnitude estimates which may be useful when comparing the fragilities of different cells at the same acoustic frequency using the same apparatus but should be used with caution when comparing results obtained at different frequencies.

Suspensions of many other cell types are sub-lethally "altered" and ultimately disrupted when they are subjected to wire-induced acoustic microstreaming fields. For example, Crowell *et al.* (1977)

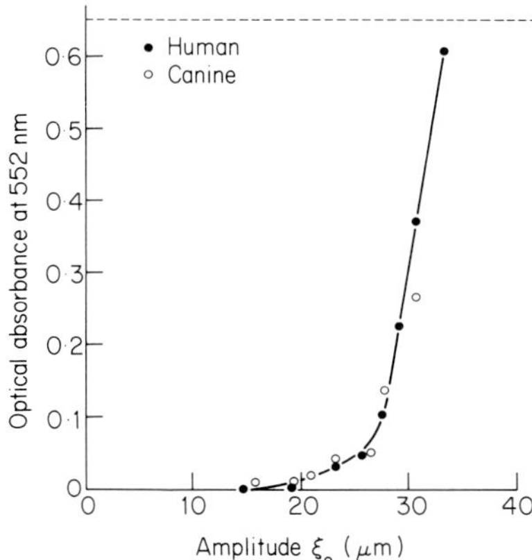

FIG. 4.7 Haemoglobin release from human and canine erythrocytes in saline suspension resulting from a 5 min exposure to a partially-submerged wire oscillating transversely at a frequency of 20 kHz. Complete osmotic haemolysis in distilled water yields an absorbance value of 0·65, as shown by the dashed line (from Williams *et al.*, 1970).

demonstrated a number of functional changes in human white blood cells which were essentially similar to the changes which could be detected after similar cells had been exposed to sub-lethal hydrodynamic shear stress in a viscometric apparatus (Dewitz *et al.*, 1977). Acoustic microstreaming can also disrupt macromolecules such as DNA (Williams, 1974b) and inactive enzymes such as malate dehydrogenase (Kashkooli *et al.*, 1980) which is again similar to the disruption of other macromolecules by non-acoustically-induced shear fields (Charm and Wong, 1970). Thus, acoustic microstreaming can produce a variety of biological effects in many biological systems (Rooney, 1972) and these effects may be duplicated by other non-acoustic hydrodynamic shear fields.

4.3.2.2 Cells in Tissues

In the preceding section the overall dimensions of the 20 kHz shear field were larger than the suspended cells which were therefore swept in close to the vibrating source, sheared, and subsequently washed away again. This cannot happen if the cells are very much larger than the dimensions of the shear field, or if the cells are held together in a solid tissue. In this case a small (*ca.* 50–100 μm diameter) source vibrating at a frequency of about 80–90 kHz may be applied to the outside of an intact cell so that an acoustic microstreaming field is generated within it. This approach has been used by many authors

(Schmitt, 1929; Nyborg and Dyer, 1960; Dyer, 1965; El'piner *et al.*, 1966) on a variety of eggs and intact animal and plant tissues. If the intracellular protoplasm has a relatively low viscosity, as is the case with most plant cells adapted to live in fresh water, then the external application of a vibrating probe or wire results in the accumulation of cytoplasmic material close to the point of contact (Gershoy and Nyborg, 1973). At slightly greater displacement amplitudes visible particles and organelles are seen to rotate and/or move about. Even higher displacement amplitudes apparently rupture some membranes and other restraining structures so that a well-defined intracellular microstreaming field is established (Nyborg *et al.*, 1977). Once it has been established, this microstreaming field will continue to flow, albeit at a reduced velocity, if the displacement amplitude of the driver is now reduced.

The long-term biological effects resulting from this perturbation of intracellular contents depend upon the severity and extent of the ultrasonic exposure. There is insufficient data available from survival studies to enable precise statements to be made but the irradiated cells are apparently able to tolerate the rupture of some intracellular membranes without being killed. At high displacement amplitudes the intracellular microstreaming can cause a variety of sub-lethal and even genetic effects. For example, Dyer (1965) applied a vibrating probe to the wall of a moss protonema while it was undergoing cell division so that the incipient or developing cross-wall was set into rotation. After sonication, he found that cell division had continued in a number of instances, but that the daughter cells were abnormal. This abnormality persisted through many cultures over several years and must be regarded as the first definitive proof that ultrasound could, under certain circumstances, produce some form of genetic or persistent teratological damage.

Human platelets are exceptionally fragile cells which are easily ruptured or induced to release their cellular contents by an acoustic microstreaming filed *in vitro* (Williams, 1974a). The micro-probe technique described above for the generation of intracellular microstreaming fields was adapted by Williams (1977) to produce a similar streaming field within the intact vascular network of anaesthetized mammals *in vivo*. In this case the 50 μm diameter tip of a Mason Horn vibrating at 85 kHz was pressed against the outside of the wall of 80–100 μm diameter mesenteric vessels which had been exposed and placed on a heated microscope stage. When the transducer was energized a symmetrical pattern of microstreaming vortices could be seen (Fig. 4.8) which closely resembled the pattern predicted from Fig. 4.2; N.B. the inner streaming vortices cannot be seen because the boundary layer thickness is only about 0·8 μm at this frequency. It should be noted that one of the outer streaming vortices contains an extremely adhesive gelatinous mass (T in Fig. 4.8) which consists of platelets which have been induced to aggregate. The microstreaming pattern disappeared as soon as the transducer was de-activated and the platelet thrombus either adhered itself to the vessel wall or was embolised downstream. These results show that an intravascular microstreaming field *in vivo* initiates thrombogenesis which may irreversibly occlude that vessel (Williams, 1977).

4.3.2.3 Microstreaming Around Solid Inclusions in vivo

Martin *et al.* (1982) reported the novel observation of rotation of blood cells within small blood vessels lying alongside the cartilaginous rods which support the optically transparent tails of tropical fish. The tails of the anaesthetized fish were viewed through a microscope while they were being exposed to therapeutic intensities of continuous wave 0·78 MHz ultrasound. Peak acoustic intensities greater than about 1 W/cm^2 (i.e. spatially averaged intensities greater

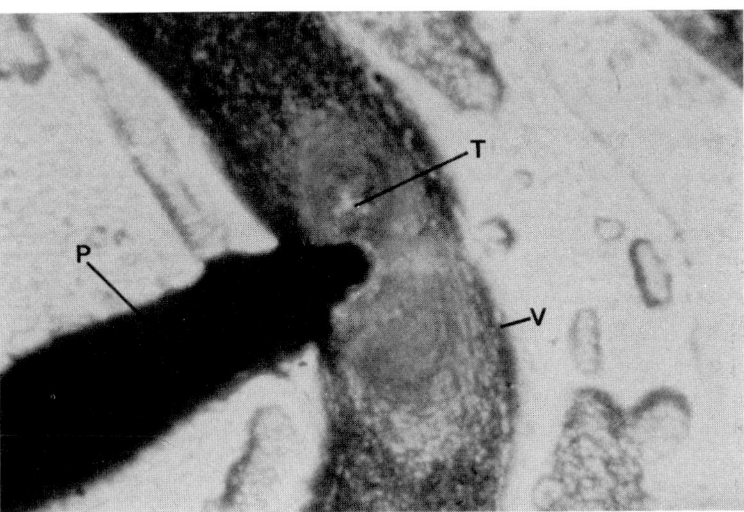

FIG. 4.8 A photomicrograph of the acoustic microstreaming field generated within an intact mammalian blood vessel using the experimental arrangement shown in Fig. 4.5. The tip of the probe (P) is being pushed against the upper surface of the blood vessel (V). It should be noted that one of the outer streaming vortices contains a gelatinous mass of aggregated platelets forming a platelet thrombus (T) (from Williams, 1981b).

than about 120 mW/cm^2) caused rotation of red blood cells within the 5–10 μm diameter capillary branches which traversed the cartilaginous rods, whereas higher intensities were required to cause rotation inside the 15–30 μm diameter vessels running paralled to the rods. The velocity of this cellular rotation increased with increasing acoustic intensity and was only observed in those portions of the vessels immediately adjacent to the edges of the cartilaginous rods.

The occurrence of cellular rotation only at the edges of the rods, together with the low rotational velocity of the streaming (only about 4 Hz even at peak intensities of the order of about 6 W/cm^2) suggests that intravascular gaseous microbubbles are not responsible for this effect. Gas bubbles would tend to be more random in their occurrence and would generate much higher velocity streaming fields which should disrupt the blood cells at these acoustic intensities.

The authors postulate the reasonable assumptions that the intravascular streaming could result from radiation torque generated by the ultrasound-induced vibration of the cartilage rods, or from complexities in the standing wave patterns set up by the exposure geometry. Another possible mechanism which they did not consider is the phase difference in particle velocity resulting from the greater inertia and/or damping within the cartilage rods (i.e. you effectively have a stationary rigid body immersed in an oscillating liquid, which should generate the same sort of microstreaming field as a vibrating solid body immersed in a stationary liquid—see section 4.3.1).

4.3.3 Stabilized Gas Bubbles

Gases are highly compressible and so the radius of a gas-filled bubble in a liquid is decreased as the acoustic pressure at that site reaches its maximum value. Conversely, the original bubble radius is increased by a similar amount as the bubble expands when the acoustic pressure attains its minimum value. This stable periodic volume oscillation in phase with the pressure changes induced by the acoustic wave is known as "breathing pulsation" (Nyborg, 1977). The displacement amplitude of the gas/liquid interface is different for bubbles of different size in the same acoustic field because of the phenomenon of resonance. It is common knowledge that a mass suspended from a spring will oscillate with maximum displacement amplitude at a certain characteristic frequency known as its resonance frequency. If you try driving that mass at a faster or slower frequency some of the energy will be absorbed within the spring and its oscillatory amplitude will be decreased. In the case of the bubble the compressible gas behaves like the spring and the surrounding liquid serves as the mass. Thus, only those gas bubbles whose dimensions are close to the resonant size for that particular driving frequency will oscillate with large displacement amplitudes.

Minnaert (1933) derived an expression for the estimation of the resonant size of a gas bubble which may be re-written as:

$$r = \frac{1}{\omega}\sqrt{\frac{3\gamma Po}{\rho}} \qquad \text{[Eq. 4.3]}$$

where r is the radius of the bubble, ω is 2π times the applied frequency, γ is the ratio of the specific heats of the gases within the bubble, Po is the pressure amplitude and ρ is the density of the surrounding liquid. This equation indicates that the resonant size for a gas bubble in water at 1 MHz is about 6·5 μm diameter.

However, as the driving frequency increases the gas can no longer operate isothermally and tends to behave adiabatically (Nyborg, 1977). This results in the bubble oscillating out of phase with the acoustic field which is driving it. This necessitates a more complex analysis (Coakley and Nyborg, 1978) which gives the resonant radius of a free gas bubble at 1 MHz at close to 3·5 μm which is in excellent agreement with the experimental value obtained by Miller (1977).

It is difficult to maintain free gas bubbles having diameters of the order of microns in a liquid because surface tension forces increase its internal pressure by an amount proportional to twice the magnitude of interfacial surface tension divided by the bubble radius so that they tend to shrink and dissolve. This can be prevented if the gas bubble is protected or stabilized against diffusion either by coating it with a film having low permeability to gases (which also tends to inhibit bubble oscillation) or by trapping it in a hole of a suitable size in a solid surface. Several early investigations drilled single small holes or arrays of holes on the driving face of metal velocity transformers and showed that the trapped gas bubbles greatly increased the rate at which that device degraded solutions of DNA (Pritchard *et al.*, 1966), eroded solid surfaces (Howkins, 1966), removed surface films (Jackson and Nyborg, 1958), or accelerated the development of an image on a photographic plate. However, the practical limitations of drilling holes smaller than a few millimetres in diameter restricted this work to the low kilohertz frequency range so that the bubbles were undergoing a variety of complex oscillatory responses and therefore the simultaneous occurrence of some form of transient cavitation-like activity must be assumed.

A novel alternative approach was adopted by Rooney (1970) who maintained a 250 μm diameter gas bubble at the end of a metal tube which was connected to a water manometer. The 20 kHz acoustic field was conducted to the bubble through the cell suspension to be irradiated. A further innovation was the inclusion of high molecular weight Dextran with the cell suspension to inhibit the development of surface waves on the gas bubble interface. His results on the haemolysis of human and canine erythrocytes are presented in Fig. 4.9 as a function of the displacement amplitude of the bubble interface. These results are virtually identical to those obtained by Williams *et al.* (1970) using a 250 μm diameter wire oscillating transversely at the same ultrasonic frequency (Fig. 4.7). These results strongly indicate that the damaging mechanism is the same in both cases and that it is the hydrodynamic shear stresses generated by the virtually identical acoustic microstreaming fields which is lysing the cells.

The development of Nuclepore® membranes (Nuclepore Corporation, Pleasonton, California) for filtration purposes has enabled these microstreaming observations to be extended into the megahertz range. The polycarbonate membranes are about 12 μm thick and are

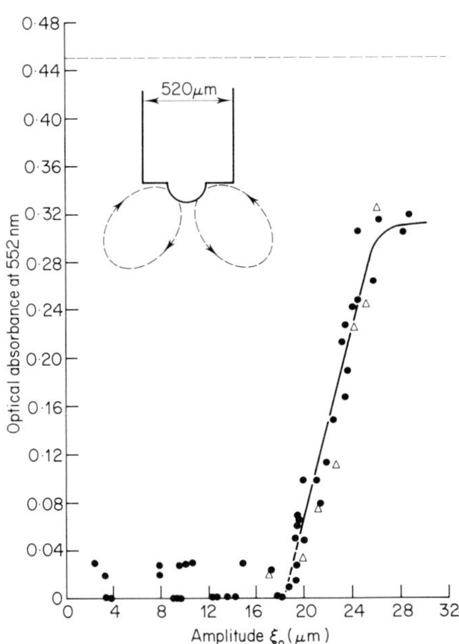

FIG. 4.9 Haemolysis of human and canine erythrocytes by the acoustic microstreaming field generated around a single 250 μm gas bubbles oscillating at 20 kHz as a function of the displacement amplitude of the gas/liquid interface (from Rooney, 1970).

randomly perforated with uniform cylindrical holes having diameters of the order of microns. Air is trapped within these holes when the membrane is made hydrophobic and is immersed in a liquid, and the gas/liquid interface at the ends of each cylindrical gas body may be driven to oscillate by a suitable ultrasonic field. These oscillating gas bodies generate acoustic microstreaming fields which can cause human platelets to aggregate together and adhere to the Nuclepore membrane in the vicinity of the gas-filled pores (Fig. 4.10 from Miller *et al.*, 1978). Unfortunately, gas bodies trapped inside the cylindrical pores of Nuclepore membranes are damped more than free bubbles of comparable size. Their resonant frequency is also changed (Nyborg *et al.*, 1977).

Despite these minor practical problems this technique has shown that stable gas bubbles can produce significant biological changes at extremely low acoustic intensities in the megahertz range. For example, Miller *et al.* (1978) have shown that fresh human platelets were induced to aggregate *in vitro* at intensities as low as 32–64 mW/cm^2 at 1 MHz. In a separate study Miller *et al.* (1979a) showed that a 2·1 MHz beam induced visible platelet aggregation at c.w.

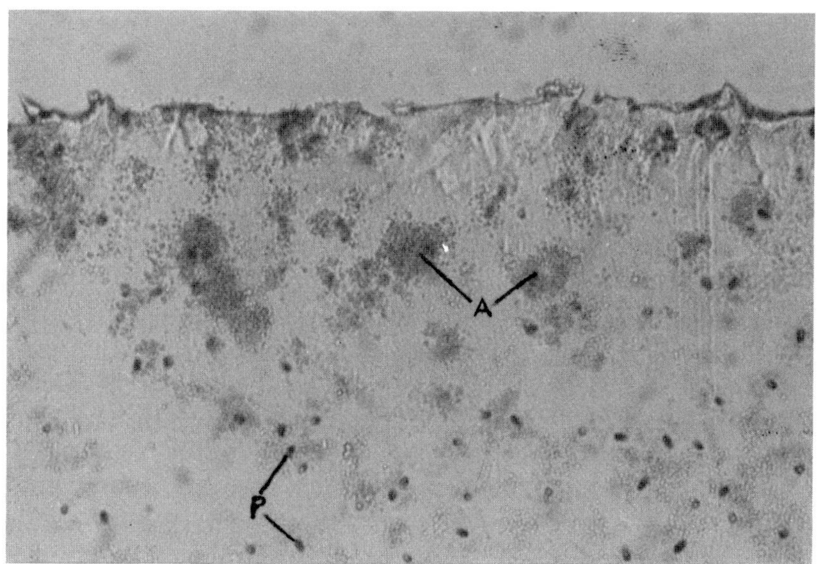

FIG. 4.10 A photomicrograph of a Nuclepore® membrane having numerous gas-filled pores (P). The membrane had been irradiated with MHz ultrasound while immersed in a suspension of human platelets and aggregates (A) have been produced around the most active pores (from Miller *et al.*, 1978).

intensities between 16 and 32 mW/cm^2. These intensities are similar to those emitted by many diagnostic instruments in current use. Miller *et al.* (1979a) also showed that the ultrasonic field emitted by a common commercial doppler diagnostic device induced platelets to aggregate in the presence of the stabilized gas bubbles. This observation must rank as one of the few unambiguous examples of a potentially hazardous biological effect produced by a commercial diagnostic device.

Platelets which have received minimal damage will aggregate to each other and adhere to surfaces, but this may be a reversible effect. More severe damage results in the platelets undergoing the release reaction and liberating the contents of their secretory granules. The platelet release reaction is usually associated with the activation of the coagulation pathway *in vivo* resulting in the formation of fibrin which binds the platelet mass together to form an irreversible clot. Adenosine triphosphate (ATP) is liberated during the release reaction and so the luciferin/luciferase enzyme system extracted from the tails of fireflies (which radiates visible light in the presence of ATP) was included with the platelet suspension and a sensitive photomultiplier tube was used to detect any emitted light (Miller *et al.*, 1979b). This technique is amazingly sensitive and can detect the release of ATP from human platelets in the presence of stabilized gas bubbles energized by 1·6 MHz ultrasound at intensities as low as about 4 mW/cm^2 (A. R. Williams and D. L. Miller, unpublished observations).

Human erythrocytes also contain ATP which is liberated when their plasma membrane is ruptured. Williams and Miller (1980) have shown that light is emitted when erythrocytes are irradiated in the presence of luciferin/luciferase and gas bodies stabilized inside hydrophobic Nuclepore membranes (Fig. 4.11).

Thus, stable gas bodies of the correct size may be driven to oscillate by extremely low intensities of ultrasound at both kilohertz and Megahertz frequencies. The hydrodynamic shear forces developed within the acoustic microstreaming fields generated by these oscillating bubbles are more than large enough to disrupt cells and would certainly constitute a potential source of hazard if these bubbles were present *in vivo*. Human platelets have been shown to be among the most fragile cells investigated to date and, if sufficient numbers of them were damaged *in vivo*, could contribute to the initiation of thrombogenesis or blood coagulation.

4.3.4 Gas Bubbles Within Plant Tissues

Most tissues of higher plants normally contain undissolved gas in

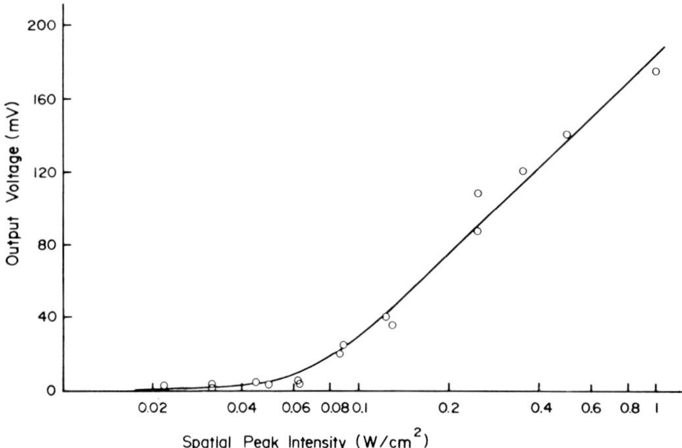

FIG. 4.11 Human erythrocytes were irradiated in the presence of stabilized gas bubbles and the luciferin/luciferase enzyme system extracted from fireflies. ATP is released from disrupted cells and is used by this enzyme system to emit photons of visible light which are detected by a photomultiplier. The maximum light output (expressed in mV) is presented as a function of the spatial peak acoustic intensity (from Williams and Miller, 1980).

intercellular channels. These gas bodies are driven to vibrate when the plant tissue is exposed to ultrasound (Miller and Kaufman, 1977; Martin *et al.*, 1978; Miller, 1979a,b). The dimensions of these gas bodies vary from plant to plant and are usually different in different tissues from the same plant. Under certain circumstances the gas bodies may also change their size even within the same tissue in response to environmental factors such as dehydration or the production of copious quantities of oxygen on a sunny day.

These intercellular gas bodies have complex geometries and may be bounded by many cells. Miller (1979a,b) has estimated the resonance characteristics of these channels and concluded that different sized gas bodies present in common plants resonate at different frequencies in the range 0·45 to 10 MHz. For example, three different cell populations within the leaves of the aquatic plant *Elodea* had differing susceptibilities to different frequencies of ultrasound and the minimum lethal threshold for each cell population generally coincided with the expected resonance frequency associated with their adjacent gas bodies (Miller, 1979b). The thresholds for cell death were as low as 75 mW/cm^2 at 0·65 MHz and 180 mW/cm^2 at 5 MHz. These intensities are remarkably low when you consider that the vibrating gas bubble is bounded by rigid walls of cellulose which tend to damp out the oscillatory motion of the gas bodies. Once a gas body has been set into oscillation the adjacent cell walls are also driven to vibrate and radiation pressure forces and acoustic microstreaming fields are developed within the adjacent cells in a manner analogous to those produced by contact with a vibrating metal source (Gershoy and Nyborg, 1973).

The presence of these gas bodies and their consequent activation by the applied ultrasonic field must therefore be borne in mind whenever a biological effect is reported using plant tissues.

4.3.5 Gas Bubble Formation and Growth

This is a difficult and nebulous topic to deal with since it depends upon so many complex independent and interdependent variables. These include the quantity of each dissolved gas and its solubility in that medium, the number of micronuclei and the magnitude of the factors enhancing nucleation, the presence or absence of solid surfaces which might contain "cracks" or cavities to stabilize bubbles, the ambient pressure and its recent history, the presence or absence of dissolved materials which might form a protective "skin" around a bubble, and the complex changes induced by an applied ultrasonic field. In general, a small bubble will tend to grow in an ultrasonic field by a process known as rectified diffusion. A simplified view of this process is that the bubble oscillates between the extreme conditions where it is compressed to its smallest radius (i.e. its internal pressure is high) and where its radius is largest (i.e. its internal pressure is low). In its compressed state the partial pressure of the gas inside the bubble will be higher than that in the surrounding liquid and so gas will diffuse out of the bubble over its entire surface area. Conversely, gas will tend to diffuse into the bubble over its entire surface area during its expanded phase. The expanded bubble has a larger surface area than the compressed bubble and so more gas molecules should diffuse in during bubble expansion than flow out during its compression.

This oscillatory gas flow is superimposed onto the normal tendency for small gas bubbles to shrink. Thus, a low power acoustic field will only slow down the rate at which a given bubble will shrink, a higher power will enable it to maintain a constant size while an even higher power will enable the bubble to grow. The fate of a bubble in an ultrasonic field of a given intensity also depends upon its size relative to the resonant size for that particular frequency. If it is far from resonance (i.e. either too small or too large) its oscillatory amplitude may be so low that it will shrink whereas a bubble close to resonant size will oscillate with a larger displacement amplitude at that same intensity and may grow. If it is supplied with abundant gas the bubble may contrive to grow beyond its resonant size so that it again oscillates with a smaller displacement amplitude in the same acoustic field.

Micronuclei must exist *in vivo* under certain conditions in mammalian tissues for these are the nucleation sites from which bubbles grow during decompression (Fulton, 1951). Blood drawn into well-wetted tubes from the veins of anaesthetized cats can withstand negative pressures of up to 713 mmHg without forming visible bubbles which suggests that no micronuclei are present (Harvey, 1951). However, numerous intravascular gas bubbles form when animals are decompressed *in vivo*. Harvey (1951) suggested that the apparent discrepancy between these observations results from the fact that micronuclei are continually being formed *in vivo* by local mechanical tensions. This is analogous to the observation that a gassy liquid in a vacuum chamber may not bubble until the walls of the vessel are given a sharp blow whereupon the liquid bubbles vigorously. An alternative method of inducing bubble formation in a liquid super-saturated with gas is to introduce a hydrophobic surface. Gas molecules which come into contact with this surface form nuclei which will continue to grow. Harvey (1951) proposes that the mechanical tensions and pressure fluctuations encountered *in vivo* may transiently expose hydrophobic regions on cellular surfaces which, if the gas conditions are favourable, will become micronuclei. This proposal is supported by numerous observations on bubble formation *in vivo* during mild experimental decompression when it was observed that anaesthetized animals at complete rest do not form visible bubbles, but that bubbles appear in profusion if muscular contraction occurs or if the animals' limbs are moved, stretched, bruised or damaged (Fulton, 1951).

An analogous situation has been observed *in vitro* in that a stationary homogeneous liquid in equilibrium with its surroundings may not exhibit any evidence of cavitation-like activity until it is mechanically disturbed. This was first observed by Hill (1972) who found that a solution of mammalian DNA was not degraded at an intensity of 1 to 2 W/cm^2 at 1 MHz until the sample chamber was mechanically rotated. It has since been found that complete rotation is not necessary to demonstrate this effect and any mechanical perturbation or even a transient fluctuation in the frequency of the ultrasonic beam is sufficient to initiate bubble activity in what was a quiescent medium (Graham *et al.*, 1980).

Similarly, it is extremely difficult to get quiescent mammalian blood to cavitate *in vivo* at MHz frequencies (Chater and Williams, 1982) and yet Gramiak and Shah (1971) could readily detect gas bubbles of microscopic size in the chambers of the beating human heart where the combination of rapid blood flow and sudden pressure changes favour nucleation.

4.3.6 Free Radical Generation and Sonochemical Effects

The nucleus of an atom is surrounded by many "shells" or orbitals at different energy levels which contain the electrons. These electrons are in their lowest energy state when they completely fill that particular shell. The innermost shell only holds a maximum of two electrons (and is full for the inert gas helium) whereas more distant shells usually hold eight electrons and are full for the other inert gases neon, argon, krypton and zenon. Those atoms having one electron more than the inert gas structure (i.e. the metals sodium, potassium, etc.) tend to lose this electron and form stable positively charged ions (Na^+, K^+, etc.). Those atoms having one electron less than the inert gases (i.e fluorine, chlorine, etc.) tend to gain an electron and form stable negatively charged ions (F^-, Cl^-, etc.). However, atoms near the centre of the periodic table of the elements (e.g. carbon) need to gain or lose four electrons to attain a stable inert gas structure. Instead of forming ions these atoms can "acquire" electrons by sharing them with other atoms and forming what are known as covalent bonds. Thus, carbon, which needs to "acquire" four electrons, can form covalent bonds with four hydrogen atoms (each of which wants to lose an electron giving the ion H^+, or to gain an electron by sharing one so that it attains the inert structure of helium) and so form methane, one of the main constituents of marsh gas.

Covalent bonds are very strong (typical energies are 60–100 kcal/mole for c-c, c-o, P-o and for O-H bonds; Cotterell, 1954) and yet each bond should not be regarded as a rigid entity but as a spring joining two masses which can therefore vibrate at a characteristic frequency. The amplitude of this oscillation increases as thermal energy is supplied to that system until at temperatures in excess of 1000°C the covalent bond may be broken and the two atoms separate, each one taking one of the shared pair of electrons. Each of the separated atoms has an unpaired electron and is therefore called a free radical (they are usually written with a dot over the radical, thus $x\text{-}y \rightarrow \dot{x}$ and $\dot{y}$). Free radicals are highly reactive species which will eagerly combine with other free radicals or a wide range of other compounds in an endeavour to regain their stable inert gas electron configuration so that their lifetime as radicals may be as short as 10^{-12} sec.

Noltingk and Neppiras (1950) deduced that the temperature rise within the cavity of a collapsing (transient) cavitation bubble could exceed several thousand degrees centigrade and would therefore result in the pyrolytic (thermal) breakdown of water vapour into $\dot{H}$ and OH radicals (Flynn, 1964). These radicals interact with each other and with undissociated water and gas molecules (such as oxygen) in the vapour phase and generate a number of longer lived reactive secondary species such as $O\dot{O}H$ and H_2O_2 (Del Duca *et al.*, 1958). Other theories have been proposed by Prudhomme (1972) and Frenkel (1940) which postulate that the initial generation of $\dot{H}$ and OH radicals within the collapsing cavity results from the generation of electrical discharges.

The quantity and chemical composition of the longer lived reactive secondary species generated within the vapour phase of an imploding cavity or bubble is critically dependent upon the composition and concentration of the gases and volatile materials dissolved within the bulk liquid. If there is too much gas or volatile vapour within the cavity it will cushion or slow down the implosion rate and so decrease the maximum temperature attained which will subsequently produce fewer $\dot{H}$ and $O\dot{H}$ radicals. If there is no oxygen dissolved in the bulk liquid, then the yield of $O\dot{O}H$ and H_2O_2 and their reaction products will be greatly reduced. Conversely, the chemical activity of a cavitating field may be greatly increased by adding some carbon tetrachloride or chloroform to the bulk liquid because this vapour is pyrolized to give radicals such as $\dot{Cl}$ and $\dot{C}Cl_3$ which may survive long enough to enter the bulk liquid and initiate an infinite number of secondary reactions (Griffing, 1952).

The plethora of chemical modifications of biological macromolecules and their constituents compiled by El'piner (1964) are mainly the result of secondary reactions occurring within the bulk liquid induced by the longer lived free radicals and agents which were formed in the vapour phase. These reactions are usually non-specific and include hydroxylation, deamination, oxidation and the cleavage of ring structures. Much attention has been devoted to these chemical effects because it has been demonstrated that the bases of DNA can be modified in an analogous manner to the modifications produced by ionizing radiation (which also generates $\dot{H}$ and $O\dot{H}$ radicals) (Wang, 1977). However, the destructive effects of the hydrodynamic shear fields associated with transient cavitation will have disrupted the cells before their DNA could have been altered (even presuming that these highly reactive molecular species could have reached the nucleus of an intact cell without encountering and reacting with other molecules en route). Those experiments where cells have been grown in a medium which had previously been exposed to ultrasound have been unable to show that long-lived reactive chemical species generated by ultrasonic cavitation can produce a genetic effect (Thacker, 1974).

Free radicals are also produced when the backbone of a macromolecule is broken by the application of a mechanical or hydrodynamic stress. In the absence of water these radicals will persist for a long time and can be detected by means of electron spin resonance (ESR). Thus, macroradicals of DNA have been detected after the anhydrous molecules have been crushed and ground (Abagyan and Butyagin, 1965). Macroradicals may also be detected in aqueous solutions providing that there is some labelled reactive molecule present to interact with the macroradicals as soon as they have been formed. Williams (1969b) used a transiently cavitating ultrasonic field to degrade isolated calf thymus DNA in aqueous solution in the presence of $^{131}I_2$ (which can form covalent bonds with free radicals) and found that some of the iodine could not be separated from the DNA fragments despite repeated chromatographic elutions.

4.3.7 General Overview of Cavitation-like Activity

There is now general agreement amongst workers in this field that stable gas bodies of the appropriate size attached to a solid surface and submerged in a liquid may be driven to oscillate by a low-intensity ultrasonic field even at megahertz frequencies. These oscillating bubbles vibrate in a linear manner and generate acoustic microstreaming fields which can damage and disrupt biological tissues. However, unlike the vibrating metal models of stable cavitation described above, the same gas bubble can generate several different patterns of acoustic microstreaming depending upon the irradiating conditions and the presence of various materials at the gas/liquid interface (Elder, 1959). The main reason for this complex behaviour is the development of a wave pattern on the gas/liquid interface known as surface waves.

Lamb (1945) has described the various modes of surface wave activity which can be produced. The simplest mode is one where the bubble surface contracts around its equator and expands in the polar direction, and vice versa; i.e. it oscillates between shapes tending towards a prolate ellipsoid and an oblate spheroid (cigar-shaped and a flattened sphere respectively). Other more complex modes are associated with shorter wavelengths so that the bubble interface changes from a smooth transparent surface to one having the appearance of frosted glass. Any given bubble may have several resonance conditions for surface waves where each resonance can be envisaged as that point where the perimeter of the bubble can just accommodate an integral number of half wavelengths (Fig. 4.12). Thus, a bubble of radius 3·3 μm which is close to the correct size for volume resonance at 1 MHz is also resonant for surface waves having frequencies of 0·78 MHz, 1·43 MHz and 2·15 MHz (Coakley and Nyborg, 1978).

These surface waves are important for three major reasons: (1) they change the microstreaming fields developed around stable oscillating bubbles so that the forces tend to be more destructive but are not amenable to precise control or quantification; (2) the bubble re-radiates some of the energy incident upon it at the resonant frequencies of the surface waves existing at that time (this may be used as a characteristic marker to detect the presence of oscillating gas bubbles as described below); and (3) at moderately high intensities the surface waves can result in the break-up of the bubble and the subsequent initiation of extremely violent (transient) cavitation-like activity.

Droplet formation by surface waves is shown schematically in Fig. 4.13. Let diagram (a) represent one half wavelength of the surface wave at the point of its maximum positive displacement, i.e. it is just about to reverse its direction of motion. Diagram (b) represents that same half wavelength an instant later where it can be seen that some of the liquid has not reversed its direction of motion. This portion of the wavelet continues to move in the same direction because its inertia (which increases as the density of the liquid is increased) is greater than the restoring force provided by the interfacial surface tension. Diagram (c) shows the droplet separated from the surface

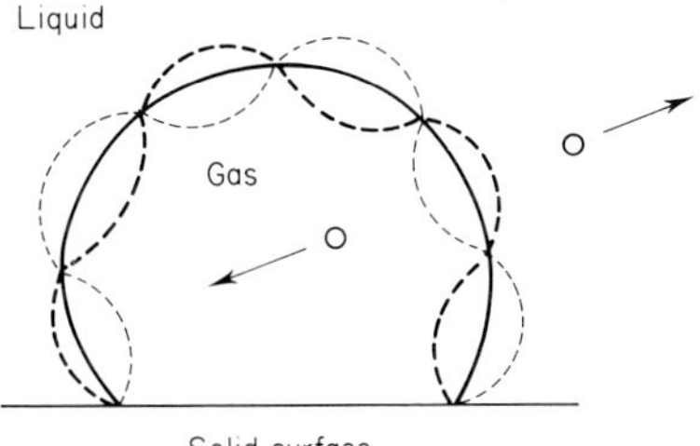

FIG. 4.12 A schematic representation of one of the possible modes of surface wave oscillatory behaviour on a gas bubble resting on a solid boundary. The free circles represent droplets of liquid or bubbles of gas formed at high displacement amplitudes.

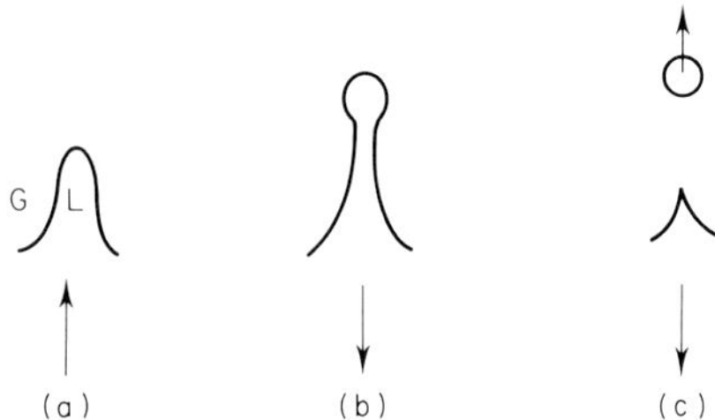

FIG. 4.13 A schematic representation of the mechanism of formation of droplets of liquid (L) in a gas (G) as a result of high amplitude surface wave activity. The arrows show the direction of motion; for details, see text.

wave and moving away from the interface. It should be noted that droplet formation only occurs when the amplitude of the surface wave is large enough to give some of the liquid enough inertial energy to overcome the surface tension forces.

A similar mechanism may also operate during the next half cycle of the surface wave which results in the injection of a droplet of gas (i.e. a small bubble) into the liquid. This process has been investigated at low frequencies by many workers including Storm (1974), Howkins (1966) and Temple *et al.* (1971) and results in the break-up of the parent gas bubble to form a cloud of microbubbles which has the appearance of a "ghost" or "shuttlecock" (if the parent bubble is moving) depending upon the acoustic field parameters and the arrangement of the viewing conditions. This microbubble cloud is a highly dynamic structure where the bubbles are attracted to and recombine with each other and with the remnants of the parent bubble, so that the size distribution and number of microbubbles present at any given instant in time is the equilibrium value between their rate of production and their rate of recombination in response to that particular combination of exposure parameters. This complex situation is further complicated if the parent bubble is moving through the liquid. In this case a "streamer" or trail of active microbubbles which may be several millimetres in length is left behind as the bubble propagates through the liquid (Hughes and Nyborg, 1962).

Microbubble formation has been extensively investigated by Willard (1954) and Nyborg and Rodgers (1967) and it has been shown that these microbubbles act as sites for new cavitational activity and give rise to the production of harmonic and wide band frequency noise emission (Neppiras and Fill, 1969). The onset of microbubble formation is also associated with the sudden onset of other phenomena indicative of transient or collapse-type cavitation such as sonoluminescence (Saksena and Nyborg, 1970) and the production of short-lived highly reactive free radicals.

At moderately high acoustic intensities a microbubble which has either been formed by surface wave induced breakdown of a large gas bubble or has grown *de novo* from a nucleation site by rectified diffusion is driven to pulsate in a non-linear manner (Coakley and Nyborg, 1978). The original theoretical work of Rayleigh (1917) was developed by Noltingk and Neppiras (1950) and predicts that a vapour and/or gas-filled bubble will expand somewhat during the negative pressure portion of the acoustic cycle and then rapidly contract to an extremely small volume when the acoustic pressure attains its maximum value. Lauterborn (1974) has confirmed the validity of a slightly modified version of this theory by showing that it adequately describes the oscillatory behaviour of a single bubble/cavity produced in a viscous silicone fluid by a pulsed ruby laser (Fig. 4.14). It can be seen that the expansion phase of the bubble cycle is broad while the contraction phase is extremely rapid. In his extensive review article, Flynn (1964) shows that the velocity of the bubble interface during contraction increases with increasing applied pressure amplitude so that it can even exceed the velocity of the sound wave in that liquid medium. This supersonic contraction generates large shock waves and develops instantaneous temperatures within the bubble contents which may be as high as several thousand degrees Kelvin.

Flynn (1964) has attempted to simplify the complex terminologies associated with acoustic cavitation by proposing that the term *stable cavitation* be applied to gas bodies which oscillate (without fragmentation) for a large number of cycles without leaving the field. Conversely, *transient cavitation* occurs if the bubble collapse is very rapid (even though it may also persist for many cycles) and corresponds to the Rayleigh prediction; it is implicit that in this

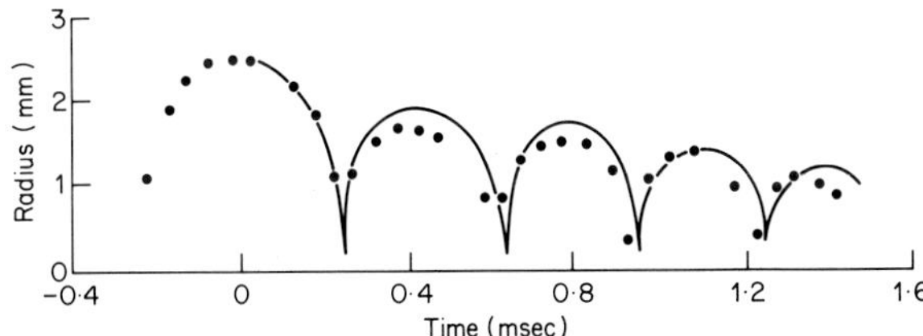

FIG. 4.14 Variation of the radius of a freely oscillating bubble with time (from Lauterborn, 1974).

collapse the bubble may disintegrate. The shock waves and surface waves associated with this rapid collapse and fragmentation suggest that it should be possible to distinguish between the various types of cavitation on the basis of their acoustic emissions. All types of cavitation will be destructive to living tissues but the transient cavitation (which tends to occur at higher acoustic intensities than stable cavitation) should be more violent and, by forcing new microbubbles into quiescent adjacent medium, exert its effects over a larger volume.

4.3.8 Physical Phenomena Associated with Cavitation-like Activity

Large stable gas bubbles are visible in transparent tissues and so the best evidence for their biological effects come from visual observation of the streaming fields they generate inside adjacent cells (Watmough, 1981). Optical techniques may also be used to detect the presence of bubbles in clear liquids; however, these are best observed from a distance in case the foreign surfaces of the microscope objectives perturb the acoustic field and/or themselves act as nucleation sites for bubble growth. Numerous small bubbles scatter incident light so that a clear liquid may appear milky. This is commonly seen when a gassy liquid such as freshly-drawn tap water is being sucked into a clear syringe—the water is cloudy while under suction and instantly becomes clear again when pressure is restored to a normal atmospheric value. Thus, visual detection of bubbles or changes in the intensity of a transilluminating light beam during sonication is good evidence supporting the presence of some form of cavitation-like activity. However, the inability to detect such changes is not necessarily conclusive proof that cavitation has not occurred, i.e. the test system may just not have been sensitive enough to detect the small amount of cavitation necessary to produce an observed biological effect (e.g. Hawley *et al.*, 1963).

In many cases the biological end point itself has been used as the method of detection of cavitational activity. For example, it is presumed that only cavitation could generate the forces which readily rupture DNA (Hill, 1972) or disrupt platelets (Williams *et al.*, 1976) or erythrocytes in suspension (Chater and Williams, 1977) or tear holes in solid tissues (Fry *et al.*, 1970).

Chemical methods are frequently employed to show that transient or collapse-type cavitation has occurred. These tests rely on the production of short-lived highly energetic chemical species called free radicals which are produced by the pyrolytic (thermal) dissociation of water vapour by the high temperatures generated within the collapsing bubbles (Noltingk and Neppiras, 1950). These free radicals may enter the surrounding liquid and chemically modify other molecules. Weissler (1959) compared the sensitivities of several different free radical scavenging systems and concluded that the production of molecular iodine from potassium iodide in the presence of carbon tetrachloride was the most sensitive. Clarke and Hill (1970) found that the sensitivity of the test was twenty times greater at 1 MHz when the carbon tetrachloride was present than when it was omitted. However, it should be noted that carbon tetrachloride is immiscible with water and has a low solubility coefficient so that some of it may be present as microdroplets in the aqueous medium having hydrophobic surfaces. These hydrophobic surfaces may encourage the formation of micronuclei for the growth of cavitation bubbles. The fact that free radical production can be detected in a given apparatus at a given combination of exposure parameters with a potassium iodide/carbon tetrachloride mixture does not automatically mean that transient cavitation will also occur in that same apparatus at those same acoustic parameters when that mixture is replaced with a biological medium having a different micronucleus and gas content.

The presence of a highly volatile material or abundant gas will inhibit the detection of free radicals by diffusing into the bubbles and "cushioning" the rate of collapse of the bubble interface. This will drastically lower the maximum temperature generated within the bubble and hence the chemical yield, but have relatively little effect on the destructive mechanical forces produced by the bubbles. This point should be borne in mind when interpreting the results of cavitation generated *in vivo* in animals because many of the gaseous anaesthetics (e.g. ether or Halothane®) are highly volatile.

In general, the "threshold" acoustic conditions necessary for the detection of free radicals corresponds with the "threshold" for the onset of the emission of visible light (sonoluminescence) (Clarke and Hill, 1970). Other workers suggest that these thresholds are close but not identical, but this may be a reflection of the different sensitivities in the detection of the various end points.

The most reliable, and yet in many ways the most complex, measurement of the presence or absence of cavitation-like activity in a medium is to detect and analyse the acoustic spectrum emitted from the irradiated tissue or sample. Inclusions differing in acoustic impedance and stable bubbles of resonant size oscillating in a linear manner will scatter or re-radiate the incident ultrasonic wave at the same frequency (fo). A bubble which is oscillating in a non-linear manner or has grown to twice the diameter of a resonant bubble will radiate a wave at half the fundamental frequency (fo/2) known as the first subharmonic (Neppiras, 1969). Bubbles which are about half the resonant size, or are at resonance and may be exhibiting low amplitude surface wave activity emit higher harmonies of the driving frequency (e.g. 2 fo and 3 fo). The development of high amplitude surface waves leading to the formation of a dynamic cloud of interacting microbubbles results in the emission of a complex spectrum containing several intense peaks superimposed onto a wide range of frequencies known as "white noise" (Coakley and Nyborg, 1978). The intensity of this white noise increases with increasing acoustic pressure amplitude and can be used as a quantitative index of the amount of "transient" cavitation that has occurred.

4.3.9 Acoustic Detection of Cavitational Activity

In its simplest form a small microphone is placed close to the sample being irradiated but positioned so that it does not interrupt or interfere with the main acoustic beam. A more sensitive technique used by Clarke and Hill (1970) and others is to position a large focused piezoelectric bowl at right angles to the axis of the main beam so that the tissue to be investigated is at the focus of the bowl. This greatly increased the collecting area and hence the sensitivity of the detection apparatus.

A novel technique has been developed by Coakley where the same focused transducer was used to generate the interrogating ultrasonic beam and to detect the acoustic emissions from oscillating bubbles close to its focus. The large impedance mismatch between a vapour or gas-filled bubble and its surrounding medium as it passes through the focal volume reflects some of the incident acoustic energy back to the transducer which emitted it. These echoes are detected as transient "spikes" or pulses superimposed onto the steady driving voltage. After filtering, these signals are a direct measure of the number and amplitude of the "transient" cavitational events occurring close to the focal volume. In general, there was a direct correlation between the number of these discrete cavitation events and the amount of degradation of molecules of DNA in solution (Coakley and Dunn, 1971) or the rupture of living cells such as the amoeba *Hartmanella castellanii* (Coakley *et al.*, 1971) (Fig. 4.15). Visual and electronic observation of these discrete cavitational events in liquids and cell suspensions has shown that a "cavitation centre" or nucleus forms and is driven towards the focal region by the "quartz wind" and radiation pressure forces. It grows as it approaches the region of highest acoustic intensity and at some critical point an "explosion" occurs resulting in the creation of an extensive and active cavitation complex. These explosive events may persist for many cycles of the sound field even at MHz frequencies and have been called "Willard events" (Coakley and Nyborg, 1978). The sphere of influence of each discrete cavitational event, i.e. that volume around each cavitational complex where the mechanical forces are so large that biological materials are disrupted or degraded, has been estimated to be about 1 mm^3 (Coakley and Dunn, 1971).

4.3.10 "Thresholds" for Cavitational Activity *in vitro*

Cavitation may appear in so many diverse forms and its presence may be detected in so many different ways that it is not surprising that different workers using techniques which vary widely in sensitivity obtain markedly different values for the "threshold" conditions necessary for its occurrence. However, even at a single frequency (e.g. 1 MHz) Iernetti (1971) has pointed out that the reported threshold values range from $1{\cdot}75 \times 10^5$ Pa to $2{\cdot}4 \times 10^7$ Pa (1·75 to 240 atmospheres) while a value as low as 7×10^4 Pa (0·7 atmospheres) has been reported by Doynon and Simont (1951). There is therefore no single "magic number" of acoustic pressure amplitude which can be given below which the absence of cavitation can be guaranteed.

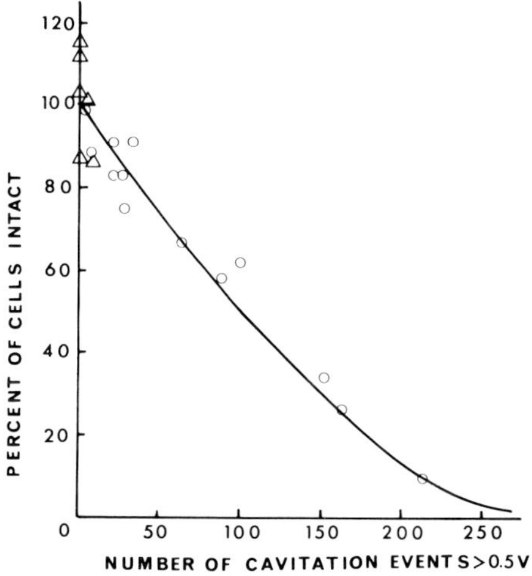

FIG. 4.15 The relationship between the number of amoeboid cells disrupted in a medium as a function of the number of discrete cavitational events which had occurred within that medium (from Coakley *et al.*, 1971).

Many of the factors influencing this "threshold" have been described earlier in this chapter; they include the number and size of micronuclei, the presence or absence of foreign surfaces (especially if they contain hydrophobic regions), the content and composition of dissolved gas, the history and pre-treatment of the solution to be irradiated and the frequency, intensity and pulsing regime of the ultrasonic beam. In general, the "threshold" for the occurrence of cavitation increases with increasing frequency in the MHz range (Fig. 4.16 from Esche, 1952). One explanation for this effect is that at high frequencies there is not enough time for gas molecules to diffuse into the micronucleus during the extremely short duration of the negative pressure portion of the acoustic cycle. A similar effect (i.e. a tendency to inhibit the development of cavitation) is obtained at lower ultrasonic frequencies if the same amount of acoustic energy is delivered in the form of many short but intense pulses separated by relatively long intervals which allow time for any micronuclei which may have begun to grow during the "on time" to spontaneously dissolve (Hill, 1972). This latter process will be accelerated if the ambient pressure in the medium is increased, i.e. the entire sample chamber is pressurized.

Thus, any biological effect produced by a given ultrasonic power which is reduced or abolished when that same amount of ultrasonic power is delivered in the form of short pulses separated by relatively long intervals, and/or when the ambient pressure has been increased, can be assumed to have been caused by ultrasonic cavitation.

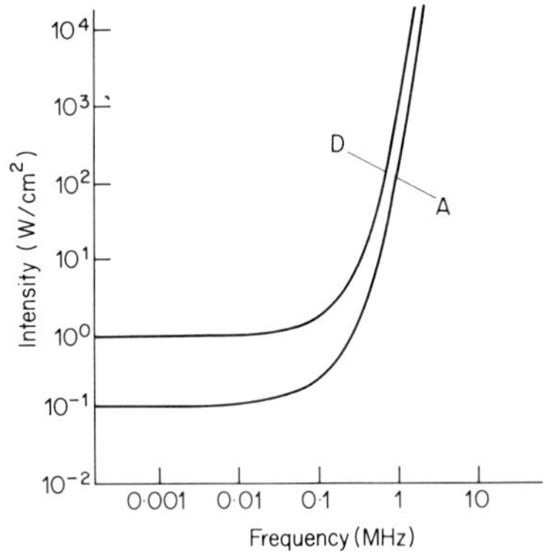

FIG. 4.16 A graphical representation of the commonly quoted "threshold" intensities which have to be exceeded before transient cavitation will occur in aerated (A) and degassed (D) aqueous media *in vitro*, as a function of the exposure frequency (modified from Esche, 1952).

Unfortunately, it is seldom this simple in practice. For example, increasing the ambient pressure may accentuate the destructive forces developed by various types of cavitational activity either by compressing bubbles which were larger than resonant size so that they now become resonant or by increasing the rate of collapse of a transient cavity. Also, delivering the same amount of acoustic energy in the form of multiple short pulses of millisecond duration (i.e. similar to the pulse lengths employed in many therapy devices) has been shown to significantly increase the amount of transient cavitational activity occurring within a medium instead of decreasing it (Ciaravino *et al.*, 1981).

Transient cavitational activity occurs most readily at low ultrasonic frequencies (*ca.* 15–40 kHz) and high acoustic power densities. However, its occurrence becomes extremely unpredictable at megahertz frequencies at intensities close to the "threshold" for its production in that medium. The major factors contributing to this unpredictability affect the generation of the nucleation sites from which the cavitational events develop. For example, Messino *et al.* (1963) have shown that the "cavitational threshold" of a liquid is markedly decreased when it is exposed to ionizing radiation *in vitro*. It is not possible to shield the media being sonicated from cosmic rays and so micro-nuclei are continually being generated within them in a random manner. Also, any liquid medium must be contained within a vessel. The walls of this containing vessel may contain microscopic gas-filled cracks or cavities or even have hydrophobic sites which will encourage the growth of small gas bubbles under the appropriate acoustic exposure conditions.

Micronuclei may also be produced within a liquid by the stirring, shaking or rotation of the sample chamber commonly employed to maintain cells in homogeneous suspension while they are being irradiated *in vitro*. Williams (1982) showed that the rate of haemolysis of human erythrocytes in saline exposed to 1 W/cm^2 of 0·75 MHz ultrasound increased as the rate of rotation of a $2 \times 2 \times 5$ mm rectangular bar magnet within the sample chamber was increased (Fig. 4.17). The samples were contained in a glass "T" piece and irradiated under exactly the same free-field acoustic exposure conditions in the near field of the transducer. The meniscus of the cell suspension was within the narrow stem of the "T" piece and was not markedly disturbed even at the highest rotation rate of 1800 r/min so that gas bubbles were not brought down from the meniscus. This additional rotation-induced haemolysis must therefore have been caused by cavitational activity resulting from the generation of micronuclei either at the edge of the stirring bar during its rotation or within the bulk of the stirred suspension (i.e. hydrodynamic nucleation). The haemolysis caused by stirring in the absence of an applied ultrasonic field was negligible, being less than 3% even after 4 min at 1800 r/min.

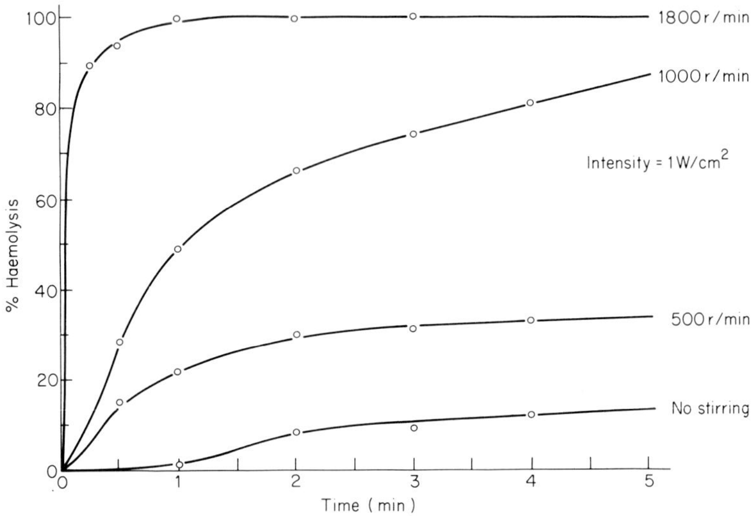

FIG. 4.17 The rate of haemolysis of human erythrocytes in isotonic saline as a function of the stirring rate of the sample under the same acoustic exposure conditions of 1 W/cm^2 at 0·75 MHz (from Williams, 1982).

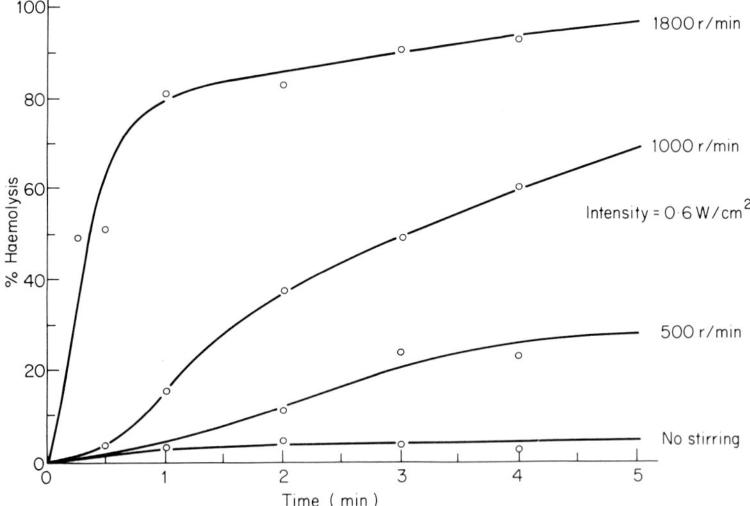

FIG. 4.18 The rate of haemolysis of human erythrocytes in isotonic saline as a function of the stirring rate of the sample under the same acoustic exposure conditions of 0·6 W/cm^2 at 0·75 MHz (from Williams, 1982).

Figure 4.18 shows a similar series of haemolysis curves at various stirring rates obtained using the same apparatus described above, but at a spatially averaged intensity of 0·6 W/cm^2 (0·75 MHz, c.w.). It can be seen that each curve at any given stirring rate has been displaced to a lower haemolysis value and that the curve obtained at 500 r/min shows essentially zero haemolysis after an exposure of one minute. Combining these results with similar sets of curves obtained at other ultrasonic intensities one can obtain a family of percentage haemolysis/intensity curves; each of which can be extrapolated to a different threshold value ranging from more than about 0·7 W/cm^2 at zero stirring rate to less than 0·1 W/cm^2 at 1800 r/min after an exposure time of two minutes. The reproducibility of the haemolysis measurement was greatest at the high-speed stirring rates and progressively decreased as the stirring rate was decreased. At zero or low-speed stirring rates it was possible to distinguish between cell suspensions which had been agitated vigorously immediately before sonication (these gave more haemolysis) and similar samples which had been mixed by gentle rotation or inversion. At high stirring speeds it was not possible to distinguish between these same two samples, presumably because of the magnitude of the effects produced by the large number of nucleation sites generated by the high-speed stirring.

Williams (1982) also investigated two other physical arrangements which may be used to prevent cells from settling during sonication. The stirring bar was removed and in one arrangement the entire glass sample chamber was clamped in a standard laboratory shaker (microid Flask Shaker) and vibrated in a vertical plane at different frequencies with a constant displacement amplitude of approximately 1·7 mm. It was found that vibration in the absence of an applied ultrasonic field did not produce cell lysis but that vibration in the presence of ultrasound (0·75 MHz, 1 W/cm^2, c.w.) resulted in a family of sigmoid curves similar to those presented in Figs 4.17 and 4.18 with more haemolysis being produced at the higher vibration frequencies. In this case it is possible that the cavitation micro-nuclei are being generated by inertial effects causing the liquid to separate from the glass walls of the sample chamber when the direction of motion is reversed.

In the other experimental arrangement, the sample chamber was subjected to low-amplitude percussive impacts at various frequencies from a hexagonal brass bar 0·5 cm in diameter attached to a variable speed motor and driven to rotate while pressed against the outside of the glass sample chamber. This arrangement did not perturb the acoustic field within the sample chamber (as determined by hydrophone measurements) or cause detectable fluid movements within the red cell suspension. Once again, no haemolysis could be detected in the absence of the ultrasonic field but the percussive impacts increased the amount of haemolysis produced when the ultrasound was switched on and the extent of this enhancement increased with increasing frequency of impacts. The mechanism of this enhancement of nucleation is presumed to be analogous to that observed during the degassing of a liquid by reduced atmospheric pressure where an apparently bubble-free liquid will abruptly form a number of visible bubbles if the containing vessel is struck or agitated (Williams, 1982).

Thus, three different physical arrangements have been shown to produce quantitatively similar results in that they increased the amount of haemolysis produced under constant acoustic exposure conditions. The magnitudes of the observed changes were so large that it is a pointless exercise to attempt to interpret the effects of an ultrasound field on cell suspensions *in vitro* without considering the effects of the pre-treatment of the sample and the effects of any stirring or agitation system on the nucleation properties of that suspension.

It should also be noted that the rate at which any given micronucleus will grow is dependent upon the instantaneous value of the acoustic pressure amplitude of a pulsed beam and its duration, rather than the time averaged intensity at that site. Williams (unpublished observations) has found that the rate of haemolysis of a stirred suspension of human erythrocytes in saline was increased when the same SATA intensity of 0·75 MHz ultrasound was delivered in the form of 2 ms pulses (generated by a commercially available therapeutic device). The magnitude of this enhanced haemolysis increased with increasing peak pulse intensity (i.e. with increasing "off" time of the transducer).

The geometry of the acoustic field also has a profound effect on the subsequent fate of micronuclei. Any interface at right angles to the path of a plane acoustic beam reflects some of the incident energy forming a partial standing wave field. The greater the mismatch in acoustic impedance, the higher the percentage of the incident energy which is reflected and therefore the greater the standing wave component of that field; a perfect reflector giving a 100% standing wave ratio. Bubbles which are smaller than resonant size are driven to regions of highest pressure amplitude (i.e. pressure antinodes) where they will rapidly grow and become highly active (Nyborg, 1974). Bubbles larger than resonant size are driven to pressure minima (i.e. pressure nodes); these will therefore be largely inactive in a "perfect" standing wave, but will still experience a significant acoustic field if it is only a partial standing wave.

Iernetti (1971) has also shown that the "threshold" ultrasonic intensity at which cavitation was first detected in air-saturated distilled water at 0·7 MHz decreased as the volume of the irradiated sample was increased. This is probably a reflection of the greater probability of finding a suitable nucleus in a larger volume of fluid. It should be noted that the volume at the focus of a focused device is small compared with that seen by a plane wave transducer. The complex geometry associated with focused ultrasonic fields also means that you are less likely to get appreciable standing wave fields associated with them. In addition the enhanced "quartz-wind" streaming through the focal region means that there is a smaller probability of a micronucleus being held in a position of high acoustic pressure amplitude long enough for it to grow to a resonant size at fairly low intensities.

In general, one would therefore expect that (in the absence of any "seeding" mechanism) a given gassy medium would exhibit cavitation-like activity at the lowest ultrasonic intensities at any given frequency in a plane-wave irradiation system having a high standing wave ratio. This "threshold" intensity would be expected to rise as the standing wave component was reduced. The highest "threshold" intensities for the same liquid at the same frequency would be obtained using the focused system, especially if it was being driven by extremely short electrical pulses.

In view of these and other considerations it is perhaps not too surprising that there is no single "threshold" value for the occurrence of cavitation. Each individual situation has to be assessed separately; for example, the "threshold" for the disruption of human erythrocytes in isotonic saline at about 1 MHz by a focused bowl arrangement is about 100 W/cm^2, this drops to about 1·5 W/cm^2 in an unstirred sample irradiated by a plane 1 MHz beam, or about 0·6 W/cm^2 in a gently stirred sample (Chater and Williams, 1977). If gas bubbles of a resonant size are deliberately introduced this "threshold" intensity is decreased to about 0·02 W/cm^2 (Williams and Miller, 1980).

4.3.11 Cavitational Activity *in vivo*

All the complications which bedevil the production and detection of cavitational activity *in vitro* also apply *in vivo* together with the added complications that experimenters are not able to significantly modify or invade the environment inside the living animal. In general, mammalian tissues *in vivo* appear to contain fewer micronuclei than solutions *in vitro*; this is probably because all fluids entering or leaving living tissues must be filtered through cell membranes. Thus, quiescent biological fluids and tissues appear to be remarkably resistant to bubble formation during decompression (Harvey, 1951) or to ultrasonically-induced cavitation at intensities similar to and greater than those commonly employed in physiotherapy (Williams *et al.*, 1981a; Chater and Williams, 1982).

Histological examination of tissues irradiated with focused fields of MHz ultrasound at intensities greater than about 2000 W/cm^2 and time durations less than about 40 μsec show lesions which are large empty-looking regions containing severely disrupted cells and cellular debris which are present immediately after the irradiation has been terminated (Fry *et al.*, 1970). Similar, but less dramatic, histological lesions are produced within other soft mammalian tissues such as liver at lower intensities (*ca.* 4–10 W/cm^2) using unfocused beams (Martin *et al.*, 1981). These lesions usually had a central core of damaged tissue surrounded by normal tissues which usually contained "blebs" of damaged tissue, i.e. small roughly spherical regions of damaged tissue joined to the central core of damaged tissue by a narrow "stalk". This histological picture of damaged tissue is remarkably similar to the break-up pattern seen when a cavitation void implodes and ejects micronuclei or microbubbles which themselves act as new sites of cavitational activity (Coakley and Nyborg, 1978). These lesions occurred near the surface of the liver at intensities of the order of 10 W/cm^2 if the body cavity was kept closed, but this "threshold" was reduced to about 4 W/cm^2 if the body wall was opened and the peritoneal cavity filled with degassed saline (Martin and Gregory, 1979). This suggests that the initiating cavitation events were occurring either within the fluids bathing the liver (peritoneal fluid or degassed saline) or at the liver/fluid interface.

The growth of micronuclei and the subsequent generation of transient cavitation occurs more readily at low ultrasonic frequencies of the order of 20–40 kHz (Fig. 4.16). These are the frequencies commonly employed to homogenize samples of biological tissue *in vitro* by means of ultrasonic disintegrators. It is therefore not too surprising that these devices may also disrupt living tissues within the intact animal if the acoustic energy is coupled into it in a suitable manner. Fishman and Willis (1977) applied the tip of the probe of one of these low frequency devices to the surface of the skin to produce a reproducible lesion which could be detected using a thermographic camera. Similar lesions were produced by Wittenzellner (1976) who found that a 25 min exposure to the skin overlying his thigh at a frequency of 800 kHz and an intensity of 1·33 W/cm^2 resulted in the formation of an acute blister, which on histological examination showed considerable loosening of the cutaneous and subcutaneous connective tissues, high grade pigmentation of the basal layers and intense accumulation and mobilization of histocytic cells and lymphocytes. These tissue lesions resulted in the formation of a scar which healed without any long-term complications.

Chater and Williams (1982) applied the tip of a 25 kHz probe to the outside of the intact vena cava of an anaesthetized rabbit and were able to demonstrate intravascular thrombus formation and cavitation-induced lysis of erythrocytes and platelets. It is highly probable that similar cavitation-induced thrombogenic effects are responsible for the reports of decreased haemorrhage observed when using ultrasonically vibrating surgical instruments (Goliamina, 1974) and ultrasonic dental equipment oscillating at 25 kHz.

Negative results have also been obtained from exposures at kHz frequencies; for example, Fishman (1968) looked for evidence of haemolysis in human blood samples taken from the antecubital vein while the hand was immersed in an ultrasonic cleaning bath for up to 45 min. Fortunately for the volunteer no intravascular haemolysis was detected. This observation should not be taken as evidence that it is safe to immerse the hand for prolonged periods in ultrasonic cleaning baths (a practice which should be discouraged).

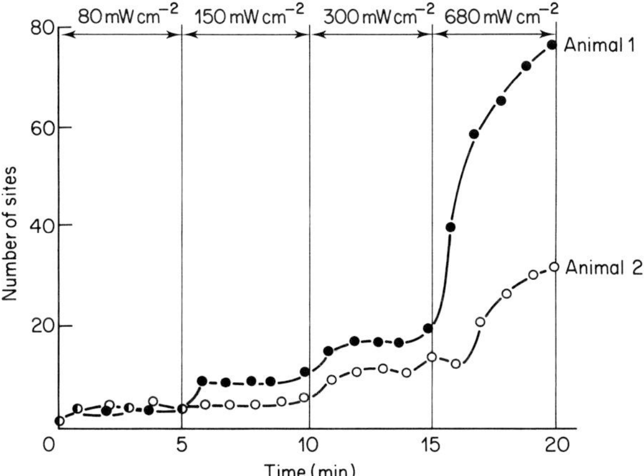

FIG. 4.19 The rate of accumulation of sites of gas bubble formation within the hind limb of guinea pigs *in vivo* during irradiation with continuous wave 0·75 MHz ultrasound (from ter Haar and Daniels, 1981).

ter Haar and Daniels (1981) and ter Haar *et al.* (1982) have shown that therapeutic intensities of MHz ultrasound can cause gas bubbles to grow within the tissues of anaesthetized guinea pigs *in vivo*. The hind limbs of the animals were depilated and submerged in a water bath where they were simultaneously exposed to two ultrasonic fields essentially at right angles to each other. One acoustic beam was emitted by a commercial therapy device at 0·75 MHz while the other was that of an 8 MHz mechanical sector scanner developed to detect gas bubble formation within tissues during decompression. This detector could visualize bubbles larger than about 10 μm in diameter; N.B. all bubbles which can be detected are significantly larger than the resonant size for free bubbles at this frequency which is of the order of about 4 μm. Figure 4.19 shows the rate of accumulation of gas bubbles as a function of time for two animals exposed to the space averaged c.w. intensities shown in the figure. It can be seen that the number of active sites increased rapidly with increasing intensity above about 100 mW/cm^2, and that the majority of the new sites were recruited within the first minute of a step increase in intensity. It should be noted that each bubble would have had to grow through its resonant size before it became large enough to have been detected by this technique. Some bubbles were only visualized on one scan frame and then disappeared, perhaps to re-appear intermittently at the same site in subsequent frames. These single frame events may have been intravascular gas bubbles passing through the plane of the scan or an extravascular bubble oscillating in size about the detection threshold of the recording system. The persistent bubbles represented extravascular bubbles which were formed mainly at the boundaries between the muscle bundles and within the fatty tissues. These *in vivo* observations by ter Haar and Daniels (1981) are the most conclusive evidence currently available that stable cavitational activity can be generated *in vivo* by therapeutic ultrasound.

Lele (1978) and his collaborators have extensively investigated the ultrasonic conditions necessary to generate the various types of ultrasonic cavitation in tissues both *in vivo* and in freshly excised tissues *in vitro*. They found that there appeared to be no distinct "threshold" intensity for the detection of sharp "spikes" or bursts of spikes of subharmonic emission when anaesthetized cat brain or freshly excised calf liver were irradiated using a focused beam of 2·7 MHz ultrasound. These spikes (which are indicative of some form of bubble oscillatory behaviour or stable cavitation) occurred only during the period of insonation, but randomly with respect to its beginning or end. They could be detected even at 100 mW/cm^2 (the lowest intensity investigated) and their time-averaged energy content increased monotonically with increasing intensity until about 1500 W/cm^2 when there was a sudden marked increase in the intensity of the emission and the concomitant generation of white noise indicative of the presence of transient or collapse-type cavitation (Lele, 1978). The attenuation coefficient of a tissue will be increased if small gas-filled bubbles or cavities grow within that tissue and scatter the incident acoustic wave. Under the exposure conditions described above, it was found that the attenuation coefficient of the irradiated tissue remained constant at 0·32 Neper/cm up to intensities of about 500 W/cm^2 and thereafter increased monotonically with increasing intensity to reach approximately 0·4 Nepers/cm at about 1500 W/cm^2. The attenuation coefficient rose abruptly to approximately 1·0 Nepers/cm at higher intensities when transient cavitation occurred.

Thus, Lele (1978) and his co-workers have demonstrated that various types of bubble oscillatory behaviour occurring within organized tissues both *in vivo* and *in vitro* may be detected at intensities as low as 100 mW/cm^2. The number of these active bubbles and the violence of their oscillatory behaviour increase with increasing intensity until at 500 W/cm^2 they cause a measurable change in the attenuation coefficient of that tissue. The irradiated tissues were examined for histological evidence of cellular damage and were divided into two groups. In the absence of the wide-band emission characteristic of transient cavitation, the observed lesions were coagulative and pan-necrotic and could not be distinguished from lesions created by heat alone. Those lesions produced when wide-band emission had been recorded had disintegrated tissue in the damaged zone and invariably contained extravasated blood cells (Lele, 1978). It should be noted that histological techniques are relatively insensitive and may not have been able to detect the more subtle or small-scale damage produced by the scattered and spasmodic stably-oscillating bubbles.

Williams (1981a) has shown that the diverse changes in platelet function and the activation of the blood coagulation system by ultrasound demonstrated *in vitro* appear to be initiated by some form of cavitational activity within the blood or plasma. In an endeavour to see if these effects could be duplicated *in vivo* an antecubital vein was cannulated in one arm of healthy adult human volunteers and sequential 2·5 ml blood samples were collected in plastic tubes. A 0·75 MHz transducer was positioned upstream of the cannulation site and activated at the highest intensity that the volunteer would tolerate (0·25–0·34 W/cm^2; continuous wave) throughout the time taken to collect the test samples (typically 30–45 sec). There was no significant elevation of the platelet specific protein β- thromboglobulin in the plasma of the irradiated samples indicating that no platelets had been lysed (Williams *et al.*, 1981).

One criticism of this *in vivo* series is that the maximum intensities which could be tolerated were too low to demonstrate a positive effect. An animal model was therefore developed whereby similar sequential blood samples could be withdrawn from the inferior vena cava of anaesthetized rabbits, before, during and after exposure to 0·75 MHz ultrasound. Rabbit platelets do not contain β-thromboglobulin but they are loaded with histamine which is liberated when the platelets are disrupted or undergo their physiological release reaction. It was found that exposure of stationary or orderly flowing blood for 30–40 sec at spatial average intensities of up to 17 W/cm^2 did not result in the release of detectable quantities of histamine from the platelets or of haemoglobin from erythrocytes *in vivo* (Chater and Williams, 1982). Contradictory results were obtained by Wong and Watmough (1981) who found that 1–2 W/cm^2 of continuous wave 0·75 MHz ultrasound directed into the beating hearts of anaesthetized rats through their diaphragms resulted in the lysis of erythrocytes (and presumably also platelets). It is highly probable that the turbulence and large pressure changes occurring within the beating heart enhanced the nucleation characteristics of the blood (i.e. hydrodynamic nucleation) and so lowered the "threshold" intensity for generating ultrasonic cavitation *in vivo*.

In summary, therapeutic intensities and frequencies of ultrasound appear to be able to induce bubble growth and oscillation (i.e. stable cavitation) *in vivo*. The effects of these oscillating bubbles on organized mammalian tissues have not been thoroughly investigated but experience with other less complex biological systems suggests that they could be potentially hazardous. The rate of growth of these bubbles would be enhanced if the same "dose" of continuous wave ultrasonic power was administered in the form of relatively long pulses (*ca.* 2 ms) but reduced if given as extremely short pulses. Those bubbles which grow within the fluid blood undergo more violent oscillatory behaviour and apparently lyse platelets and erythrocytes which activate the coagulation system (this may or may not lead to the development of an intravascular thrombus depending on the "sensitivity" of that person's coagulation system at that time). Fortunately, quiescent or orderly flowing blood cannot readily be driven to cavitate under the exposure conditions employed in diagnosis or therapy, but it would be a wise precaution to avoid conditions which enhance nucleation such as the beating heart and artificial valves or stenoses, and to investigate the additional potential hazard which might result from the deliberate introduction of minute gas bubbles as contrast agents to enhance diagnostic imaging techniques (Ziskin *et al.*, 1972).

The effects of other non-acoustic factors which may enhance nucleation *in vivo* have also not been adequately investigated. The decompression experiments reported in Fulton (1951) show that more gas bubbles are formed within limbs which are undergoing muscular activity than within resting limbs. This mechanical enhancement of bubble nucleation may also apply to ultrasound exposures. If it does, it could make a significant difference in physiotherapy because one might expect different cavitation-induced effects whether a limb or joint was irradiated while it was resting or while it was being actively exercised.

Some physiotherapy devices permit the simultaneous exposure of a tissue to therapeutic ultrasound and to pulses of electrical current (i.e. simultaneous acoustic and Faradic stimulation). The electrical stimulation would cause muscular contraction and possibly result in the mechanical enhancement of nucleation as described above. In addition, the electrical currents may directly generate cavitational nuclei within the tissues by electrolysis. These combination treatments therefore ought to be used with extreme care either until they are shown to be no more hazardous than conventional single-modality ultrasonic therapy, or there is clear experimental evidence that the combination therapy is significantly more effective.

REFERENCES

Abagyan, G. V. and Butyagin, P. Yu. (1965). Electron paramagnetic resonance spectra observed on mechanical treatment of DNA preparations. *Biofizika* **10**, 763–770.

Carstensen, E. L. and Schwan, H. P. (1959). Acoustic properties of haemoglobin solutions. *J. Acoust. Soc. Amer.* **31**, 305–311.

Charm, S. E. and Wong, B. L. (1970). Shear degradation of fibrinogen in the circulation. *Science* **170**, 466–468.

Chater, B. V. and Williams, A. R. (1977). Platelet aggregation *in vitro* by therapeutic ultrasound. *Thrombosis and Haemostasis* **38**, 640–651.

Chater, B. V. and Williams, A. R. (1982). Absence of platelet damage *in vivo* following the exposure of non-turbulent blood to therapeutic ultrasound. *Ultrasound Med. Biol.* **8**, 85–87.

Ciaravino, V., Flynn, H. G. and Miller, M. W. (1981). Pulsed enhancement of acoustic cavitation; a postulated model. *Ultrasound Med. Biol.* **7**, 159–166.

Clarke, P. R. and Hill, C. R. (1970). Physical and chemical aspects of ultrasonic disruption of cells. *J. Acoust. Soc. Amer.* **47**, 649–653.

Coakley, W. T. and Dunn, F. (1971). Degradation of DNA in high-intensity focused ultrasonic fields at 1 MHz. *J. Acoust. Soc. Amer.* **50**, 1539–1545.

Coakley, W. T. and Nyborg, W. L. (1978). Cavitation; dynamics of gas bubbles; applications. *In* "Ultrasound: Its application in Medicine and Biology" (Ed. F. J. Fry), pp. 77–159, Chap. 4. Elsevier, Holland.

Coakley, W. T., Hampton, D. and Dunn, F. (1971). Quantitative relationships between ultrasonic cavitation and effects upon amoebae at 1 MHz. *J. Acoust. Soc. Amer.* **50**, 1546–1553.

Cotterell, T. L. (1954). "The Strengths of Chemical Bonds". Butterworths, Sevenoaks.

Crowell, J. A., Kusserow, B. K. and Nyborg, W. L. (1977). Functional changes in white blood cells after microsonation. *Ultrasound Med. Biol.* **3**, 185–190.

Del-Duca, M., Yeager, E., Davies, M. O. and Hovorka, F. H. (1958). Isotopic techniques in the study of the sonochemical formation of hydrogen peroxide. *J. Acoust. Soc. Amer.* **30**, 301–307.

Dewitz, T. S., McIntire, L. V., Martin, R. R. and Sybers, H. D. (1979). Enzyme release and morphological changes in leukocytes induced by mechanical trauma. *Blood Cells* **5**, 499.

Doynon, D. and Simont, Y. (1951). *Comp. Rend.* **232**, 2011, 2411.

Dunn, F. and Pond, J. B. (1978). Selected non-thermal mechanisms of interaction of ultrasound and biological media. *In* "Ultrasound: its Applications in Medicine and Biology" (Ed. F. J. Fry), pp. 539–560, Chap. 9. Elsevier, Holland.

Dyer, H. J. (1965). Changes in the behaviour of mosses treated with ultrasound. *J. Acoust. Soc. Amer.* **37**, 1195.

Dyson, M., Woodward, B. and Pond, J. B. (1971). Flow of red blood cells stopped by ultrasound. *Nature (Lond.)* **232**, 572–573.

Elder, S. A. (1959). Cavitation microstreaming. *J. Acoust. Soc. Amer.* **31**, 54–64.

El'piner, I. E. (1964). "Ultrasound: Physical, Chemical and Biological Effects". Consultants Bureau, New York.

El'piner, I. E., Faikin, I. M. and Basurmanova, O. K. (1966). Intracellular microcurrents caused by ultrasound waves. *Fed. Proc.* **25**, T716–720.

Esche, R. (1952). Untersuchungen der Schwingungskavitation in Flüssigkeiten. *Acustica* **2**, 208–218.

Ewen, S. J. (1969). Devices for calculus removal. *J. Amer. Dent. Assoc.* **78**, 795–798.

Fishman, S. S. (1968). Biological effects of ultrasound: *in vivo* and *in vitro* haemolysis. *Proc. Western Pharmacol. Soc.* **11**, 149–150.

Fishman, S. S. and Willis, J. N. (1977). Development of a stress test by exposing the arm to cavitating ultrasound: an application for thermography. *Proc. Western Pharmacol. Soc.* **20**, 221–226.

Flynn, H. G. (1964). *In* "Physical Acoustics", Vol. IB. (Ed. W. P. Mason), pp. 57–172. Academic Press, New York.

Frenkel, Ya. I. (1940). On the electrical effects connected with cavitation caused by ultrasonic oscillations in a liquid. *J. Phys. Chem. (U.S.S.R.)* **14**, 305–308.

Fry, F. J., Kossoff, G., Eggleton, R. C. and Dunn, F. (1970). "Threshold" ultrasonic dosages for structural changes in the mammalian brain. *J. Acoust. Soc. Amer.* **48**, 1413–1417.

Fulton, J. F. (1951). "Decompression sickness: Caisson Sickness, Diver's and Flier's Bends and Related Syndromes". W. B. Saunders, Philadelphia.

Gershoy, A. and Nyborg, W. L. (1973). Perturbation of plant-cell contents by ultrasonic microirradiation. *J. Acoust. Soc. Amer.* **54**, 1356–1367.

Goliamina, I. P. (1974). Ultrasonic surgery. Proc. Eighth Intern. Congress on Acoustics, London. pp. 63–69.

Graham, E., Hedges, M., Leeman, S. and Vaughan, P. (1980). Cavitational bio-effects at 1·5 MHz. *Ultrasonics* **18**, 224–228.

Gramiak, R. and Shah, P. M. (1971). Detection of intracardiac blood flow by pulsed echo-ranging ultrasound. *Radiology* **100**, 415–418.

Griffing, V. (1952). The chemical effects of ultrasonics. *J. Chem. Phys.* **20**, 939–942.

Harvey, E. N. (1951). Physical factors in bubble formation. *In* "Decompression Sickness' (Ed. J. F. Fulton), pp. 90–114, Chap. 4. W. B. Saunders, Philadelphia.

Hawley, S. A., MacLeod, R. M. and Dunn, F. (1963). Degradation of DNA by intense, non-cavitating ultrasound. *J. Acoust. Soc. Amer.* **35**, 1285–1287.

Hill, C. R. (1972). Ultrasonic exposure thresholds for changes in cells and tissues. *J. Acoust. Soc. Amer.* **52**, 667–672.

Holtzmark, J., Johnsen, I., Sikkeland, T. and Skavlem, S. (1954). Boundary layer flow near a cylindrical obstacle in an oscillating incompressible fluid. *J. Acoust. Soc. Amer.* **26**, 26–39.

Howkins, S. D. (1966). Solid erosion in low-amplitude sound fields. *J. Acoust. Soc. Amer.* **39**, 55–61.

Hughes, D. E. and Nyborg, W. L. (1962). Cell disruption by ultrasound. *Science* **138**, 108–114.

Iernetti, G. (1971). Cavitation threshold dependence on volume. *Acustica* **24**, 191–196.

Jackson, F. J. and Nyborg, W. L. (1958). Small-scale acoustic streaming near a locally excited membrane. *J. Acoust. Soc. Amer.* **30**, 614–619.

Kashkooli, H. A., Rooney, J. A. and Roxby, R. (1980). Effects of ultrasound on catalase and malate dehydrogenase. *J. Acoust. Soc. Amer.* **67**, 1798–1801.

Lamb, H. (1945). *In* "Hydrodynamics". Dover Publications, New York.

Lauterborn, W. (1974). *In* "Finite-Amplitude Wave Effects in Fluids" (Ed. L. Bjørnø), pp. 195–202. IPC Science and Technology Press, Guildford.

Lele, P. P. (1978). Cavitation and its effects on organised mammalian tissues—a summary. Appendix I. *In* "Ultrasound: Its Application in Medicine and Biology" (Ed. F. J. Fry), pp. 737–741. Elsevier, Holland.

Martin, C. J. and Gregory, D. W. (1979). A microscopic investigation of changes in mouse liver produced by ultrasound. Paper presented at the British Bioacoustical Discussion Group Meeting, British Institute of Radiology, London, July 6th, 1979.

Martin, C. J., Gemmell, H. G. and Watmough, D. J. (1978). A study of streaming in plant tissue induced by a Doppler fetal heart detector. *Ultrasound Med. Biol.* **4**, 131–138.

Martin, C. J., Gregory, D. W. and Hodgkiss, M. (1981). The effects of ultrasound *in vivo* on mouse liver in contact with an aqueous coupling medium. *Ultrasound Med. Biol.* **7**, 253–265.

Martin, C. J., Pratt, B. M. and Watmough, D. J. (1982). A study of ultrasound-induced microstreaming in blood vessels of tropical fish. *Brit. J. Cancer* **45**, 161–164.

Messino, D., Sette, D. and Wanderlingh, F. (1963). Statistical approach to ultrasonic cavitation. *J. Acoust. Soc. Amer.* **35**, 1575–1583.

Miller, D. L. (1977). Stable arrays of resonant bubbles in a 1 MHz standing-wave acoustic field. *J. Acoust. Soc. Amer.* **62**, 12–19.

Miller, D. L. (1979a). A cylindrical-bubble model for the response of plant-tissue gas bodies to ultrasound. *J. Acoust. Soc. Amer.* **65**, 1313–1321.

Miller, D. L. (1979b). Cell death thresholds in Elodea for 0·45–10 MHz ultrasound compared to gas-body resonance theory. *Ultrasound Med. Biol.* **5**, 351–357.

Miller, M. W. and Kaufman, G. E. (1977). Effects of short-duration exposures to 2 MHz ultrasound on growth and mitotic index of Pisum sativum roots. *Ultrasound Med. Biol.* **3**, 27–29.

Miller, D. L., Nyborg, W. L. and Whitcomb, C. C. (1978). *In vitro* clumping of platelets exposed to low intensity ultrasound. *In* "Ultrasound in Medicine" Vol. 4 (Eds D. White and E. A. Lyons), pp. 545–553. Plenum, New York.

Miller, D. L., Nyborg, W. L. and Whitcomb, C. C. (1979a). Platelet aggregation induced by ultrasound under specialised conditions *in vitro*. *Science* **205**, 505–507.

Miller, D. L., Williams, A. R. and Nyborg, W. L. (1979b). Photochemical detection of platelet damage induced by low intensity ultrasound. Proc. 24th Ann. Meeting of American Institute Ultrasound in Medicine, Oklahoma City.

Minnaert, M. (1933). On musical air bubbles and the sounds of running water. *Phil. Mag.* **16**, 235.

Morris, D. R. and Williams, A. R. (1978). Membrane fatigue as a parameter in shear-induced lysis of erythrocytes. *Amer. Inst. Chem. Engn., Symp. Series* **74**, 27–30.

Neppiras, E. A. (1969). Subharmonic and other low-frequency emission from bubbles in sound-irradiated liquids. *J. Acoust. Soc. Amer.* **46**, 587–601.

Neppiras, E. and Fill, E. E. (1969). A cyclic cavitation process. *J. Acoust. Soc. Amer.* **46**, 1264–1271.

Noltingk, B. E. and Neppiras, E. A. (1950). Cavitation produced by ultrasonics. *Proc. Phys. Soc. (Lond.)* **B63**, 674–685.

Nyborg, W. L. (1965). Acoustic streaming. *In* "Physical Acoustics", Vol. 2B (Ed. W. P. Mason), pp. 265–383, Chap. 11. Academic Press, New York.

Nyborg, W. L. (1975). "Intermediate Biophysical Mechanics". Cummings, Menlo Park, California.

Nyborg, W. L. (1977). "Physical Mechanisms for Biological Effects of Ultrasound", pp. 78–8062. HEW Publication (FDA).

Nyborg, W. L. and Dyer, H. J. (1960). Ultrasonically-induced motions in single plant cells. Proc. 2nd Intern. Conf. Med. Electron., Paris, June 24–27th, 1959. pp. 391–396.

Nyborg, W. L. and Rodgers, A. (1967). Motion of liquid inside a closed vibrating vessel. *Biotech. Bioeng.* **9**, 235–241.

Nyborg, W. L., Gershoy, A. and Miller, D. L. (1977). Interaction of ultrasound with simple biological systems. *Proc. Ultrasonics Int.* 19–27.

Parry, J. S., Cleary, B. K., Williams, A. R. and Evans, D. M. D. (1971). Ultrasonic dispersal of cervical cell aggregates. *Acta Cytol.* **15**, 163–166.

Pritchard, N. J., Hughes, D. E. and Peacocke, A. R. (1966). The ultrasonic degradation of biological macromolecules under conditions of stable cavitation: I. Theory, methods and application to DNA. *Biopolymers* **4**, 259.

Prudhomme, R. D. (1972). *Vijnana Parishad Anusandhan Patrika* **15**, 3.

Lord Rayleigh (1917). On the pressure developed in a liquid during the collapse of a spherical cavity. *Phil. Mag.* **34**, 94–98.

Rooney, J. A. (1970). Hemolysis near an ultrasonically pulsating gas bubble. *Science* **169**, 869–871.

Rooney, J. A. (1972). Shear as a mechanism for sonically induced biological effects. *J. Acoust. Soc. Amer.* **52**, 1718–1724.

Saksena, T. K. and Nyborg, W. L. (1970). Sonoluminescence from stable cavitation. *J. Chem. Phys.* **53**, 1722–1734.

Sanders, M. F. and Coakley, W. T. (1972). Factors affecting biodegradation by a transversely oscillating wire at 20 kHz. *Exp. Cell Res.* **73**, 410–414.

Schmid-Schönbein, H. and Wells, R. E. (1969). Fluid drop-like transition of erythrocytes under shear. *Science* **165**, 288–290.

Schmitt, F. O. (1929). Ultrasonic manipulation. *Protoplasma* **7**, 332–340.

Storm, D. L. (1974). Interfacial distortions of a pulsating gas bubble. *In* "Finite-

Amplitude Wave Effects in Fluids" (Ed. L. Bjørnø), pp. 234–239. IPC Science and Technology Press, Guildford.

Temple, P. R., Detenbeck, R. W. and Nyborg, W. L. (1971). Sonoluminescence from a "shuttlecock" associated with a single gas bubble. *J. Acoust. Soc. Amer.* **50**, 112.

ter Haar, G. R. and Daniels, S. (1981). Evidence for ultrasonically induced cavitation *in vivo*. *Phys. Med. Biol.* **26**, 1145–1149.

ter Haar, G. R., Daniels, S., Eastaugh, K. C. and Hill, C. R. (1982). Ultrasonically induced cavitation *in vivo*. *Brit. J. Cancer* **45**, 151–155.

Thacker, J. (1974). An assessment of ultrasonic radiation hazard using yeast genetic systems. *Brit. J. Radiol.* **47**, 130–138.

Wang, S. Y. (1977). Ultrasonic radiation of nucleic acids and components. *In* "Symposium on Biological Effects and Characterizations of Ultrasound Sources", pp. 196–205. HEW Publication (FDA) 78–8048.

Watmough, D. J. (1981). Cavitation in biological objects at low and medium intensities of ultrasound. Paper presented at UBIOMED-V (Symposium on Ultrasound in Biology and Medicine), Pushchino, U.S.S.R., Sept. 7–11.

Weissler, A. (1959). Some sonochemical reaction rates. *J. Acoust. Soc. Amer.* **32**, 283–284.

Willard, G. W. (1954). Vibrating liquid surfaces as generators of bubbles and drops. *J. Acoust. Soc. Amer.* **26**, 933.

Williams, A. R. (1969a). An electromagnetic modification of the Zimm-Crothers viscometer. *J. Sci. Instr. (J. Phys. E.)* **2**, 279–281.

Williams, A. R. (1969b). DNA macroradicals produced by ultrasonic irradiation. *Biopolymers* **8**, 555–558.

Williams, A. R. (1971). Hydrodynamic disruption of human erythrocytes near a transversely oscillating wire. *Rheol. Acta* **10**, 67–70.

Williams, A. R. (1974a). Release of serotonin from human platelets by acoustic microstreaming. *J. Acoust. Soc. Amer.* **56**, 1640–1643.

Williams, A. R. (1974b). DNA degradation by acoustic microstreaming. *J. Acoust. Soc. Amer.* **55**, S17.

Williams, A. R. (1977). Intravascular mural thrombi produced by acoustic microstreaming. *Ultrasound Med. Biol.* **3**, 191–203.

Williams, A. R. (1981a). Interactions of ultrasound with platelets and the blood coagulation system. Proc. Ultrasound Interactions in Medicine and Biol. Symposium, Reinhardsbrunn, East Germany, Nov. 10–14.

Williams, A. R. (1981b). The induction of intravascular thrombi *in vivo* by means of localised hydrodynamic shear stresses. *In* "Basic Aspects of Blood Trauma" (Eds H. Schmid-Schönbein and P. Teitel), pp. 63–73. Martinus Nijhof, The Hague.

Williams, A. R. (1982). Absence of meaningful thresholds for bioeffect studies on cell suspensions *in vitro*. *Brit. J. Cancer* **45**, 192–195.

Williams, A. R. and Slade, J. S. (1971). Ultrasonic dispersal of aggregates of *Sarcina lutea*. *Ultrasonics* **8**, 85–87.

Williams, A. R. and Miller, D. L. (1980). Photometric detection of ATP release from human erythrocytes exposed to ultrasonically activated gas-filled pores. *Ultrasound Med. Biol.* **6**, 251–256.

Williams, A. R., Hughes, D. E. and Nyborg, W. L. (1970). Hemolysis near a transversely oscillating wire. *Science* **169**, 871–873.

Williams, A. R., Sykes, S. M. and O'Brien, W. D. Jr (1976). Ultrasonic exposure modifies platelet morphology and function *in vitro*. *Ultrasound Med. Biol.* **2**, 311–317.

Williams, A. R., Chater, B. V., Allen, K. A. and Sanderson, J. H. (1981). The use of β-thromboglobulin to detect platelet damage by the therapeutic ultrasound *in vivo*. *J. Clin. Ultrasound* **9**, 145–151.

Wittenzellner, R. (1976). Tissue-damaging effects of ultrasound. *Ultrasonics* **14**, 281–282.

Wong, Y. S. and Watmough, D. J. (1981). Haemolysis of red blood cells *in vitro* and *in vivo* caused by therapeutic ultrasound at 0·75 MHz. Proc. Ultrasound Interactions in Medicine and Biology Symposium, Reinhardsbrunn, East Germany, Nov. 10–14.

Ziskin, M. C., Bonakdarpour, A., Weinstein, D. P. and Lynch, P. R. (1972). Contrast agents for diagnostic ultrasound. *Invest. Radiol.* **7**, 500–505.

5. BIOEFFECTS OF ULTRASOUND

If enough ultrasonic energy is directed into any biological material then it will eventually be heated (Chapter 3) or disrupted by cavitation (Chapter 4), or both. Thus a biological effect (defined as a detectable change from "normal") can be produced in virtually every living organism or structure if it is supplied with enough acoustic energy. The problem is to condense the vast literature reporting bioeffect experiments into one cohesive but concise chapter without degenerating into a mere catalogue of reports. It has therefore been impossible to include many of the publications in this field and to accord others the prominence they deserve.

My approach has been to briefly summarize the reported interactions between megahertz ultrasound and (a) isolated macromolecules, (b) membranes and sub-cellular organelles, (c) cells in suspension and (d) cells in tissues, both *in vitro* and *in vivo*. The articles which are referred to within each section are those which are discussed and quoted most frequently or best illustrate a particular point. This brief review will be followed (Chapter 6) by a survey of the bioeffects obtained at the lowest ultrasonic intensities and the attempts to delineate the "boundary region" between positive and negative effects as well as the arguments for and against a statutory upper limit for the power output of diagnostic devices.

5.1 ISOLATED MACROMOLECULES

5.1.1 DNA

Deoxyribonucleic acid (DNA) may be extracted and purified from a number of biological sources and exists in saline solution as a very long semi-rigid coil consisting of two chains wound in an α-helix (Watson and Crick, 1953). Any long molecule may be stretched by an applied hydrodynamic shear field so that a stress is developed within that molecule which (if it is large enough) tends to break both chains of that molecule at the same time (Ryabchenko *et al.*, 1964) preferentially near its centre (Frenkel, 1944). These hydrodynamic shear fields are developed around solid bodies vibrating at ultrasonic frequencies while submerged in a liquid (Williams *et al.*, 1970) or around gas bodies driven to oscillate by an applied acoustic field (i.e. stable cavitation as described in Chapter 4).

The British Bioacoustic Discussion Group Meetings held in the early 1970s were the scene of interminable controversy concerning the "threshold" intensity needed to disrupt isolated DNA in solution at about 1 MHz. On the one hand Dr C. R. Hill and his associates (Hill *et al.*, 1970) were quoting "threshold" values of about 0·4 W/cm^2 for a 3 min exposure to degrade DNA isolated from calf-thymus and salmon sperm, while on the other Dr W. T. Coakley was obtaining "threshold" values of the order of about 1000 W/cm^2. In retrospect it can be seen that both groups of workers were correct and that the different "thresholds" were due to fundamental differences in their exposure systems. Hill *et al.* (1970) used a plane wave transducer and rotated the sample chamber which contained a relatively large volume of DNA solution (this procedure enhanced the growth and development of stable cavitation bubbles). However, Dr Coakley used a focused bowl in which only a small quantity of liquid was present within the focal volume at any instant (which therefore only contained a small number of cavitation nuclei) compounded by the fact that any nuclei which did begin to grow at moderately low intensities were washed out of the focal volume by the quartz wind streaming before they had attained their resonant size.

All bioeffect fields are plagued by articles which either report work which cannot be repeated in other laboratories or misinterpret the significance of a repeatable observation and report it in such a way as to cause unnecessary concern to clinical users of that modality. A classic example of the latter was the article by Galperin-Lemaitre *et al.* (1975) who reported the not unreasonable finding of the degradation of isolated DNA in solution at intensities as low as 200 mW/cm^2 of MHz ultrasound. Unfortunately, these authors incorrectly referred to this degradation as a mutagenic effect and may be cited as evidence of one by someone who has not seen the swift rebuttals by Thacker (1975) and Coakley and Dunn (1975). DNA degradation *in vivo* is not likely to be a significant potential biological hazard at low ultrasonic intensities because it is much more difficult to generate any form of cavitational activity *in vivo* than it is *in vitro*. Also, the shear forces generated by the cavitational activity *in vivo* would be so large that the entire cell is likely to be disrupted before its DNA is degraded.

It is nevertheless possible that long-lived free radicals or other highly reactive chemical species generated within a transiently cavitating field could chemically modify DNA and produce single strand breaks or genetically significant changes. It should be borne in mind that there is only a small probability of any given highly reactive molecule being able to penetrate the nucleus of an intact cell without reacting with another biomolecule en route. Combes (1975) irradiated transforming DNA and Thacker (1974) subjected yeasts and bacteria to media which had been driven to cavitate and neither author was able to demonstrate any mutagenic effects. It is also worthy of note that several studies have exposed bacterial and mammalian cells to ultrasound and isolated the DNA from the intact survivors. After allowing for procedural artifacts there was no evidence for single or double stranded breaks in the DNA from any of the experimental series (Graham *et al.*, 1980; Treton *et al.*, 1977).

5.1.2 Enzymes

The rate of a reaction catalysed by an enzyme usually increases as the temperature is increased until the temperature becomes so high that it begins to change the three-dimensional conformation of the enzyme itself (i.e. thermal denaturation). Thus, over a limited range of temperature enzymic reactions tend to be accelerated by exposure to MHz ultrasound.

The three-dimensional conformation of an enzyme is crucial to its functioning and so enzymes may be inactivated by large hydrodynamic shear fields which can break inter- and intra-chain bonds and change the shape of the molecule (Tirrell and Middleman, 1978). Kashkooli *et al.* (1980) have shown that the microstreaming fields generated around a wire oscillating transversely at 20 kHz can inactivate the enzyme malate dehydrogenase but not the enzyme catalase. An extensive series of experiments were conducted by Dunn and Macleod (1968) who exposed the enzymes α-chymotrypsin, trypsin, aldolase, lactate dehydrogenase and ribonuclease to ultrasonic frequencies ranging from 1 to 27 MHz at intensities as high as 10^4 W/cm^2. This study showed that the enzymes were inactivated only if the suspension was driven to cavitate.

Stefanovic *et al.* (1959, 1960) reported the controversial observations that enzymes in solution were inactivated by 3 MHz ultrasound within the intensity range 1 to 3 W/cm^2. Investigations in other laboratories showed that the reported effects were not due to the direct interaction between the ultrasonic waves and the protein molecules, but rather to a reaction between the suspension medium and the rubber material used as part of the containing vessel (Dunn, 1971).

Thus, as in the case of isolated DNA, the adverse interactions between ultrasound and isolated enzymes appear to be mainly caused by the hydrodynamic shear forces associated with ultrasonic cavitation and are therefore not likely to be hazardous in the case of the intact cell (which will be disrupted by these same forces). Both DNA and other biological macromolecules in solution may possibly be disrupted by hydrodynamic forces resulting from the quartz wind-streaming through the focal volume of a focused ultrasonic field (Coakley and Dunn, 1972) but again the intact cell will almost certainly be destroyed before its intracellular molecules are degraded or modified.

5.2 MEMBRANES AND SUB-CELLULAR ORGANELLES

5.2.1 Artificial Membranes

The simplest artificial membranes are bi-molecular leaflets of water-insoluble lipids such as oxidized cholesterol which can model some of the behaviour of biological membranes when appropriate modifiers are added (Mueller and Rudin, 1968; Tien, 1974). Ochs and Burton (1974) and Pasechnik and Sokolov (1973) observed that the capacitance of their artificial membranes changed when exposed to sonic waves of frequency 50–150 Hz which they interpreted as stretching of the lipid film. Rohr and Rooney (1978) performed similar capacitance and electrical conductivity experiments on bilayer lipid membranes exposed to 1 MHz ultrasound and could detect no change in the conductance, the capacitance, or the dependence of each on the voltage applied across the membrane until the intensity exceeded about 1·5 W/cm^2 and ruptured the membrane.

Lakshminarayanaiah and Siddiqi (1972) and Mendez *et al.* (1976) investigated the permeability of water through artificial membranes composed of collodion and cellulose respectively. It was found that ultrasonic intensities comparable with those used in therapy (0·18 W/cm^2 in Mendez *et al.*, 1976) increased the apparent permeability coefficient for water from $5{\cdot}03 \times 10^{-5}$ cm/sec to $11{\cdot}62 \times 10^{-5}$ cm/sec. This effect may be relevant in phonophoresis, i.e. that process whereby ultrasound facilitates the penetration of chemicals through membranes such as the skin.

5.2.2 Biological Membranes

The simplest and most thoroughly investigated biological membrane is the plasma membrane of the mammalian erythrocyte. It has the disadvantage of a small size (the entire cell at rest is a bi-concave disk about 8 μm diameter and about 2 μm thick) but has the advantage that it is the sole remaining structural element within that cell.

Lota and Darling (1955) observed that intracellular potassium ions were released when human erythrocytes were irradiated with 0·5 to 3 W/cm^2 of 1 MHz ultrasound. The amount of free K^+ increased both with increasing intensity and time of exposure. The authors concluded that since they could not detect "appreciable" loss of haemoglobin from the irradiated cells, the mechanism of K^+ release was a change in the permeability of the membrane perhaps resulting from the net relative motion between the intact cell and the fluid in which it was suspended. An alternative and much more likely explanation is that some of the erythrocytes were being disrupted by acoustic cavitation liberating both K^+ ions and haemoglobin. The relatively greater sensitivity of the assay system for K^+ would then indicate a significant change whereas the less sensitive spectrophotometric assay for haemoglobin would still be close to its control value.

Bundy *et al.* (1978) exposed avian erythrocytes for 30 min to 1 MHz ultrasound at a spatial average intensity of 0·6 W/cm^2 and measured the rate of uptake of labelled leucine. They found that the total uptake by sonicated cells in 10 min was approximately 5% less than that by the control cells. This small but statistically significant effect was mainly due to a change in the nonmediated (passive) component of transport and not with the mediated (active) transport of this amino acid.

Adler and Hrazdira (1980) have shown that the electrophoretic mobility of human erythrocytes and of mouse bone marrow cells was reduced after the cells were exposed for 20 min to 0·8 MHz ultrasound at intensities of 0·5 to 1·5 W/cm^2. Similarly, Taylor and Newman (1972) and Joshi *et al.* (1973) observed that the electrophoretic mobility of Ehrlich ascites cells was decreased by ultrasound and that the magnitude of this effect increased with increasing frequency. These measurements indicate that ultrasound is apparently able to change the distribution of charged groups on the outside of the cell membrane.

The survival rate of a population of cells immersed for a given time in a medium containing a toxic substance is determined (at least in part) by the rate at which that substance can enter the cell. Kremkau *et al.* (1977) have shown that the rate of cell killing by some anti-tumour drugs was increased if those cells were exposed to MHz ultrasound while in the presence of the drug. This observation apparently shows that ultrasound has caused more drug to enter each cell (i.e. a phonophoretic effect) or has caused some other intracellular damage which potentiates or complements the toxic effects of the drug.

The abdominal skin of the frog *Rana pipens* is commonly employed as a model membrane because its transmembrane potential is determined by the sodium concentration outside and the potassium content inside the skin. When bathed in standard frog Ringer's solution this membrane potential is sensitive to changes in pH, temperature and membrane metabolic state. If the membrane potential is short circuited a current flows which is proportional to the flux of sodium ions through that membrane (Lehmann and Biegler, 1954). Coble and Dunn (1976) irradiated a small portion of an isolated frog skin preparation with a focused beam of 1 MHz ultrasound and observed reversible intensity-dependent changes in both membrane potential and the short circuit current. It is interesting to note that the membrane potential was decreased by the ultrasound if the internal surface of the skin faced the transducer whereas it was increased by the same amount if the membrane was reversed. Lehmann and Krusen (1954) proposed a thermal mechanism for the membrane alterations they observed following the exposure of isolated frog skin to both continuous wave (1·8 and 3 W/cm^2) and pulsed (1·8 W/cm^2 time average with 30 ms on time and 30 ms off) 1 MHz ultrasound. They found an irreversible increase in membrane permeability, a decrease in the active transport of sodium ions and an increase in the isoelectric point of the membrane. These effects could be duplicated by increasing the temperature of the membrane to 41°C in the absence of ultrasound.

Identification and interpretation of possible specific ultrasound-induced changes in membrane potential or function becomes even more difficult as the complexity of the cells or tissues being irradiated is increased. There are many reports of changes in the permeability of the membranes of muscle (e.g. Hughes *et al.*, 1963) or of nerve (e.g. Fry *et al.*, 1958) but so many different types of cells are being irradiated and the ultrasonic intensities are usually so high that both thermal and cavitational effects could be occurring and generating a plethora of effects including a limited amount of cellular destruction.

Thus, there appears to be evidence which indicates that the electrophoretic and transport properties of the membranes of some cells in suspension is altered during or following exposure to ultrasound. In general, the mechanisms proposed to explain these changes involve some form of hydrodynamic streaming either generated by stable cavitation bubbles or arising from relative motion between the cell and its suspending medium. These mechanisms are less likely to occur within solid tissues *in vivo* and so it may not be possible to extrapolate these *in vitro* observations to explain the complex histological and morphological changes seen within solid tissues *in vivo*.

5.2.3 Isolated Organelles

There have been relatively few publications on the effects of Megahertz ultrasound on isolated subcellular organelles. Campbell and Kernot (1962) and Van der Decken and Campbell (1964) irradiated preparations of microsomes and ribosomes isolated from rat liver for 1 min within the transiently cavitating field generated by an ultrasonic disintegrator. They observed the release of protein and RNA, a decrease in the ability of the suspensions to synthesize proteins from amino acids and an increase in phosphodiesterase activity. Those organelles which survived this drastic exposure showed no obvious structural changes when examined by electron microscopy. Similarly, Hughes (1972) has shown that "stable" cavitation bubbles can uncouple respiration and the phosphorylation of ADP in mitochondria isolated from the protozoan *Tetrahymena pyriformis*. Higher intensities of ultrasound result in the onset of transient cavitation which is associated with leakage of matrix enzymes and impairment of respiration.

5.3 CELLS IN SUSPENSION

There are many practical advantages to be gained by irradiating living cells in suspension *in vitro*. The experimentalist has complete control over the ultrasonic exposure conditions and can choose to use a standing wave field or a plane travelling wave, or to irradiate in the near or far field of the transducer. The chamber which holds the cell sample can be designed so that it does not unduly perturb the acoustic beam and so the dosimetry is relatively straightforward.

There are also many biological advantages to be gained by irradiating cells in suspension. One can obtain a homogeneous population of cells which can be asynchronous or a variety of techniques can be employed to induce synchrony so that the cells may be irradiated at any stage within their life cycle. The cells are free to move in suspension and so the quartz wind-streaming aided by any additional stirring ensures that the irregularities in exposure due to the non-uniform nature of the ultrasonic beam (see Chapter 2) are evened out. The techniques for maintaining living cells in suspension are well established and the wide variety of available prokaryotic (i.e. no separate nucleus) or eukaryotic (nucleated) cells ensure that these are amongst the most common bioeffect exposure systems.

The greatest advantage of irradiating living cells in suspension under near ideal exposure conditions is that there are so many sensitive techniques which can be applied to the surviving cells to determine what effects, if any, have resulted from this exposure. For example, the parameters most commonly measured include: (1) viability; (2) morphological changes; (3) functional changes; (4) biochemical changes; (5) behavioural changes; (6) genetic changes. Ultrasound-induced modification of each of these parameters will be discussed separately below.

5.3.1 Disadvantages of Using Cells in Suspension

The major disadvantage as far as working with cells in suspension is concerned is that the mechanism of interaction of those cells with the acoustic beam will almost certainly be different to that of those same cells embedded within a solid tissue when irradiated with the same acoustic beam. For example, the amount of heat absorbed from an ultrasonic beam by a given cell will be the same whether that cell is within a tissue or free in suspension. Within a tissue that cell is surrounded by other similar cells all of whom have absorbed similar quantities of heat and so all of the absorbed heat goes to increase the temperature of that cell. A cell in suspension is usually surrounded by a medium which has a much lower acoustic absorption coefficient (this is a function of the protein content of that medium which is about 70 g% for living cells but only about 3–4 g% even for complex growth media). The surrounding medium therefore absorbs less heat from the acoustic beam resulting in a smaller rise in temperature. Because of its small volume and relatively large surface area, heat loss from a single cell into an infinitely large non-absorbing medium is extremely rapid so that the steady-state temperature rise within that cell is negligible even at acoustic intensities which would have thermally denatured that same cell had it been within a solid tissue (Nyborg, 1977; Love and Kremkau, 1980).

Another equally significant change in the interaction mechanism is that the preparative procedures and even the walls of the sample chamber itself greatly increase the probability of some form of cavitational activity occurring during the irradiation period (see Chapter 4). The fluid nature of the medium suspending the cells means that once a stably oscillating bubble is established it may move through the suspension and so subject a large number of the cells to hydrodynamic shear stresses. Even if that same oscillating bubble was to be formed within a solid tissue its oscillatory behaviour would be damped by the surrounding cells and it would not be able to wander freely through that tissue and so far fewer cells would be affected by it.

Thus, irradiation of living cells in suspension provides a convenient model system after which the cells are amenable to a battery of sophisticated tests to evaluate subtle changes in most of that cell's vital functions. The disadvantage of this procedure is that any observed changes may not be an accurate reflection of the changes which that same cell would have undergone had it been irradiated within a solid tissue because the cell suspension technique minimizes the role of heating and maximizes the role of cavitation and its associated streaming.

5.3.2 The Cell Cycle

Another major difference between cells in a tissue and cells in suspension is that the cells within the tissue are contact inhibited and are therefore not actively preparing for their next cell division. Cells in suspension are not subject to the complex feedback mechanisms which prevent unrestrained growth within a tissue and are therefore actively growing and replicating themselves at the maximum rate permitted by environmental factors such as temperature, pH, and oxygen and the availability of suitable nutrient materials and the absence of toxic waste products.

If viable non-dividing single cells are introduced into a fresh growth medium, then there is a characteristic delay (called the lag phase) before those cells begin to divide (Fig. 5.1). This delay is followed by a period of regular growth where the number of cells within that population is found to have doubled after a characteristic regular interval of time. When plotted on a semi-logarithmic scale as shown in Fig. 5.1 this portion of the growth curve is a straight line and is commonly called the log phase of growth. This unrestrained

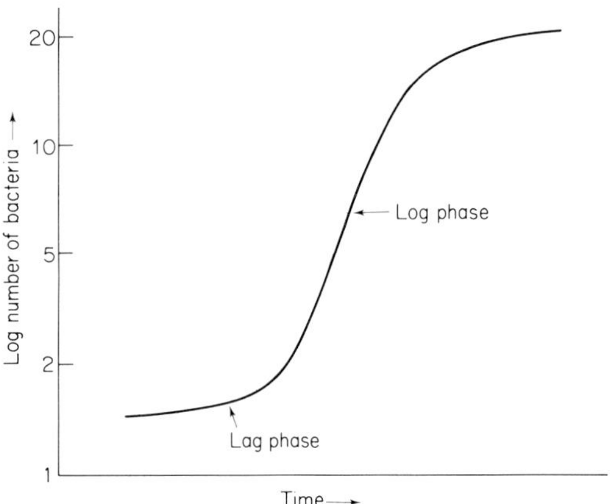

FIG. 5.1 A diagrammatic representation of the growth rate of a typical bacterial suspension. After an initial delay or lag phase, the organism grows at a constant rate with a characteristic doubling time for each growth medium and incubation temperature.

rate of growth cannot continue indefinitely and so either one of the nutrients becomes limiting or the cell population is inhibited by its own toxic waste products and so the growth rate begins to decrease.

Thus, each cell type has a characteristic replication time which can be found from the slope of the log phase of the growth curve obtained under the ideal growing conditions for that cell. This replication time is the interval between the end of one cell division and the end of the next cell division for each of the daughter cells, i.e. one complete cell cycle. This replication time can be as short as 20 min for some procaryots growing under ideal conditions, but is typically about 18–24 hours for mammalian cell lines growing in culture at 37°C. This cell cycle is divided into two visually distinct phases; one is from the start of the mitotic process itself until the two daughter cells separate (this is called the M phase and usually comprises 5–10% of the division cycle), the other is called the interphase (i.e. the remaining 90–95% of the cell cycle) when the cell exhibits few if any visible changes.

Despite the absence of visible changes a cell in interphase is extremely active and undergoes a regular progression of biochemical changes. For example, a cell will only incorporate labelled thymidine triphosphate while it is actively synthesizing DNA; this is called the S phase and lasts for about 30–50% of the cell cycle depending upon the cell type (Fig. 5.2). There is a period of active growth before the S phase begins which is called the G_1 phase and occupies 30–40% of the cell cycle, and another called the G_2 phase which occurs after the S phase and occupies the remaining 10–20% of interphase. These phases are indicated in Fig. 5.2 which also shows the relative duration of each phase within the life cycle of a mouse hepatoma cell growing in tissue culture and dividing once every 24 hours (Watson, 1976).

5.3.3 Cell Viability

Moore and Coakley (1977) found that Chinese Hamster cells grown as a monolayer *in vitro* remained viable even though they had been exposed to 30 times greater ultrasonic intensities and up to 1000 times longer exposure times than those which have been shown to produce pathological lesions within mammalian tissues *in vivo*. This strongly indicates that heating and not some other physical property of the wave is the dominant mechanism causing tissue cell damage *in vivo*, and that this heating can be greatly reduced or eliminated *in vitro*. Similarly, Bleany *et al.* (1972) exposed CHLF hamster cells to 1·5 MHz continuous wave at intensities of up to 8·8 W/cm^2 and at higher intensities using a variety of pulsed regimes for one hour and found no reduction in cell survival providing that the temperature of the cell suspension did not exceed 45°C.

There is general agreement that virtually all types of living cells,

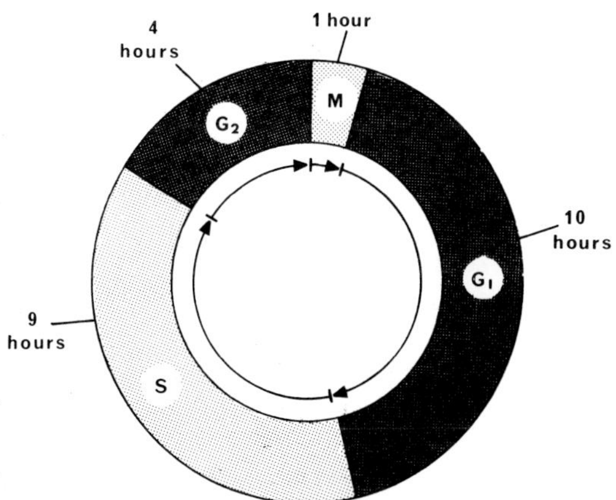

FIG. 5.2 A diagrammatic representation of the life cycle of a typical cell growing under optimal growth conditions. After mitosis (M) the cell grows but does not synthesize DNA (G_1 phase); this is followed by a period of cell growth together with the synthesis of DNA (S phase). After a further period of growth without DNA synthesis (G_2 phase), the cell undergoes mitosis once again (M phase). The times refer to the typical duration of each phase in a mammalian cell growing in tissue culture.

including those having extremely strong cell walls, will be disrupted by the streaming forces associated with transient cavitation events (Neppiras and Hughes, 1964). Some form of cavitational activity is also presumed to be the dominant destructive mechanism for cells in suspension or growing as monolayers when they are subjected to MHz ultrasound (Armour and Corry, 1982; Moore and Coakley, 1977; Morton *et al.*, 1982). Clarke and Hill (1969) found that cells undergoing mitosis were most susceptible to ultrasound and suggested that the intact cells which survived exposure to MHz ultrasound remained unaffected in terms of their subsequent growth and proliferation rate. This suggestion has not been substantiated by other workers who have found that some of the intact cells irradiated for 1–15 min with 1 MHz ultrasound at intensities of 1–10 W/cm^2 had a decreased viability as determined both by trypan blue dye exclusion and colony-forming ability (Kaufman *et al.*, 1977). Watmough *et al.* (1977) observed intracellular damage within some of their intact surviving cells (this is described in the next sub-section) and Ciaravino and Miller (1978) showed that about 16% of the intact post-sonicated cells underwent spontaneous lysis within 24 hours (3 W/cm^2, 2 min, 1 MHz). Kaufman and Miller (1978) exposed Chinese Hamster V-79 cells to 1 MHz ultrasound (1 min, 2·5 W/cm^2 spatial peak) and found that there was little if any apparent growth of the surviving intact cells for 24 hours. This apparent cessation of growth during the first 24 hours appeared to be due to an approximate balance between the proliferation of viable cells and the corresponding loss of non-viable cells into the growth medium.

Li *et al.* (1977) found that increasing the temperature of the sonication vessel from 37°C to 43°C had no effect on the number of surviving intact cells (1 MHz, 2 W/cm^2 pulsed) but reduced the colony-forming ability of the survivors by 10–50%. This indicates that the mechanism of cell disruption (presumably cavitation) is temperature-independent over this range whereas the incompetence of the repair mechanisms (which results in the reduction in the colony-forming ability) was aggravated by the increased temperature, i.e. cell inactivation is temperature-dependent (Li *et al.*, 1977).

Most publications indicate that ultrasound damages cells in suspension, but some apparently show a hormetic effect, i.e. a beneficial or stimulating effect resulting from the application of a small "dose" of a harmful or irritating agent. For example, Hrazdira *et al.* (1974) irradiated mouse bone marrow cells *in vitro* at 800 kHz at various intensities (0·3–0·7 W/cm^2) for 10 or 20 min and then injected them into mice which had been irradiated with 600 rad of ionizing radiation. It was found that the number of bone marrow colonies that grew within the spleen of the host mouse was significantly increased if the bone marrow cells had been irradiated with ultrasound *in vitro*. The magnitude of this effect increased as both the intensity and the time of exposure to that ultrasound was increased even though the number of intact cells progressively decreased.

5.3.4 Morphological Changes

Many authors have described a multiplicity of ultrastructural defects or lesions as seen by the electron microscope within cells which had been subjected to various intensities of MHz ultrasound. Unfortunately, the most commonly reported findings are enlargement of the mitochondria and swelling of the lacunae of the endoplasmic reticulum which are characteristic responses of these organelles to a wide range of unfavourable stimuli including heat, ionizing radiation, lack of oxygen, etc. (Hrazdira *et al.*, 1974).

The electron micrographs presented by Watmough *et al.* (1977) are reasonably representative of the changes commonly observed in suspensions of irradiated mammalian cells. HeLa cells (originally human epithelial cells derived from an ovarian tumour) were irradiated in their growth medium at 0·75 MHz for various times and intensities under conditions where there was a high standing wave component. The control cells shown in Fig. 5.3 have a normal healthy appearance and small quantities of cell debris are found in the spaces between the cells as is usual for a suspension culture. Figure 5.4 shows the remains of a similar cell culture irradiated for 5 min at a spatial average intensity of 0·9 W/cm^2 where it can be seen that the plasma membranes have been ruptured and largely removed and the organelles have been grossly distended by the influx of fluid. These grossly abnormal cells usually co-exist with the cells which have received little or no damage. Figure 5.5 shows cells irradiated for only 20 sec at 0·9 W/cm^2; they can be seen to be intact but more "condensed" than their controls and to have numerous "slits" (S) within their cytoplasm which can be seen on higher magnification to be bounded on one side by endoplasmic reticulum (Watmough *et al.*, 1977).

These electron micrographic changes produced at 0·75 MHz are remarkably similar to the changes observed by Williams (1973) after he exposed mouse Ehrlich ascites tumour cells to the acoustic microstreaming field generated around an oscillating steel wire. Figure 5.6 summarizes these changes and shows: an undamaged cell (A) coexisting with two types of "sublethally" damaged cells, one of

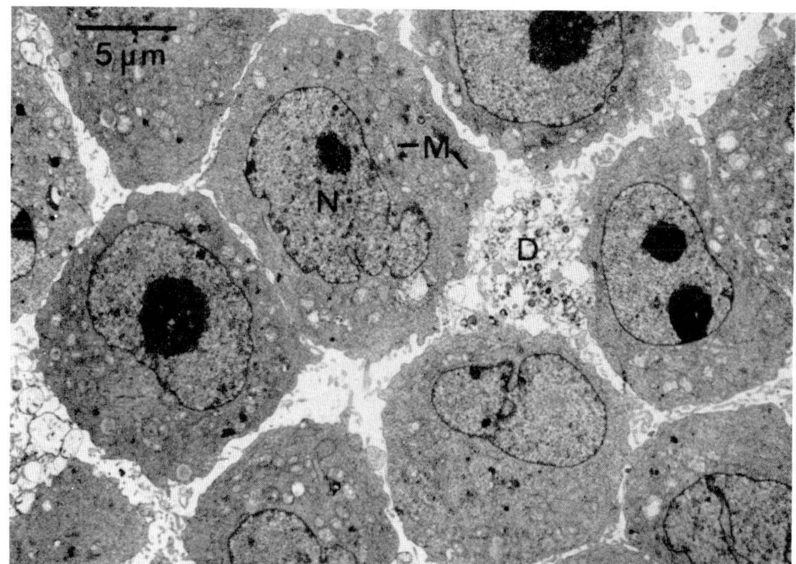

FIG. 5.3 An electron micrograph of untreated (Control) HeLa cells showing the nucleus (N), mitochondria (M) and some cell debris (D) (from Watmough *et al.*, 1977).

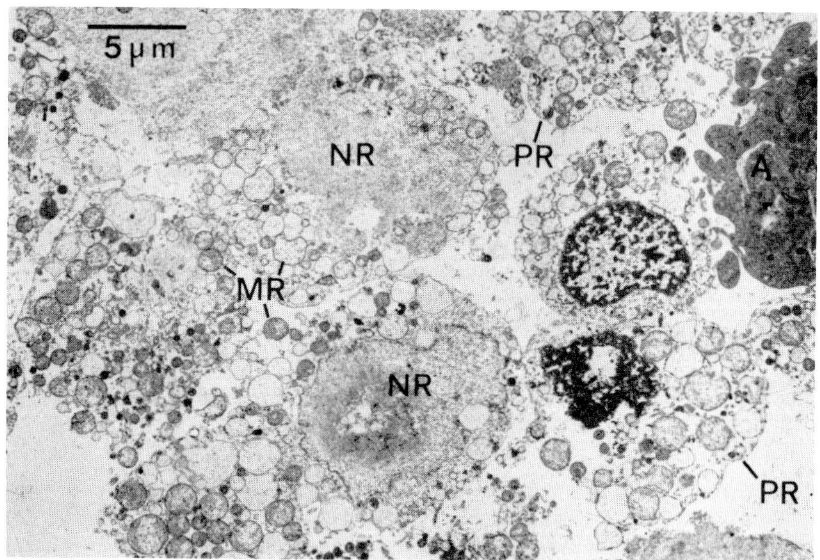

FIG. 5.4 An electron micrograph of HeLa cells exposed to 0·9 W/cm^2 of 0·75 MHz ultrasound for 5 min. An occasional cell (A) appears to be relatively normal, but the majority of the cells have been stripped of their plasma membrane so that their nuclear remnants (NR), mitochondria (MR) and other cytoplasmic vesicles have swollen and ruptured (from Watmough *et al.*, 1977). PR refers to the remains of the plasma membrane.

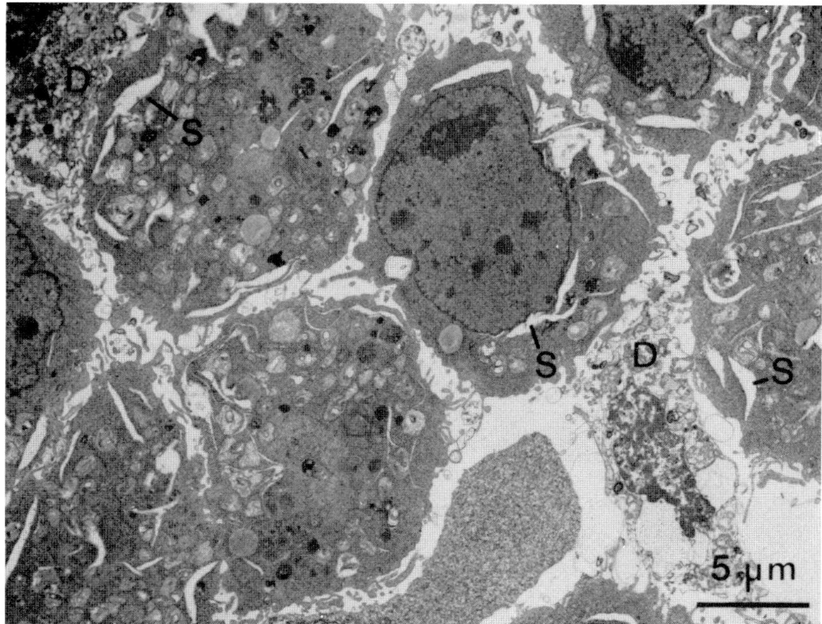

FIG. 5.5 An electron micrograph of HeLa cells exposed to 0·9 W/cm^2 of 0·75 MHz ultrasound for 20 sec showing highly condensed cells with numerous slits (S) within their cytoplasm (from Watmough *et al.*, 1977).

which has swollen (C) as shown by the enlarged mitochondria while another cell has "condensed" (B) and has "slits" similar to those shown in Fig. 5.5; (D) is a cell whose nuclear membrane has been ruptured even though it is still surrounded by cytoplasm, and (E) shows cells whose plasma membranes have been partially or completely removed so that the organelles have swollen and burst and are only being held together by the cell's microtubular and microfibrillar systems (Williams, 1973). These similarities suggest that the ultrasound-induced changes observed by Watmough *et al.* (1977) might also have been caused by acoustic microstreaming fields which could have been generated by some form of stable acoustic cavitation bubbles.

Figure 5.7 is also from Watmough *et al.* (1977) and shows where the nuclear membrane of a HeLa cell irradiated for 120 sec at 0·9 W/cm^2 has been ruptured releasing the nucleoplasm into the cytoplasm. It was proposed that this damage could have been caused by an intracellular or even intranuclear microbubble. However, the nuclear membrane can be damaged within an intact cell if that cell is mechanically deformed such as when it is suddenly sucked into a narrow capillary or pore and so the nuclear membrane damage shown

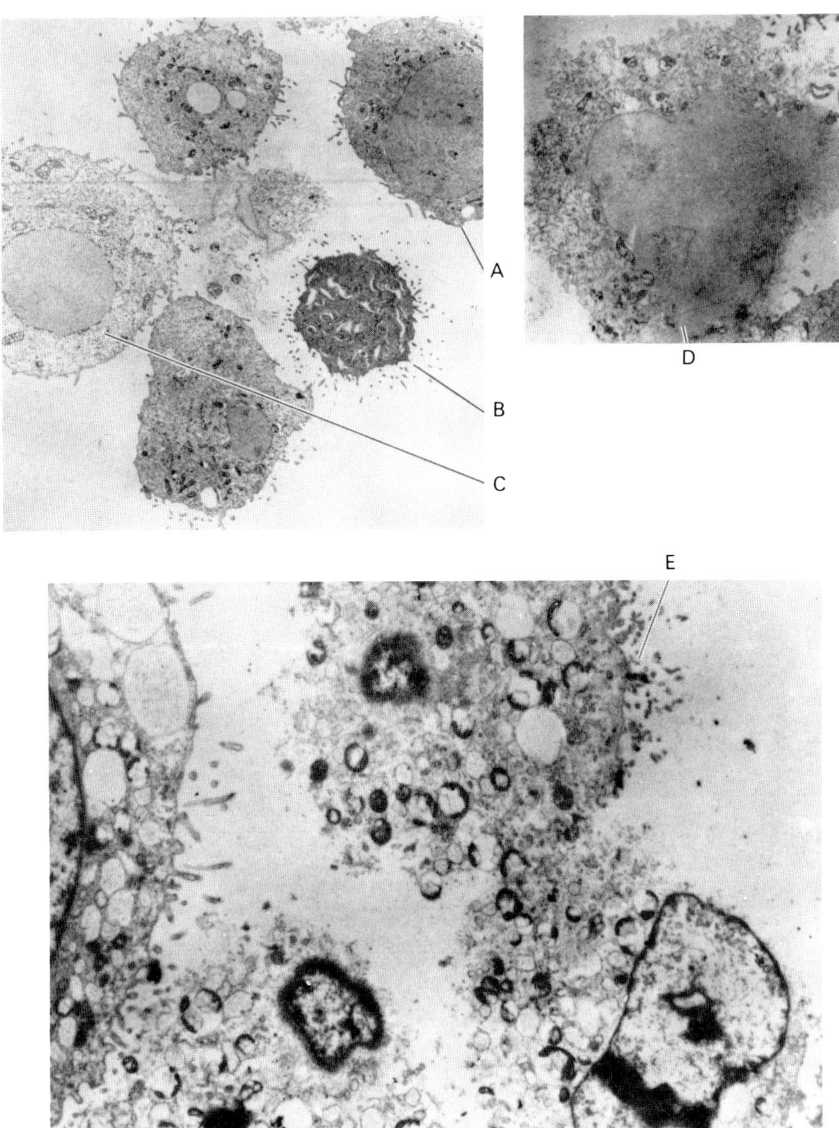

FIG. 5.6 Three electron micrographs of Ehrlich Ascites tumour cells which have been subjected to the acoustic microstreaming field generated around a transversely oscillating wire (from Williams, 1973). An undamaged cell (A) coexists with a swollen cell (C) and one which has condensed (B) and developed slits similar to those shown in Fig. 5.5. The plasma membranes have been partially removed from the cells in micrograph (E) and the subcellular organelles have swollen and burst. The nuclear membrane of cell (D) has been ruptured even though the cell has retained most of its cytoplasm.

in Fig. 5.7 could perhaps result from large-scale cellular distortions produced by the microstreaming field as shown in Fig. 5.6(D).

Liebeskind *et al.* (1982) exposed suspensions of a fibroblast cell line (3T3) and normal rat peritoneal cells to pulsed ultrasound from a commercially available diagnostic instrument *in vitro*. They observed a variety of post-sonication ultrastructural changes including an increase in the number and distribution of perichromatin granules within the nucleus and an increase in the microfibrillar and microtubular content of the cytoplasm. These changes may have been associated with the increased cell motility which was reported to persist through at least 10 generations in culture. Other morphological changes which could be seen 5 min after ultrasound exposure include bulk invagination of the cytoplasm into the nuclear domain and separation of the inner and outer nuclear membranes (Liebeskind *et al.*, 1982).

One unique morphological change which might have been induced by ultrasound is the formation of giant cells of chinese hamster ovary (CHO) cells grown in culture. Miller *et al.* (1977) exposed CHO cells to continuous wave 1 MHz ultrasound at 5 W/cm^2 for 5 min, spread them on plates and allowed them to grow for 7 days. Despite the

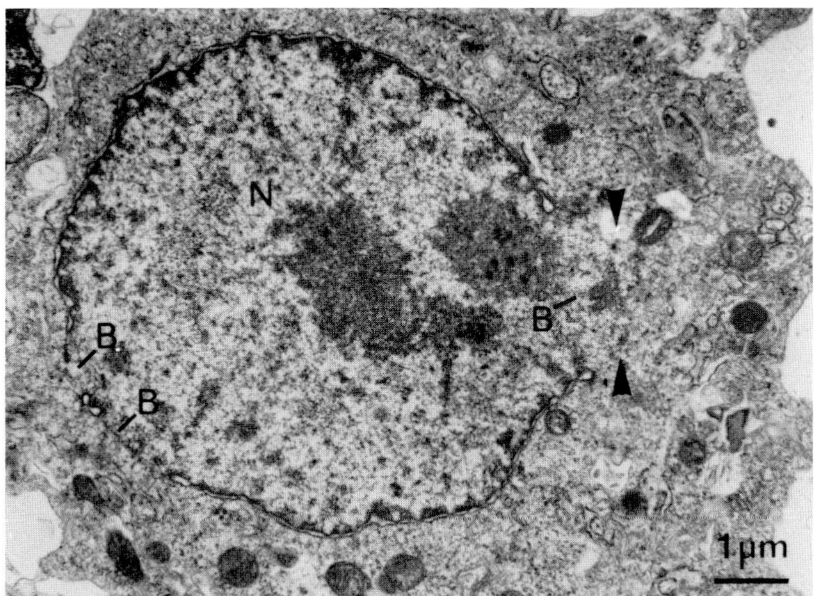

FIG. 5.7 An electron micrograph of a HeLa cell exposed to 0·75 MHz ultrasound showing where the membrane of the nucleus (N) has burst (B) in several places (arrowed) releasing nucleoplasm into the cytoplasm (from Watmough *et al.*, 1977).

average decrease in the number of surviving cells in the sonicated samples they contained almost 2·5 times as many giant cells per cell as their controls ($P = 0{\cdot}001$). To my knowledge there have been no other reports of giant cell production associated with exposure to ultrasound.

5.3.5 Functional Changes

In general, all the viable cells within a mammal have many common biochemical functions in that they all respire, excrete, actively transport nutriments, and, under certain circumstances, are potentially capable of reproducing themselves. As well as performing these common but necessary functions, most mammalian cells have become differentiated which means that they have been specifically modified to perform a limited number of additional highly specialized tasks. Some cells, especially those commonly found in blood, continue to perform various aspects of their specialized tasks in suspension *in vitro*.

Crowell *et al.* (1977) subjected white cell enriched canine plasma to the acoustic microstreaming field generated around a tungsten wire oscillating transversely at 20 kHz. They observed cell lysis at wire displacement amplitudes greater than 8 μm but also measured changes in phagocytic and bacteriocidal indices and metabolic activities at amplitudes significantly less than the value required to disrupt the cells. Visual observation of the number of bacteria engulfed by (and/or adhering to the surface of) the polymorphonuclear neutrophilic leucocytes showed that the average cell's phagocytic index increased with increasing displacement amplitude up to about 7 μm and thereafter decreased with increasing displacement amplitude. Other measurements showed that the number of bacteria killed by each leucocyte (i.e. its relative bacteriocidal capacity) did not exhibit this same trend but declined progressively with increasing displacement amplitude (as did the rate of oxygen utilization by those same cells). Crowell *et al.* (1977) proposed that this discrepancy in their observations could have arisen if more bacteria were attached to the outside of the cells sheared at displacement amplitudes less than about 7 μm, but that these bacteria were not engulfed by the "damaged" cells. This increased adhesion could be the result of changes in the distribution of the charges on the cell surface resulting from exposure to the microstreaming field.

Several other authors have observed similar changes after the leucocytes had been exposed to megahertz ultrasound *in vitro*. For example, Saggio and Sommer (1980) report an increase in the phagocytic activity of macrophages, whereas Fung *et al.* (1977) report a decrease in the rate of uptake of labelled thymidine by human lymphocytes after they had been exposed to therapeutic intensities and frequencies of ultrasound. One criticism which can be levelled against most of the experiments performed using leucocytes is that the authors have usually not attempted to remove most of the platelets. These platelets are extremely fragile cells which are readily disrupted and/or activated by ultrasound *in vitro* and release chemically active agents which can subsequently alter both their own functioning and that of other adjacent cells (Williams *et al.*, 1976a).

Platelets are anucleate cell fragments of the megakaryocytes of the bone marrow whose function is to adhere to each other and to materials like collagen and to release potent chemicals which initiate and accelerate various aspects of the blood coagulation system. Irradiation of anticoagulated human whole blood *in vitro* at 1 MHz for 5 min at spatial average intensities as low as 65 mW/cm^2 caused a time-dependent decrease in its recalcification time (i.e. the time taken to form strands of fibrin following the addition of enough calcium ions to overcome the effects of the anticoagulant) (Williams *et al.*, 1976b). This effect was subsequently shown to be due to damage and activation of the blood platelets. Electron micrographic investigations were performed on the thrombi produced by the recalcification of human platelet-rich plasma (PRP; i.e. anticoagulated whole blood with the erythrocytes and most of the white cells removed by centrifugation). Platelet thrombi from irradiated samples contained some debris from disrupted platelets and the remaining intact platelets were abnormal in that (a) they were more vacuolated, (b) they formed a larger number of smaller aggregates, and (c) their clot retraction mechanism was impaired. These anatomical and functional changes could be duplicated by incubating normal platelets with small amounts of platelet debris from homogenized platelet suspensions (Williams *et al.*, 1976a).

A suspension of PRP is turbid because the large number of individual small platelets scatter the incident light. However, once they are stimulated to undergo their normal physiological function or are subjected to an adverse physical or chemical environment, they adhere to each other until they form a smaller number of large aggregates which permit more light to pass through the sample. Chater and Williams (1977) found that under certain conditions ultrasonic irradiation alone could initiate aggregation of the platelets and that the magnitude of this effect decreased with increasing frequency at the constant intensity of 1·8 W/cm^2 (Fig. 5.8).

For any one human donor the amount of aggregation produced by

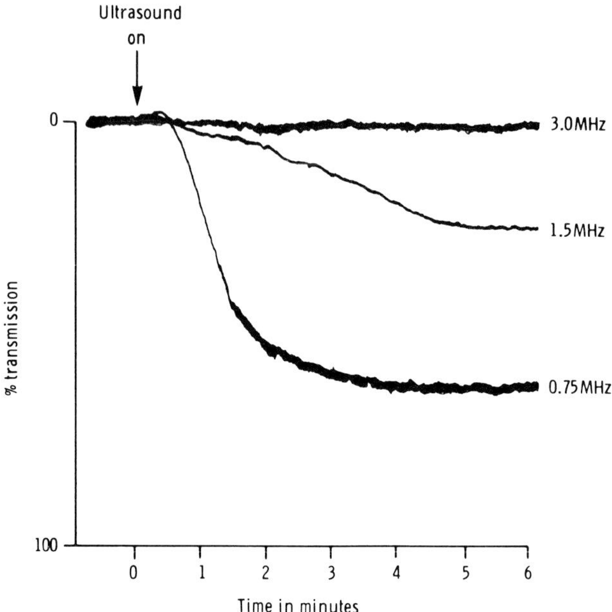

FIG. 5.8 Representative changes in light transmission through a sample of human platelet rich plasma resulting from the aggregation of the platelets following exposure to 1·8 W/cm^2 of ultrasound at three different frequencies (from Chater and Williams, 1977).

any given ultrasonic exposure at 0·75 MHz was remarkably constant. However, PRP from different donors sometimes exhibited wide variation (Fig. 5.9). It is known that PRP from different donors may have a different sensitivity to aggregation by ADP and so in a separate series of experiments aliquots of PRP from the same donors used to compile Fig. 5.9 were exposed to near threshold levels of 4 μm and 10 μm adenosine diphosphate (ADP). The circled number on each curve in Fig. 5.9 refers to the sensitivities of those platelets to ADP-induced aggregation with No. 1 being the most sensitive and No. 5 being the least sensitive. It can be seen that the most sensitive platelets required a lower ultrasonic intensity to induce aggregation, and that the maximal extent of this aggregation was always greater than that obtained with less sensitive platelets (Chater and Williams, 1977).

β-Thromboglobulin (β-TG) is a protein found only within human platelets which is liberated when the platelets are disrupted or undergo their normal physiological release reaction (Ludlam *et al.*, 1975). Its concentration can be measured with an accuracy of about

±2–3 ng/ml by means of a specific radioimmunoassay and normal plasma values range from 20–40 ng/ml rising to more than 5000 ng/ml in serum where all the platelets have released. The uppermost curve in Fig. 5.10 (−release inhibitor) shows the additional amount of β-TG released into the plasma when anti-coagulated blood was irradiated at 0·75 MHz. This shows that above about 0·8 W/cm² there was a steep intensity-dependent increase in

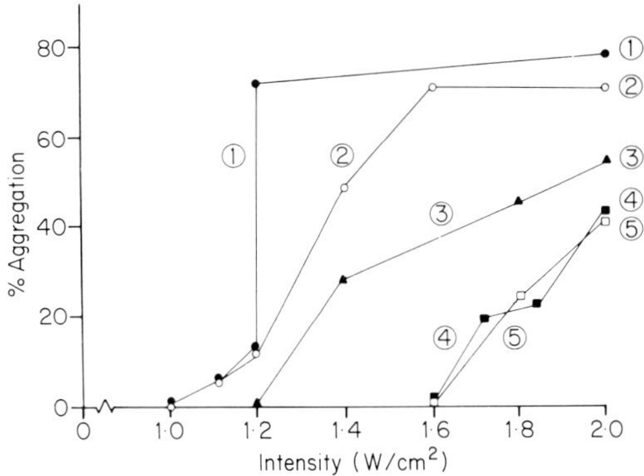

FIG. 5.9 The percentage change in light transmission (i.e. percentage aggregation) varied both with the ultrasonic intensity and with the donor of the platelet rich plasma. Each curve was obtained with a different human donor and the encircled number beside each curve refers to the inherent aggregability of that sample with No. 1 producing the most aggregation and No. 5 the least in response to a near-threshold challenge of ADP (from Chater and Williams, 1977).

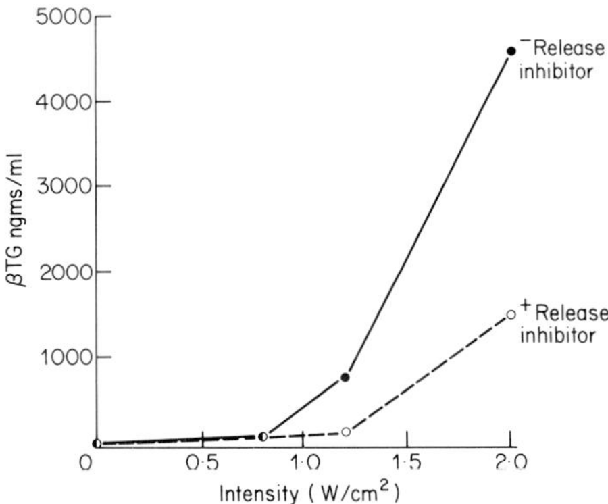

FIG. 5.10 The release of β-Thromboglobulin from human platelets *in vitro* by 0·75 MHz ultrasound as a function of the space averaged c.w. intensity (from Chater and Williams, 1977). Only about one third of the amount of β-TG was released if the samples were irradiated in the presence of an inhibitor of the platelet release reaction (for details see text).

the amount of β-TG released into the plasma (Williams *et al.*, 1978). There was no sharp cut-off point or "threshold" as the ultrasonic intensity was reduced below 0·8 W/cm^2. Other (unreported) measurements showed that the curve of β-TG release began to merge with the baseline at intensities below about 0·6 W/cm^2, but the precise value varied from day to day.

The lower curve in Fig. 5.10 (+release inhibitor) was obtained using the same blood sample as the upper trace except that it contained a mixture of EDTA and theophylline which prevented the platelets undergoing their normal release reaction. It can be seen that this inhibitor decreased the amount of β-TG released by ultrasound to about 30% of the control value. The only plausible mechanism for liberating β-TG from these functionally-inert platelets is by mechanically disrupting them. This argument in favour of cavitation is supported by the fact that high levels of β-TG release were always accompanied by a corresponding increase in the levels of plasma haemoglobin from disrupted erythrocytes (Williams *et al.*, 1978).

All of the above results indicate that some form of cavitational activity is being generated at "suprathreshold" intensities of ultrasound which subsequently disrupts a small proportion of the platelet and (if they are also present) erythrocyte populations. When enough cells have been disrupted, materials (including ADP) liberated from these disrupted cells will induce other intact functional platelets to undergo their release reaction (liberating β-TG and more ADP) and to aggregate.

The "threshold" intensity for any one of the effects reported above is increased if the sample to be irradiated is degassed or is not stirred during the sonication procedure. Conversely, the "threshold" intensity is decreased if the samples to be irradiated are bubbled, agitated, stirred or aerated. The lowest "threshold" values obtainable *in vitro* are found when microscopic gas bodies are deliberately introduced into the sonication medium. These micron-sized gas bubbles tend to dissolve spontaneously, but they may be stabilized within the uniformly sized pores of hydrophobic Nuclepore® membranes. When these stable gas bubbles are present, human platelets can be disrupted at average intensities as low as 8 mW/cm^2 at 1·6 MHz (A. R. Williams and D. L. Miller, unpublished observations) and can be induced to aggregate by the ultrasonic field emitted by a conventional doppler diagnostic instrument (Miller *et al.*, 1979).

There are other recent but as yet unconfirmed reports of functional changes produced in cells in suspension at extremely low (diagnostic level) intensities of MHz ultrasound. Liebeskind *et al.* (1979)

reported that 20–30 min exposures of 2·5 MHz ultrasound (200 Hz pulse repetition rate, spatial peak intensity 35·4 W/cm^2, 3 μsec "on" time) increased the immunoreactivity of HeLa cell nuclei in their G_1 phase to antinucleoside antibodies (i.e. this suggests the formation of single-strand breaks in the DNA and/or low levels of additional repair synthesis). These same authors also claimed an increased rate of transformation of a C3H mouse cell line which, if it can be independently verified, could be a significant potential hazard of medical ultrasound. Siegel *et al.* (1978a,b,c) have reported that diagnostic intensities of MHz ultrasound under conditions which encouraged wave reflection decreased the number of cells which could attach themselves to a flat plastic surface. Two independent laboratories have so far been unable to duplicate these observations. In view of their potential significance, the experiments by Liebeskind *et al.* (1979) and Siegel *et al.* (1978a,b,c) will be discussed further in Chapter 6.

5.3.6 Biochemical Changes

Both increases and decreases in the rates of DNA and protein synthesis and in the intracellular concentrations of various ions and metabolites have been reported to occur following the exposure of mammalian cells grown in tissue culture to MHz ultrasound. Prasad *et al.* (1976) report a 14% decrease in DNA synthesis in HeLa cells exposed for 10 min to 1 MHz pulsed ultrasound at an average intensity of 4 mW/cm^2. Conversely, Fung *et al.* (1977) reported that some cells within a population of human lymphocytes irradiated for 3–12 min with a commercial (Sonicaid) Fetal Doppler unit showed a significant stimulation of DNA synthesis. This effect did not occur if the same cell suspensions were irradiated for 15–30 min which suggests that the above result may have been an artifact. Kaufmann and Kremkau (1978) irradiated L-1210 leukaemia cells with 2·22 MHz ultrasound for 10 min at a spatial average intensity of 10 W/cm^2 and observed a 54% reduction in the uptake of thymidine and its incorporation into new DNA. This observation may reflect the number of intact cells which have been irreversibly damaged and are destined to undergo spontaneous lysis.

Webster *et al.* (1978, 1980) found that the rate of protein synthesis in general, and collagen synthesis in particular, was significantly stimulated in human embryonic fibroblasts grown in tissue culture after they had been exposed to 3 MHz ultrasound (spatial and temporal peak intensity 0·5 W/cm^2, pulsed 2 ms on and 8 ms off for 5 min). They also observed that this stimulatory effect was virtually abolished if the entire sample exposure system was pressurized with helium to twice the ambient atmospheric pressure. The only parameters or effects of an ultrasonic wave which would be changed by this relatively small increase in the ambient pressure would be a reduction in the rate of generation of cavitational events and a decrease in the magnitude of the damage which they would inflict. This observation, coupled with their electron micrographic detection of cellular damage similar to that presented in Figs 5.3 to 5.6, strongly indicates that this stimulation of protein synthesis is the direct consequence of some form of cellular damage or stimulation caused by the streaming associated with some form of acoustic cavitation.

This stimulation of non-specific protein synthesis may reflect a general increase in the metabolic activity of that cell, possibly in an endeavour to repair any damage caused by the cavitation-induced microstreaming (i.e. Hormesis). This proposal is supported by many observations of increased enzymic activity within tissues which have been irradiated under conditions where bubble-induced streaming would be expected (Istomina and Ostrovskij, 1936; Majewski *et al.*, 1966; Ruban and Dolgopolov, 1952). Alternatively, all of this increased cellular activity could merely have been stimulated by the microstreaming-induced influx of sodium ions and/or the loss of potassium ions from the cell as has been demonstrated in rat thymocytes irradiated at 1·8 MHz for 30 min at about 1 W/cm^2 (Chapman, 1974; and his students McNally, 1977; Tucker, 1976).

In contrast to the majority of published reports which describe ultrasound-induced increases or decreases in the intracellular concentrations of various metabolites, some articles have appeared in press which report that they have been unable to detect any significant change. For example, Glick *et al.* (1979) irradiated both mouse peritoneal cells and human amniotic fluid cells with 2 MHz ultrasound (1 W/cm^2, 2000 sec, 34°C) and were unable to detect changes in the intracellular levels of the cyclic mono-nucleotides cAMP and cGMP. From my own laboratory experience and that of my colleagues it can be observed that obtaining negative results in bioeffect experiments is far more common than would be led to expect from a survey of the literature. It is a sad reflection on the criteria used to select bioeffect articles for publication in a specialized journal that a sloppy investigation which produces a spurious but "startling" positive result will usually be accepted whereas a well-conducted study which yields a negative result may not even be written up because of the high probability of rejection. In fairness it must be pointed out that if one did not have a reasonably rigorous selection process then the scientific literature would rapidly become swamped with poorly conducted studies and a multitude of negative studies.

5.3.7 Behavioural Effects

A behavioural effect occurs when a living organism is able to detect some parameter of the acoustic wave (either directly or indirectly) which causes that organism to change the magnitude, rate or direction of some active process over which it has control. Typical examples on the macroscopic scale are the avoidance by fish of a beam of therapeutic ultrasound which is propagating through their tank (Wood and Loomis, 1927) or the pain-induced response of a person subjected to intolerably high intensity during therapy. On a smaller scale, Hawley and Dunn (1964) demonstrated significant changes in the behaviour of rotifers at extremely low intensities of the order of 1 mW/cm^2 at 270 and 510 MHz, but not at intermediate frequencies. This latter observation is especially interesting in that these two frequencies are nearly harmonically related which implies that a resonance effect might be implicated in the detection mechanism.

Mummery (1978) cultured human fibroblast cells *in vitro* and observed their growth rate and motility by means of time-lapse photography. It was demonstrated that the spontaneous mobility of the fibroblasts on their two-dimensional substrate was increased following exposure to therapeutic intensities and frequencies of ultrasound, but that this increased mobility resulted from a decrease in the number of changes of direction by each cell rather than an increase in their mean speed. Normal motility was recovered 24 hours after treatment.

Liebeskind *et al.* (1982) exposed 3T3 fibroblasts grown in tissue culture to the pulsed ultrasonic field emitted by a commercial diagnostic instrument (SATA intensity 15 mW/cm^2, SPTP intensity 35·4 W/cm^2; pulse length 3 μs and pulse repetition rate 200 Hz). They observed that the insonated cells continued to divide and grow normally, but that they showed striking differences in cell motility and in the behaviour of the cell surface. The ultrasound-exposed cells did not spread as well over their substrate and showed less contact inhibition as well as a marked time-dependent "bubbling" over their surface reminiscent of that seen at the poles of untreated cells during cytokinesis. This behaviour was observed throughout the cell cycle and was reported to persist for at least 10 generations in culture (Liebeskind *et al.*, 1982).

5.3.8 Genetic Changes

If ultrasound is able to modify the DNA within a viable cell or to otherwise alter the way in which that DNA will be distributed between its daughter cells, then it could result in the malformation of a growing organism or structure (i.e. a teratogenic effect) or even induce a neoplastic change with possibly disastrous consequences for that organism. If these nuclear modifications were to occur within the gametes or the fertilized ovum then these changes might be passed on to future generations. In view of the widespread use of ultrasonic techniques in obstetrics and gynaecology with the concomitant exposure of the ovaries and developing embryo, there has naturally been an extensive programme of investigations in this field.

In August 1963 Dr Hubert J. Dyer induced an acoustic microstreaming field inside the apical protonema cells of the moss *Physcomotrium hyriforme* by the external application of a fine-pointed metal probe vibrating at 80 kHz (i.e. simulating the effects of an external stable cavitation bubble). These cells were all undergoing mitotic divisions of the nucleus at the time (i.e. in their M phase) and 22 of these treated cells survived this and the subsequent handling and subculturing techniques and continued to grow. Two visibly detectable morphological mutants were found amongst the irradiated clones whereas none were found in the controls or in the many hundreds of subcultures made of the original wild-type strain (Dyer, 1972). The subcultures derived from these abnormal mosses were also abnormal and some were still alive 18 years later (after 25–30 subcultures). Preliminary biochemical investigations confirmed that there were many differences in chemical composition between the wild type and mutant strains indicating a widespread genetic "reorganization" rather than a single point mutation (H. J. Dyer, personal communication). One mutant contained more DNA than the original wild type whereas the other contained less DNA than the wild type. It is tempting to speculate that these mutants could have arisen either by the inclusion of an additional chromosome or by the loss of one complete chromosome during cell division. There was no evidence of mutant clones from cells irradiated during interphase.

Spencer (1952) reported a transmissible morphological change in plants grown from peas (*Pisum sativum*) which had been exposed to ultrasound before germination had become apparent. Despite the high acoustic intensities (20 W/cm^2, c.w., 0·5 MHz for 20 sec) most seeds were able to germinate and produced plants which were almost twice as tall as their unsonicated controls. Cross-breeding experiments showed that only the offspring of the female sonicated plants continued to grow as abnormally tall plants. This indicates that the ultrasound has not produced a dominant genetic mutation, but instead has resulted in a change of some cytoplasmic factor which is transmitted through the cytoplasm of the seeds produced by the female plants. The effects of these non-nuclear phenotype modifying factors (called dauermodifiers) may be transmitted through a number of successive generations before gradually disappearing.

Suspensions of single cells have been widely used to demonstrate the genetic changes induced by ionizing radiation. They offer many practical advantages in that they are readily available and are easy to culture, have a short cell cycle so that many generations of the organism can be studied within a few days or weeks and they are amenable to a number of well established techniques to pinpoint the nature and exact locus of any ultrasound-induced mutation. These advantages are of course countered by some disadvantages in that the cells in suspension are more likely to be affected by cavitation-induced streaming than cells in a solid tissue whereas the role of heating will tend to be diminished. Another potential experimental disadvantage is that if the damaging mechanism is a rotational torque arising from sudden discontinuities in intensity within the acoustic beam, then this would tend to rotate the entire cell if it is free in suspension whereas it would tend to rotate only the cellular contents if that cell was embedded within a tissue (this latter situation is closer to the conditions produced by Dyer, 1972).

5.3.8.1 Studies with Microorganisms

Combes (1975) could not detect an increase in the back-mutation of an auxotropic strain of *Baccillus subtilis* after it had been irradiated for 5 min at 2 MHz at temporal peak intensities of up to 60 W/cm^2 (20 μs pulses, duty cycle 0·004). He was also unable to detect mutagenic lesions *in vivo* after transforming DNA had been irradiated *in vitro*. Thacker (1974) used the yeast *Saccharomyces cerivisiae* to test for ultrasound-induced mutations in nuclear genes and in mitochondrial DNA as well as for recombination of a nuclear gene. He used a variety of exposure conditions simulating diagnostic (20 μsec pulses with 5 μsec spaces and a peak pulse power of 10 W/cm^2), and therapeutic conditions (5 W/cm^2 continuous for up to 30 min), and also even more powerful exposure regimes. These tests yielded no evidence for increased mutations or recombinations, even when only about 0·1% of the cells remained viable, provided that chemical agents such as hydrogen peroxide were not allowed to accumulate and that excessive heating was avoided.

5.3.8.2 Studies using Human Lymphocytes

Lymphocytes in human blood are roughly spherical cells (with many microvilli) about 10 μm in diameter and comprise about 25% of the leukocyte or white cell population. They contain relatively little cytoplasm and are therefore almost completely filled with their large nucleus. These cells may be induced to divide if they are cultured in a suitable growth medium *in vitro* and stimulated with certain chemical agents such as phytohaemagglutinin. Other agents may be added to arrest cell division at a particular stage and the nuclear contents smeared on a slide, stained and examined.

Macintosh and Davey (1970, 1972) and Macintosh (1971) started one of the biggest wild goose chases the ultrasound biological effects field had ever seen when they reported that the ultrasonic field from a commercial diagnostic Doppler instrument increased the incidence of chromosome aberrations in irradiated human lymphocytes. These papers stimulated many other laboratories to attempt to duplicate these findings using a variety of frequencies, intensities, exposure systems and irradiation media. Some reports began to show positive results which apparently confirmed Macintosh and Davey, e.g. Kunze-Mühl and Golob (1972; 2 MHz, 1 h, 20 mW/cm^2) and Fischman (1973; Marrow cells 1·3 MHz, 2 min, 22 and 40 W/cm^2), while others reported inconclusive data which was not statistically significant but indicated that the effect might be real, e.g. Bugnon *et al.* (1972; 2·5 MHz, 3–5 min, 10 mW/cm^2), one of the several series of experiments by Coakley *et al.* (1971 and 1972) and Serr *et al.* (1971; human amniotic fluid cells, 6 MHz, 10 h, 22 mW/cm^2).

However, the majority of the articles describing attempts to duplicate these studies found no increased incidence of chromosome abnormalities (e.g. Abdulla *et al.*, 1971; Bobrow *et al.*, 1971; Boyd *et al.*, 1972; Braeman *et al.*, 1974; Brock *et al.*, 1973; Buckton and Baker, 1972; Coakley *et al.*, 1971, 1972; Mermut *et al.*, 1973; Rott *et al.*, 1972; Watts *et al.*, 1972; Watts and Stewart, 1972) even though the ultrasonic exposure conditions ranged from the diagnostic through the therapeutic zones and into the region where transient cavitation would be expected. Thus, after many years of controversy the greatest weight of evidence seems to indicate that ultrasound does not cause chromosome abnormalities. But, how could the positive results have been obtained in the first place? The answer appears to lie in the experimental complexities inherent in the preparation of the chromosome spreads and in the difficulties the operator experiences in quantifying the aberrations. On a "real" slide prepared as part of any experiment the individual chromosomes are seldom completely spread out as shown in the "classical" photographs in textbooks. In practice, many of the chromosomes may overlap each other and the operator is frequently called upon to decide whether or not they are normal or abnormal. It is therefore imperative that the operator does not know if the slide that he or she is scoring is from a sonicated or a control series because it is a normal human tendency in the case of a marginal decision to say it is normal if you know that it is a control and to say abnormal if you know it has been sonicated and you believe that ultrasound can cause aberrations. The slides on which the Macintosh and Davey (1970) and (1972) articles were based were scored by Macintosh in the full knowledge of which samples had been sonicated and which were the control. When he repeated this work using the same ultrasonic Doppler instrument and exposure system, but this time scoring the slides without knowing which was which (i.e. the experiment was performed double-blind) he found no increase in chromosomal abnormalities (Macintosh *et al.*, 1975).

5.3.8.3 Studies using other Mammalian Cells

The search for ultrasound-induced chromosome abnormalities in lymphocytes also spread to other mammalian cell types, some of which have already been described above. Hill *et al.* (1972) found no significant increase in the chromosome aberration rate in Chinese hamster cells following 1 or 2 hour exposure to 1 MHz (50 μs pulse, duty cycle 0·05, peak pulse intensity 150 W/cm^2) and Galperin-Lemaitre *et al.* (1973) who found that 2–5 min of 0·87 MHz ultrasound at 1–1·5 W/cm^2 did not damage bone marrow chromosomes. Similar negative results were also obtained *in vivo* by Ikeuchi *et al.* (1973) who found no evidence of embryonic chromosome damage in foeti irradiated at 40 mW/cm^2 for 5 min at 2 MHz before a therapeutic abortion and by Hara (1980) who irradiated foetal fibroblasts *in vitro* for 1 hour with 2 W/cm^2 of 2 MHz ultrasound.

In retrospect, the whole topic of chromosome aberrations probably received far more attention than it deserved, because as Thacker (1973) and Wells (1977) point out, if these chromatid and chromosome breaks were to occur *in vivo* (instead of during the preparation of cell smears *in vitro*) then they would most probably result in the death of that cell.

In a recent report, Liebeskind *et al.* (1979a) found that cultured mammalian cells exposed to the ultrasonic field from a diagnostic instrument (2·5 MHz, pulsed, spatial and temporal peak intensity 34·4 W/cm^2 for 20–30 min) bound more antinucleoside antibodies during their G_1 phase. These antibodies are specific for single stranded or denatured DNA and are normally only bound during the S phase. In the same report Liebeskind *et al.* (1979a) reported the incorporation of labelled thymidine into non-S phase cells which again suggests DNA synthesis (possibly repair) at an abnormal time during the cell cycle. However, these same authors found no evidence of DNA strand breakage as indicated by alkaline-sucrose density gradient ultracentrifugation. These observations have not been duplicated in other laboratories.

In a separate paper, Liebeskind *et al.* (1979b) report that diagnostic levels of ultrasound induced a small but significant increase in the rate of sister chromatid exchange in fresh human lymphocytes as well as in a human lymphoblast line. The significance of sister chromatid exchange (i.e. the exchange of discrete sections of chromosomal material between two essentially identical chromatids) is unknown, but it does appear to reflect some aspects of chromosome damage (Latt and Schreck, 1980). Contradictory results were obtained by Morris *et al.* (1978) who found no evidence of sister chromatid exchange in human leukocytes after a 10 min exposure to 1 MHz ultrasound at intensities between 15 and 36 W/cm^2. Similar negative results were reported by Barrass *et al.* (1982) who exposed fibroblast cells in culture to therapeutic intensities of 3·15 MHz ultrasound (SATA intensity 3 W/cm^2 continuous wave).

Negative effects of pulsed ultrasonic exposures on sister chromatid exchange in CHO cells have also been reported by Barnett (1982) and Wegner and Meyenburg (1982). However, Barnett (1982) indicated that he apparently obtained positive results at a pulse repetition rate of 1 kHz (but not at 10 kHz) when he used SPTP intensities of the order of 2·5 kW/cm^2.

5.4 CELLS IN TISSUES

Cells within solid tissues are less liable to be modified by cavitation-induced microstreaming than cells in suspension, but they are more susceptible to thermal damage. Thus, the rate at which ultrasonic energy is delivered to a tissue, the absorption characteristics of that tissue at that frequency, the spatial distribution of energy within the beam and the rate at which the deposited heat is removed by conduction and convection (i.e. blood flow) are crucial parameters which will determine the nature and magnitude of any observed lesion. Unfortunately, the problems of dosimetry are magnified when one considers intact tissues, especially if they are still within the living organism since multiple reflections occur from interfaces differing in acoustic impedance (this is the fundamental basis of the use of ultrasound as an imaging technique, e.g. B-scans). Most authors choose to avoid many of these problems by performing their dosimetry in an echo-free water bath and reporting the peak and/or average intensity which exists at the point in space which would be occupied by the organ to be irradiated. This measured intensity must therefore be regarded as the maximum upper limit of intensity which could be found within that organ in the absence of reflection. Absorption and reflections occurring before the incident wave reached that organ would significantly reduce the intensity of the exposure whereas normal reflections from behind the irradiated organ would tend to increase the acoustic power density at that site and also change the nature of the irradiation system by developing a standing wave field.

5.4.1 Effects of Standing Wave Fields

The ultrasonic exposure conditions are markedly different for a tissue or organ subjected to an acoustic field having a strong standing wave component as compared to that same organ subjected to that same incident intensity beam in the absence of the reflected wave. One obvious difference which has been extensively investigated by Dyson *et al.* (1974) is that a standing wave can prevent the normal flow of blood cells through small blood vessels and cause them to accumulate in discrete stationary bands separated by zones of clear plasma as shown in Fig. 5.11. This phenomenon which usually disappears as soon as the acoustic power is turned off is commonly called blood cell stasis or blood cell banding instead of the original less specific term parathrombosis (Summer and Patrick, 1964). The spacing between alternate bands of cells is one half wavelength of the compressional wave in that medium and ter Haar and Wyard (1978) have shown that it is the radiation pressure forces developed by the standing wave field which are responsible for the position and orientation of each band. Other forces accelerate the rate of formation of the band, the most significant being: (1) the inter-particle force resulting from the interaction of each oscillating body with the acoustic field re-radiated by the other, and (2) the attractive radiation force developed by pulsating gas bubbles (ter Haar and Wyard, 1978).

The second major change in the acoustic exposure conditions produced by a strong standing wave field is that it greatly enhances the development of cavitational activity within the intact tissue. Crum and Eller (1970) have shown that gas bubbles of different sizes

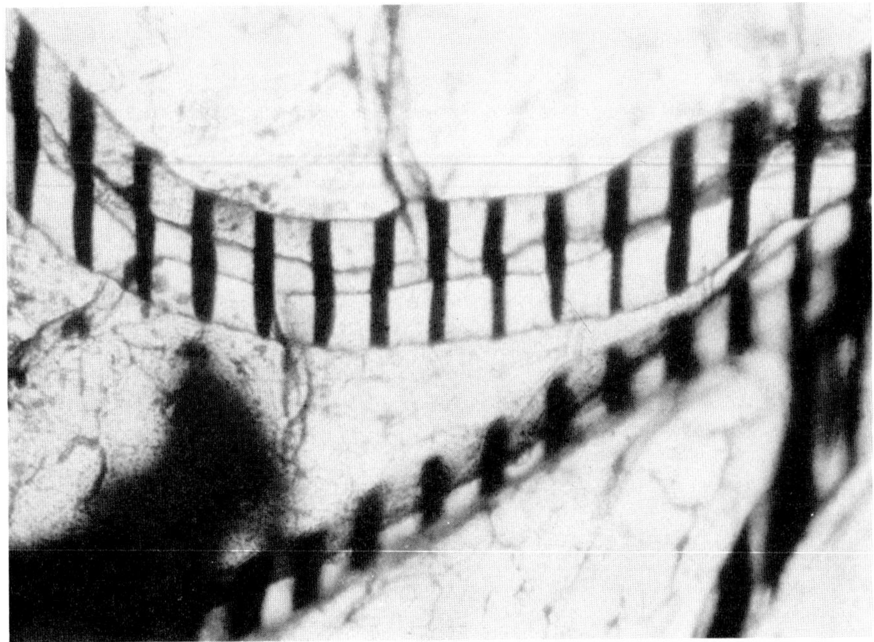

FIG. 5.11 A photomicrograph of the banding pattern produced in the vitelline membrane of the developing chick embryo by blood cell stasis in a MHz standing wave field (photograph supplied by Dr M. Dyson).

are separated by the radiation pressure forces generated by a standing wave field so that bubbles larger than the resonant size for that frequency are driven to the pressure node where they will not be particularly active. However, bubbles close to, or smaller than, their resonant size will be driven to the pressure antinodes (i.e. the regions of maximum varying acoustic pressure amplitude) where they will rapidly grow to resonant size and be driven to oscillate with maximum amplitude. Goldman and Lepeshkin (1952) have shown that cell damage *in vitro* occurs predominantly at these pressure antinodes when these gas bubbles are present. Thus, the formation of a standing wave *in vivo* will collect sub-resonant gas bubbles or micronuclei at the point of maximum acoustic pressure amplitude and hold them there until they fuse together and/or grow by rectified diffusion to the point where they oscillate with maximum amplitude and generate microstreaming forces which mechanically damage the endothelial cells lining blood vessels and the blood cells themselves (ter Haar *et al.*, 1979). This process frequently results in the consolidation of the red blood cell "plug" with fibrin so that it does not break down as soon as the acoustic field is turned off.

There is no single "threshold" intensity for the production of red cell stasis since it depends upon the ultrasonic frequency and the magnitude of the standing wave component, the animal's blood pressure and the velocity of blood flow within that vessel, the orientation and dimensions of that blood vessel and even the difference in density between the blood cells and the surrounding plasma. Under laboratory conditions this effect cannot usually be observed at intensities less than about 0·6 W/cm^2 at 3 MHz, and can be avoided at even higher intensities if the irradiating transducer is kept moving or if short pulses of irradiation are used.

5.4.2 Physiological Effects of Ultrasound

Most of the articles in the vast literature describing the diverse physiological effects of ultrasound on tissues, organs and whole animals can be divided into one of three main categories, namely those describing ultrasound-induced changes in: (1) morphology, (2) biochemical composition and activity and (3) function. The section describing functional changes is so large that in the interests of producing a compact yet cohesive guide to this field it has been necessary to subdivide it into a number of related sub-sections each devoted to one particular biological property (e.g. cell growth) or to one particular tissue or organ system (e.g. muscular tissues, or nervous or reproductive systems). In the interests of brevity and clarity the number of references and the details of the experimental systems described within each section have had to be reduced to a minimum. The approach adopted below has been chosen to complement the numerous existing reviews of the biological effects of ultrasound including those by Edmonds (1980), Fry (1979), Hussey (1975), Lele (1979), Stratmeyer (1977), and Wells (1977).

5.4.2.1 Morphological Changes

The most readily apparent changes which can be produced within intact organs by ultrasonic irradiation are those which result from the death of the irradiated cells. In some tissues such as the kidney, large necrotic lesions may be visible to the naked eye (e.g. Fridd *et al.*, 1977) but in general the tissues have to be stained whereupon the dead or dying cells can be differentiated from the surrounding normal, viable tissues (Fry *et al.*, 1970; Johnston and Dunn, 1976). There is general agreement that small discrete lesions developed within the focal volume of a focused transducer are thermal in origin (Chapter 3) and that their boundaries are determined by the balance between the rate of deposition of heat and its rate of removal by conduction and convection. Thus, the intensity times exposure time combination required to produce a lesion having dimensions of the order of millimetres (where heat loss by diffusion can be appreciable) may be more than that required to produce a larger uniform lesion (where diffusion of heat from the periphery will help to heat the interior) (Nyborg, 1977).

An additional complication associated with the use of a broad unfocused beam of continuous wave ultrasound (similar to that sometimes used in physiotherapy) is that one is more likely to obtain cavitation-induced effects at spatially averaged intensities as low as about 4–10 W/cm^2. For example, Martin and Gregory (1979) found morphological lesions suggestive of cavitational activity in the exposed livers of mice irradiated with 4 W/cm^2 of 0·8 MHz ultrasound. Similarly, O'Brien *et al.* (1977) irradiated mice testes *in vivo* with 10 W/cm^2 of 1 MHz ultrasound and observed a variety of progressive degenerative changes even though the spatial distribution of this damage was highly variable with severely damaged tissues adjacent to apparently normal tissue. This damage pattern was not related to the distribution of intensity across the testis (which was uniform to within 5%) which suggests that the cellular damage was produced by some form of stable cavitational activity (O'Brien *et al.*, 1977). The kidney lesions reported by Fridd *et al.* (1977) after 2 min of 5 MHz irradiation at about 10 W/cm^2 tended to be wedge-shaped which is typical of vascular infarction. This again implies that cavitational as well as thermal mechanisms may have been contributing to the observed lesion by acting in concert in the development of an intravascular blood clot.

Numerous ultrasound-induced changes in cellular ultrastructure have been reported and some of these are briefly summarized in Table 5.1. The most commonly observed changes occurring at the lowest intensities are swelling of the mitochondria (Valtonen, 1967; Dvorak and Hrazdira, 1966) and dilation of the endoplasmic reticulum (Valtonen, 1967). These changes are similar to the changes observed within single cells which have been irradiated while in suspension *in vitro*. In general, these changes are the typical response of any cell to an unfavourable environment and could simply be the response of that tissue to a sub-lethal rise in temperature. The ultrasound-induced changes found in muscle tissues, for example, are very similar to the changes observed in muscle biopsies taken from animals which have just completed a strenuous exercise programme where it is presumed that the elevated temperature and acid environment constitute a temporarily unfavourable environment. It is therefore extremely difficult to draw any meaningful conclusions regarding the mechanism of the interaction of ultrasound with tissues from electron-micrographic observations alone.

Webster *et al.* (1978, 1980) correlated the dilation of the endoplasmic reticulum seen in irradiated human fibroblasts with an increased rate of protein synthesis *in vitro*. These observations apparently imply that the protein synthetic pathways within the cell have been stimulated and that increased quantities of tropocollagen (the precursor of collagen fibres) are being temporarily stored within the endoplasmic reticulum prior to their being processed by the Golgi body and secreted outside the cell.

Despite the number of investigators who have reported positive or negative changes in the appearance of the mitochondria within irradiated cells, only Stephens *et al.* (1978) mention the consistent finding of an intra-mitochondrial amorphous electron-dense material. The quantity of this material seemed to be greater in those mitochondria which had suffered the most disruption of their cristae and so it may simply be a non-specific interaction of the mitochondrial contents with the fixatives and/or heavy metal stains used in the preparatory procedures which must be undertaken before cells can be examined using the transmission electron microscope.

5.4.2.2 *Biochemical Changes*

Table 5.2 summarizes some of the reports describing positive or negative effects of ultrasound on the biochemical activity or chemical composition of mammalian tissues. The relative inaccessibility of cells within a solid tissue means that the assay systems used to detect any change are usually less sensitive than those used to look for similar changes within cells in suspension. For example Robinson *et al.* (1972) exposed human placental tissue to diagnostic and therapeutic intensities of ultrasound *in vitro* for 8 hours and used a variety of histochemical staining techniques to look for changes in the intracellular levels of various enzymes including diaphorases, hydroxysteroid dehydrogenases, acid phosphatase, alkaline phosphatase and adenosine triphosphatase. In view of the limited sensitivity of histochemical staining techniques and the prolonged (8 hour) ultrasonic exposure in a medium where the oxygen had been replaced by carbon dioxide, it is perhaps not surprising that the authors were unable to demonstrate any ultrasound-induced changes

TABLE 5.1

Electron Micrographic changes produced by Ultrasonic Irradiation

Effect	Animal	Tissue or organ	Intensity	Frequency	Duration	Reference
Rupture of myofibrils	chick	muscle myofibrils	2 W/cm^2	800 kHz	2·5 min	Samosudova and El'piner (1966)
Abnormal densities within mitochondria	mouse	liver, pancreas, kidney	1 W/cm^2	2 MHz	500 sec (continuous)	Stephens *et al.* (1978)
Swelling and disruption of mitochondria	mouse	placenta, corpus luteum	1 W/cm^2	2 MHz	400 sec	Stolzenberg *et al.* (1978)
Mitochondrial swelling and their progressive loss of internal structure	rabbit	Brown-Pearce carcinoma	150 W/cm^2	1·5 MHz	1·3 sec	Dimitrieva (1964)
Degenerative changes and disorientation of mitochondria	rat dog	kidney	20–30 W/cm^2	6 MHz	15 min to 3 hours	Bernstine and Dickson (1972)

Changes in internal structure of mitochondria and swelling of endo-plastic reticulum	mouse	liver	1 W/cm^2	1 MHz	2 min	Valtonen (1967)
Number of lysosomes increased	mouse	liver kidney	3 W/cm^2	1 MHz	5 or 10 exposures each 5 min	Majewski *et al*. (1966) Majewski and Jankowiak (1967)
Dilation of endo-plasmic reticulum	chick human	blood vessels fibroblasts (*in vitro*)	0·5 W/cm^2	3 MHz	15 min	Dyson *et al*. 1974
Disturbed continuity of plasma membrane mitochondrial damage	rat	bone marrow	1–5 W/cm^2	0·8 MHz	1 min	Dvorak and Hrazdira (1966)
Inner membrane of mitochondria separated from the outer membrane	mouse	sartorius muscle	10 W/cm^2	0·56 MHz	1 min	Borovyagin and El'piner (1964)
Swelling of basal labyrinth, microvilli of ciliated epithelium and mitochondria	dog	kidney	1 W/cm^2 c.w.	0·880 MHz	20 min	Pinchuk *et al*. (1971)

TABLE 5.2

Biochemical Changes produced in Tissues by Ultrasonic Irradiation

Effect	Animal	Tissue or organ	Intensity	Frequency	Duration	Reference
Decrease in glutathione and increase in ascorbic acid	guinea-pig	muscle, thorax	4 W/cm^2	0·8 MHz	10 min	Straburzynski *et al.* (1965)
Increase in saccharides and DNA content	rat	granulomatous tissue	1 W/cm^2	0·8 MHz	10 exposures each 2 min	Pospisilova and Rottova (1977)
Increased activity of lysosomal enzymes	mouse	liver, kidney	3 W/cm^2	1 MHz	5 or 10 exposures each 5 min	Majewski *et al.* (1966) Majewski and Jankowiak (1967)
Increase in collagen synthesis	rat	granulomatous tissue	1 W/cm^2	0·8 MHz	10 exposures each 2 min	Pospisilova *et al.* (1971) Pospisilova and Rottova (1977)
No significant changes in cAMP, cGMP and histamine levels	mouse	skin, lung and peritoneal cells	1 W/cm^2	2 MHz continuous wave	100 and 200 sec	Glick *et al.* (1979)

Increased serum and CSF levels of GOT and GPT	dog	brain	1·5 mW/cm^2 (average) pulsed 4 μs on and 15 ms off. (Peak = 5·8 W/cm^2)	2 MHz	6 to 10 hours	Tsutsumi *et al.* (1964)
No detectable changes in histochemical enzyme levels	human	placental tissue (*in vitro*)	24·3 mW/cm^2 (average) 79 W/cm^2 (peak) pulsed 4 μs on 1·4 ms off	2 MHz	8 hours	Robinson *et al.* (1972)
No changes in histochemical enzyme levels	human	placental tissue (*in vitro*)	6·3 W/cm^2 (continuous)	2 MHz	8 hours	Robinson *et al.* (1972)
Altered sulphate metabolism	guinea-pig	skin	1·7 W/cm^2	3 MHz	30 sec	Wells (1974)
Transient change in protein content and metabolism	rabbit	eye (*in vivo*)	0·2 to 1 W/cm^2	0·88 MHz	5 min/day for 10 days	Marmur and Plevinskis (1978)
Decreased levels of Thyronin	rat	thyroid	0·2 to 1 W/cm^2 0·2 W/cm^2	0·88 MHz	5 min 5 min/day for 10 days	Stereva and Beleva-Staikova (1976)

which might have been superimposed onto those changes resulting from tissue anoxia (Robinson *et al.*, 1972).

Tsutsumi *et al.* (1964) reported that the levels of the enzymes glutamate oxaloacetate transaminase (GOT) and glutamate pyruvate transaminase (GPT) in the serum and cerebrospinal fluid of dogs were increased after the animals had been exposed to 2 MHz ultrasound (average intensity 1·5 W/cm^2, pulsed 4 μs on and 15 ms off, peak intensity 5·8 W/cm^2) for 6–10 hours. The animal's brain was irradiated through the intact skull and so the intensity of the incident ultrasonic beam was attenuated by reflection at the tissue–bone interfaces and by absorption within the bone itself. Their results show that despite a marked variation from animal to animal, there appeared to be a statistically significant increase in the levels of these enzymes which reached a peak about 6 hours after irradiation before returning to normal. It is difficult to extract details of the experimental arrangement from this article, but the dogs were apparently anaesthetized and then immobilized in gypsum so that they could be awake but restrained during the long exposure period. In view of the psychological trauma associated with immobilizing active animals such as dogs for long periods (over 9 hours for 7 of them) it is not surprising that there were changes in the level of these (and presumably many other) enzymes and that these levels returned to normal a few days after the animals had been released from the gypsum. The authors did not include adequate control series (i.e. animals which had been immobilized for similar periods under the same conditions but not exposed to ultrasound) and so the results obtained by Tsutsumi *et al.* (1964) should be viewed with some scepticism. If this work should be repeated, it would be more humane, and the results more interpretable, if the animals were to be kept anaesthetized throughout the sonication period. A better alternative would be to attach a less cumbersome version of the transducer to a more docile animal which could be conscious and unrestrained during the exposure period. Ideally informed human volunteers could be used even though it would not be advisable to sample the cerebrospinal fluid in this case.

A large number of articles describing ultrasound-induced changes in the biochemical composition of mammalian tissues have not been included in Table 5.2 for the sake of clarity. Some of these omitted reports include a decreased activity of lysosomal enzymes in liver (Keller and Tanka, 1977); inhibition of uptake of calcium ions by smooth muscle (Hu *et al.*, 1978); a decrease in the permeability of synovial membranes (Engel, 1971); interruption of nervous conduction (Herrick, 1953; Tippe, 1979); changes in skin permeability

(Carney *et al.*, 1972; Sarvazyan and Pashovkin, 1979); and many others. Unfortunately, many of the publications in this area can be criticized on the grounds that the intensities used were so high, and the exposure times so long, that the amount of ultrasonic power introduced into a small mammal must have caused so much thermal damage that in some cases it is surprising that the observed changes were relatively small.

5.4.2.3 Functional Changes

Tissues and organs are specialized to perform a limited number of specific functions and it is usually relatively easy to detect changes in the rate and/or magnitude of one or more of these functions after that organ or tissue has been exposed to ultrasonic radiation. Most adult organs are in a steady-state condition where there is a continuous turnover of cells to replace dying or damaged cells but there is usually no net increase in the size of that organ. However, if a small portion of an organ is damaged, the constituent tissues will respond by initiating various repair processes; in some organs such as the liver, this repair process may include regeneration and replacement of the damaged portion. The first two parts of this sub-section will therefore be devoted to the effects of megahertz ultrasound on normal growing tissues (e.g. the developing embryo) (Section 5.4.2.3.1) and on the wound healing and regenerative systems in soft and hard tissues and the implications of these studies in physiotherapy (Section 5.4.2.3.2).

The remaining parts of this sub-section will briefly review some of the effects of megahertz ultrasound on: the mammalian nervous system (Section 5.4.2.3.3); the organs of special sense (i.e. the eye and ear (Section 5.4.2.3.4); the muscular system (Section 5.4.2.3.5); the blood coagulation system (Section 5.4.2.3.6); the immunological system (Section 5.4.2.3.7); the hormonal system (Section 5.4.2.3.8); and the reproductive system (Section 5.4.2.3.9).

5.4.2.3.1 Effects of ultrasound on Foetal Growth. The errors involved in measuring the growth rate of mammalian foeti *in vivo* are relatively large and so it is not surprising that those studies which have investigated the effect of diagnostic intensities on foetal growth rate have nearly all yielded negative results (e.g. McClain *et al.*, 1972; Shoji *et al.*, 1971; Warwick *et al.*, 1970). One notable exception is the study by Pizzarello *et al.* (1978) who exposed rats *in utero* for 5 min to 1·5 mW/cm^2 of pulsed 2·25 MHz ultrasound. The pregnant rats were irradiated 3, 5, 6 or 15 days post-conception and the embryos were all examined on day 17. The foeti irradiated 3, 5 or 6 days post-conception were reputed to be consistently smaller than their controls whereas irradiation on day 15 produced normal sized foeti. This report should be viewed with scepticism on several counts: the number of animals used in this study were not given and the work was not subjected to statistical analysis. The magnitude of the reported weight reduction was too large to have been missed by the earlier investigators. Also, it should be remembered that diagnostic intensities of ultrasound had no detectable effect on the growth rate of human cells in suspension *in vitro* (Loch *et al.*, 1971; Serr *et al.*, 1971). It therefore appears as though the article by Pizzarello *et al.* (1978) may be another example of a poorly executed investigation which has produced a spurious result and generated unnecessary anxiety concerning the safety of diagnostic ultrasound.

O'Brien (1976) exposed the foeti of CFI mice *in utero* to therapeutic intensities of continuous wave 1 MHz ultrasound (0·5–5·5 W/cm^2 with exposures of 10–300 sec). The foeti were insonated on the 8th day of gestation and were removed by laparotomy on the 18th day and examined for visible malformations and body weights. In this study 2866 foeti were examined from 273 litters and the pooled results showed a statistically significant reduction in the weight of the insonated foeti. A similar dose-dependent reduction in foetal weight was also found by Stolzenberg *et al.* (1980) who irradiated CFW Swiss-Webster mice with 1 W/cm^2 of 2 MHz ultrasound at various postconception times between 1 and 13 days. Intrauterine temperature rises were measured as part of this latter study and it was found that the mice irradiated with 1 W/cm^2 of 2 MHz ultrasound in a water-bath at 34°C experienced a 6°C rise in temperature after 3 min (i.e. a final temperature of 40°C) while the mice irradiated with the same "dose" of ultrasound while immersed in a water bath at 37°C experienced a 10°C rise in temperature to 47°C. These large temperature elevations would almost certainly have a deleterious effect on the developing embryo as was confirmed by Hara (1980).

Sikov and Hildebrand (1976) found no significant change in the weight of foetal rats exposed *in utero* 9 days post conception to 2·8–32·7 W/cm^2 of a focused beam of continuous wave 3·2 MHz ultrasound. Similarly, Sikov *et al.* (1977) also found no change in foetal weights after the rat foeti had been irradiated with 0·01–1 W/cm^2 of 0·93 MHz ultrasound for 5 min on their 15th day of gestation. Stratmeyer *et al.* (1977) also found no weight loss in mouse foeti delivered by laparotomy after they had been exposed *in utero* on the 10th day post-conception (0·25 and 0·8 W/cm^2, 1 MHz, continuous) but observed a statistically significant weight increase in some of the pups irradiated *in utero* and allowed to survive for 36 or 51 days before being sacrificed.

If the article by Pizzarello *et al.* (1978) is discounted, it may be concluded that diagnostic intensities of megahertz ultrasound appear to have no detectable effect on the growth rate of mammalian foeti *in utero* under laboratory conditions. However, therapeutic intensities can produce deleterious effects which appear to be related to both the magnitude and the duration of the temperature rise induced within the test animal. Another factor which has not been systematically investigated is the effect of the high therapeutic intensities on the well-being of the pregnant animal. The maternal spinal cord will also be exposed to high intensities of ultrasound if a broad beam is used to irradiate the foeti, and this may result in hind limb paralysis or bladder distension caused by the retention of urine. These unfortunate sequelae would undoubtedly change that pregnant animal's rate of consumption of food and water, but it has not been determined whether or not this could contribute to the foetal weight reduction which has been observed after high intensity ultrasonic exposures.

5.4.2.3.2 Effects of ultrasound on wound healing. (a) Soft Tissues. Dyson *et al.* (1968) demonstrated the remarkable ability of ultrasound to increase the rate of repair of 1 cm square wounds cut out of the ears of rabbits. Identical holes were punched in both ears and one ear was irradiated with 3·6 MHz ultrasound (5 min treatment 3 times per week, 0·1 W/cm^2 continuous, or 0·25–8 W/cm^2 peak intensity pulsed at various duty cycles) and the rate of closure of the hole compared with that of the other, untreated, ear. The irradiated wounds healed significantly more rapidly than their controls, the maximum enhanced growth rate being 1·3 times that of the controls and was obtained after 21 days using a treatment 0·5 W/cm^2 peak intensity pulsed 2 ms on and 8 ms off (i.e. a timer averaged intensity of 0·1 W/cm^2). The temperature elevation caused by this ultrasound exposure was 1·5°C which was presumed to have been too small to account for the observed effects and so it was proposed that some form of acoustic microstreaming may have played an important part in this process.

Dyson *et al.* (1976) also showed that the rate of healing of varicose ulcers caused by venous stasis was increased if the apparently healthy tissues at the margins of the ulcer were irradiated with 3 MHz ultrasound (up to 10 min treatments three times per week, peak intensity 1 W/cm^2 pulsed 2 ms on and 8 ms off giving a time averaged intensity of 0·2 W/cm^2). Figure 5.12 shows that 28 days after commencing the sonication treatment there was a significant ($0{\cdot}025 < P < 0{\cdot}05$) decrease in the areas of unepithelialized granulation tissue.

These well designed studies complement the numerous unproven claims that ultrasound has a beneficial therapeutic action in the treatment of various soft tissue lesions (Summer and Patrick, 1964). Other studies which claim therapeutic effectiveness on soft tissue injuries or lesions include Tuchman (1956) who relieved the joint contracture associated with scleroderma; Markham and Wood (1980) who produced an improvement in the mobility of the tissues in cases of Dupuytren's contracture; Hustler *et al.* (1978) who found an increased rate of fading of some experimentally induced bruises in the pinnae of guinea pigs; and an increased rate of healing of trophic ulcers (Galitsky and Levina, 1964) and pressure sores (Paul *et al.*, 1960). Similarly, Bierman (1954) reported that ultrasonic treatment increased the mobility of scar tissue resulting from burn injury which

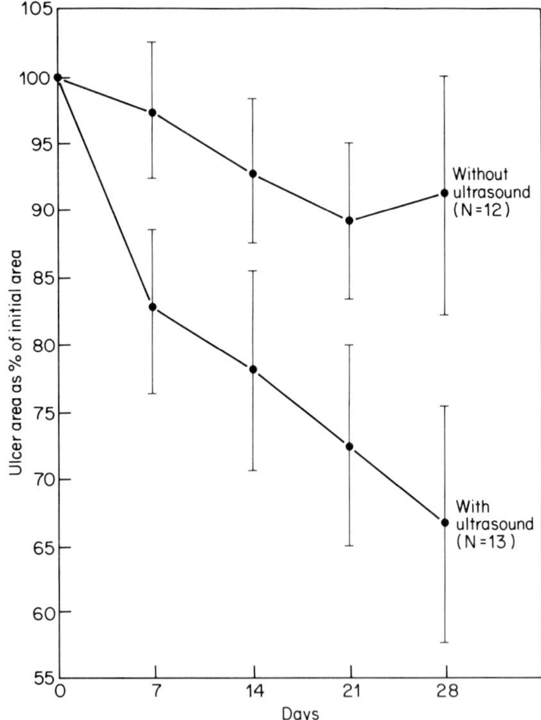

FIG. 5.12 The effect of pulsed 3 MHz ultrasound at a S.A.T.A. intensity of $0{\cdot}2\ W/cm^2$ on the rate of decrease of the areas of varicose ulcers (from Dyson *et al.*, 1976).

had contracted and severely limited movement, and Drastichova *et al.* (1973) and Webster (1980) demonstrated an increase in the tensile strength of newly-formed scar tissue.

The repair processes which occur after a soft tissue has been damaged can be divided into three consecutive phases. These are: (1) the inflammatory phase where the haemostatic plug is invaded by leucocytes which break down and engulf tissue debris and invading pathogens and (2) the proliferative phase which is characterized by the invasion of fibroblast cells and the formation of new granulation tissue. Collagen fibres are laid down and are attached to the healthy tissues at the margins of the lesion and to the myofibroblast cells which subsequently contract and pull the edges of the wound together. The third phase is the remodelling or maturation phase which is a continuous process of collagen degradation and synthesis in an attempt to modify the mechanical properties of the scar tissue so as to resemble that of the healthy tissue which it has replaced. Dyson and her collaborators have shown that therapeutic intensities and frequencies of ultrasound are able to modify certain aspects of these three phases and that ultrasound does appear to be beneficial if it is administered during the early stages of wound healing.

The wound healing scheme outlined above applies to skin and many muscular tissues, but it does not apply to specialized internal organs such as the liver. If the liver is extensively damaged then the surviving cells will be stimulated to divide in a near synchronous manner so that it may regrow to its original size (Bucher, 1963). Kremkau and Witcofski (1974) exposed the livers of anaesthetized Charles River CD rats to 50 mW/cm^2 (spatial average, continuous wave) of 1·9 MHz ultrasound and then surgically removed about 70% of the lobulated liver mass. They subsequently observed a 20–80% reduction in the number of mitotic figures within the remaining portion of the liver. Two independent laboratories have attempted to duplicate these observations but have reported negative effects. Miller *et al.* (1976) exposed rat livers to 2·2 MHz ultrasound at intensities ranging from 60 to 16 000 mW/cm^2 while Barnett and Kossoff (1977) used 2·5 MHz pulsed ultrasound at time averaged intensities of 15 and 75 mW/cm^2. Both groups reported that there was no difference between the number of mitotic figures found in the remaining portion of the irradiated livers and that of their controls.

Some amphibia are able to regenerate a completely new limb after the original one has been lost in the struggle to escape from a predator. The regenerating limb-bud provides a convenient experimental model of rapidly dividing and differentiating cells which in many ways resembles the situation found in the developing foetus.

Pizzarello *et al.* (1975) have used this system and claim that adverse growth changes (roughly equivalent to teratological effects) were produced after low power ultrasonic exposures from a commercial diagnostic device. However, some doubts have been expressed as to the experimental validity of the reported observations.

(b) Hard Tissues. In general, the bony or calcified tissues of the body have received less attention than the surrounding soft tissues. The absorption coefficient of these hard tissues is greater than that of soft tissues and so therapeutic intensities of ultrasound which are able to enter bone can produce a marked local increase in temperature. Buchtala (1948) shaved the legs of dogs and subjected them to 0·5–1 W/cm^2 of 800 kHz ultrasound in a water bath using a stationary transducer. He found that the bones of young animals were more susceptible than those of older animals, but that gross morphological damage could be observed in all cases. Consequently, he proposed that growing bones should not be exposed to therapeutic intensities of ultrasound. Barth and Wachsmann repeated these observations in 1949 and again kept the transducer stationary but moved the animal's legs slightly to simulate the situation found in physiotherapy. They found that average intensities of the order of 3 W/cm^2 were required to produce thickening and then removal of the periosteum about 1–2 weeks post-exposure (N.B. it is not known how they performed their dosimetry and so the value of 3 W/cm^2 should be viewed as an approximate figure). Even higher intensities of 800 kHz ultrasound led to the formation of what they call 'slowly creeping fractures". They therefore concluded that growing bones may be exposed to low therapeutic intensities of ultrasound provided that a moving applicator technique is used (which presumably reduces the temperature rise within the bone).

The effects of low therapeutic intensities of ultrasound on the rate of healing of fractures has attracted sporadic interest. Goldblat (1969) exposed rabbits with fractures of the third metatarsal to 0·4 and 8 W/cm^2 for 8 min per day for 15 treatments commencing on the third day post fracture. The rates of healing of the sonicated and control fractures were compared from X-ray photographs. On the basis of histological examination alone it was concluded that the lower dose of ultrasound enhanced the process of regeneration and differentiation. There were no detectable differences between the healed fractures of either group after 45 days. Duarte (1976) punched holes in rabbit's femurs and exposed them to 5 and 10 MHz pulsed ultrasound for 10 min per day for up to 15 days. Histological examination showed that after 4 days the exposed limbs showed a high degree of cellular activity (indicative of osteosynthesis) compared with their controls. After 15 days the irradiated holes were reputed to have healed completely while the controls were still in the preliminary stages of healing. Knokh and Knaut (1975) and Petrov *et al.* (1975) have also concluded that ultrasound increases the rate of osteosynthesis even though Gornia *et al.* (1974) refer to certain complications which may have been caused by excessive temperature elevations. Some of the morphological effects of ultrasound on bony tissues have been described by Abramovich (1970) and James and Halliwell (1970) while Krenztlin and Cattaneo (1968) investigated the histochemical changes produced in the dental pulp by ultrasonic treatment. These studies together with the various ongoing and unpublished observations on the rates of healing of sonicated versus control fractures suggest that low intensities of therapeutic ultrasound do not have a detectable adverse effect upon bone growth and regeneration and in fact it may even have a beneficial effect.

Dyson and Brookes (1982) fractured both of the non-weight-bearing fibulae in anaesthetized Wistar rats and left them to heal without fixation. One fracture was exposed to ultrasound (1·5 or 3 MHz, SATA intensity 0·5 W/cm^2 pulsed 2 ms on and 8 ms off for 5 min on 4 consecutive days) while the other was left as its control. Ultrasonic treatment within the first two weeks following fracture apparently increased the rate of healing and improved the quality of the repair whereas treatment only during the third or fourth week following fracture resulted in the formation of more collagen which may not be beneficial in fracture repair.

Low frequency ultrasound (*ca.* 20–30 kHz) may also be used as part of the procedure used in what is commonly called bone welding. This technique is used to fuse the pieces of a broken bone together under conditions where it is not possible to mechanically hold them together by means of a splint or a plaster cast. The bone is exposed in the vicinity of the fracture and a fluid paste composed of a mixture of bone powder and a monomer which polymerizes by a free radical mechanism is forced into the fissure of the fracture and around a splint of bone or an inert solid material. The region is then subjected to low frequency ultrasonic vibration from a hand-held probe whose tip is vibrating longitudinally with displacement amplitudes as high as 45 μm (Poliakov, 1970; Neumann, 1980). The free radicals generated by transient acoustic cavitation and the high local temperatures produced by frictional contact with the vibrating tip cause the monomer to polymerize. Preliminary clinical trials using this technique are claimed to be successful (e.g. Brug *et al.*, 1976) but the long-term effects of the permanent polymeric mass on the strength of the bone and the local and systemic effects of any non-polymerized monomer have not been fully investigated. In view of the many potential sources of biological damage associated with this technique it should not be used unless there is no other practicable means of holding the broken edges of the bone in apposition.

(c) Implications in Physiotherapy. The observations on the effects of megahertz ultrasound on foetal and adult tissues reported above apparently indicate that there is a fairly narrow range of exposure conditions (lying between those which produce no detectable change and those which produce adverse or harmful effects) which seem to stimulate tissue growth and the healing processes (Oakley, 1982). This supports the firm conviction held by many physiotherapists that ultrasound does have a beneficial therapeutic effect even though there are remarkably few well-designed clinical trials to convince the more sceptical that it is efficacious. It is not possible to pinpoint the mechanisms which might be responsible for any beneficial effect other than the local heating and the presumed increase in microvascular blood flow associated with this heating. It has been proposed that it is the piezoelectric properties of bone which result in its being strengthened after exercise. If this is true, then ultrasonic exposure may induce similar effects at the fracture site and so stimulate osteosynthesis.

One reason why some therapists are unable to obtain beneficial results with ultrasound may be that they are not using enough sound energy. The intensities (in W/cm^2) quoted on the controls of the early machines (which are still in common use) refer to the average intensity emitted in the continuous mode. If the machine is now set to a pulsed mode then the peak intensity remains the same but the total ultrasonic power emitted by the machine is decreased with a consequent reduction in the heating effect, etc. For example, a machine set to deliver 1 W/cm^2 in the continuous mode only emits 0·5 W/cm^2 (time averaged power) when pulsed 1 : 1 or 0·2 W/cm^2 when pulsed 1 : 4. Thus, the use of low intensity settings in a pulsed mode emits so little acoustic energy that it is unlikely to be hazardous to the patient but it may also be too low to be effective as a treatment. While it can be argued that the patients appear to benefit from these low-power treatments (possibly due to a placebo effect) it would seem to be a more effective use of the therapist's and the patient's time if the emitted ultrasonic power was increased to the point where it became efficacious. This is the main reason why physiotherapists ought to be encouraged to design and participate in clinical trials to identify the optimal dosage parameters for their ultrasonic treatment regimes.

5.4.2.3.3 The Nervous System. One of the first surgical applications proposed for ultrasound was the selective destruction of small well-defined regions within the brain without damaging the surrounding tissues, i.e. the formation of trackless lesions. Consequently, there has been an intensive series of investigations to determine the range of ultrasonic exposure parameters which will destroy nerve tissue within the central nervous system (see Chapter 3; Dunn and Fry, 1971). The moderately high attenuation coefficient of nervous tissue means that it is one of the most susceptible of the so-called solid tissues to the damaging effects of ultrasonic irradiation. For example, Herrick (1953) and Anderson *et al.* (1951) report that they were able to destroy the sciatic nerves of experimental animals using high therapeutic intensities of ultrasound without affecting the histological structure of the surrounding muscular tissue. The spinal cord is particularly at risk because of the intimate association between the spinal nerves and the sympathetic nerve ganglia and the highly absorbing bony vertebral column. Consequently, the exposure conditions commonly used to irradiate the foeti of pregnant mice (e.g. 80–200 sec exposure of 2 MHz continuous wave ultrasound using a broad beam transducer at an average intensity of 1 W/cm^2) frequently results in maternal hind limb paralysis which is associated with grossly distended urine-filled bladders and impacted masses of faecal material within the flaccid intestines indicative of inactivation of the autonomic system (Stolzenburg *et al.*, 1980b).

Less attention has been paid to the functional effects of low-intensity ultrasound on the nervous system. Stuhlfauth (1952) demonstrated that the temperature within the sciatic nerve of a rabbit irradiated with MHz ultrasound was higher than that of the surrounding tissues. Studies on the peripheral nerves of frogs and cats have shown that focused ultrasound can reversibly block axonal conduction with the small fibres being more susceptible than the large fibres (Young and Henneman, 1961). Takagi *et al.* (1960) reported that reflex discharges were stimulated and that spontaneous discharges appeared on the ventral root of the spinal cord at an intensity of 3·2 W/cm^2. Doubling the intensity resulted in additional spontaneous discharges but a decrease in reflex discharges. Borrelli *et al.* (1981) report that synapses are damaged within CNS tissues irradiated with short intense pulses of ultrasound even when the tissues appear morphologically normal under the light microscope. Madsen and Gersten (1961) and Esmat (1975) also report ultrasound-induced changes in the conduction velocities of peripheral nerves whereas other laboratories (including my own) have been unable to demonstrate any changes at therapeutic intensities and frequencies (unpublished observations).

Hu and Ulrich (1976) exposed the brains of anaesthetized squirrel monkeys to 2·25 MHz ultrasound at an intensity of 3·4 mW/cm^2 and detected evoked potentials on electrodes chronically implanted in the mid-line parietal region. These evoked potentials progressively disappeared within about 3 min even though the ultrasound was still being administered. This possible direct stimulation of the CNS has been proposed as the mechanism responsible for the increased foetal activity reputed to occur during investigations using Doppler foetal heart monitors (David *et al.*, 1975). These low intensity bioeffects will be discussed in more detail in Chapter 6, together with the various reports of changes in the rate of maturity of diverse neural functions in neonatal animals which had been exposed to diagnostic intensities of ultrasound *in utero* (Murai *et al.*, 1975).

5.4.2.3.4 Organs of Special Sense. (a) Eye. The visual accessibility of a "naked" portion of the nervous and vascular system behind the retina as seen through the optically transparent cornea and lens, together with the extensive ophthalmological uses of ultrasound for both diagnostic and therapeutic applications have resulted in the eye being selected for a large number of bioeffect investigations. Lizzi *et al.* (1978a) have shown that high intensities of focused ultrasound can produce opaque prolate elipsoidal cataracts within the lens or cornea of anaesthetized mammals *in vivo*. The threshold intensity needed to produce a barely detectable lesion had to be increased as the time of ultrasound exposure was decreased so that a relatively constant amount of energy was required for cataract production with exposures less than 100 ms in duration. Longer exposures required a progressively greater input of total acoustic power (Fig. 5.13). These observations are consistent with a thermal mechanism for cataract production (see Chapter 3) where more energy had to be supplied for exposures longer than 100 ms to allow for the heat lost by diffusion to the tissues outside the focal volume. The lens and cornea are particularly suitable tissues for demonstrating thermally-induced lesions because they do not have an extensive vascular supply to carry away heat. Somewhat similar lesions can also be produced within the retina, choroid and sclera (Lizzi *et al.*, 1978b) at lower intensities but requiring longer exposure times. This apparently indicates that the cellular structures within the sensory portion of the eye are more susceptible to thermal damage than those of the lens and cornea, but they are also being afforded a certain measure of protection by their active hlood supply.

Lizzi *et al.* (1978b) also report a novel "subthreshold" effect in that the retinal blood supply in an albino rabbit can be seen to blanch within the focal region when it is subjected to an intensity/time combination just below the level required to produce a chorioretinal lesion. This blanched region disappears when the ultrasound is turned off. At even lower subthreshold intensities the blanching may disappear even though the 10 MHz ultrasonic beam has not been switched off. It is not known if this transient blanching is due to microvascular constriction (i.e. direct neural or neuromuscular stimulation by the ultrasound) or is a mechanical effect resulting from the compression of the nutrient capillaries by the radiation pressure forces emanating from the transducer.

Barnett and Kossoff (1977) irradiated the retina of cats for 30 min with a focused beam of pulsed 7·5 MHz ultrasound. The peak intensity during each pulse was estimated to be about 100 W/cm^2 but each pulse was extremely brief (0·25 μs) and was repeated at a pulse repetition rate of 15 kHz giving a time averaged power output of 1·3 mW (i.e. simulating many of the characteristics of ophthalmic diagnostic devices except for a greatly increased P.R.F. and exposure duration). The total exposure energy passing through the focal volume at the retina was estimated to be 2·3 J compared with about 7 μJ for a typical diagnostic exposure. No histological changes could be detected within the retina which is in agreement with Lizzi *et al.* (1978a) who found that this same amount of energy would have to be

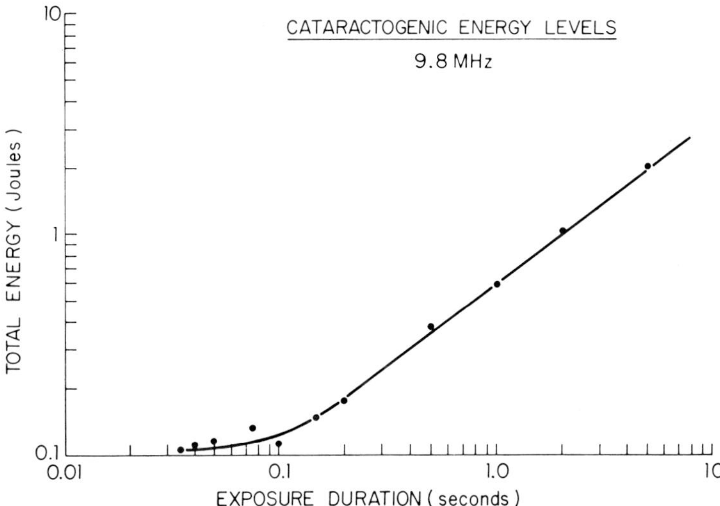

FIG. 5.13 The relationship between the total ultrasonic energy (J) which has to be delivered to the lens or cornea of an anaesthetized mammal *in vivo* to produce a barely detectable lesion, as a function of the duration of the exposure (from Lizzi *et al.*, 1978a).

delivered in about 5 sec instead of 30 sec to produce a thermal lesion in the eye of a rabbit (Fig. 5.13).

Numerous articles have claimed that ultrasound has a beneficial therapeutic effect on the course of many diverse ophthalmological diseases. Tsokh (1974) investigated the effects of ultrasound on the course of chemical burns of the eyes while Greguss (1974) determined its effect on the pupillary response. Yamamoto and Baba (1963) claimed a 75% success rate for the improvement in visual acuity caused by ultrasound in patients who had had myopia for less than 2 years. They even made the surprising claim that the visual acuity improved in one eye when the other eye was treated with ultrasound. This study was repeated by Greguss and Bertenyi (1976) who found that therapeutic ultrasound (850 kHz, peak intensity 1·5 W/cm^2 pulsed at an unspecified mark/space ratio, administered for 5 min every second day over a 20-day period) had a beneficial effect upon the progression of myopia in patients less than 20 years of age even though the results were not as spectacular as those indicated by Yamamoto and Baba (1963).

(b) Ear. Ultrasound has been used with varying degrees of success over the last 30 years to surgically destroy the sensory structures within the vestibular portion of the inner ear. This somewhat drastic procedure is used in an attempt to alleviate the distressing symptoms of Meniére's disease (a condition whereby the patients are subjected to recurrent episodes of severe vertigo). Relatively large amounts of acoustic energy (up to about 800 J) may be directed into the vestibular system by a variety of approaches (as described in Wells, 1977) and the resulting temperature rise causes a variety of histological damage to the sensory structures (James *et al.*, 1963; Barnett *et al.*, 1973). The various clinical reports indicate that most patients appear to obtain at least some relief from their symptoms without damage to their hearing. However, some patients have suffered severe hearing losses without any improvement in their symptoms which has been interpreted in terms of a badly aligned ultrasonic beam which has entered the cochlea instead of the utricle.

Molinari (1968) has shown that much lower intensities of ultrasound can change both the cochlear microphonic potentials and the action potentials generated by the functional cochlea. Gavrilov *et al.* (1975) showed that irradiation of the cochlea of a frog with a beam of focused ultrasound resulted in the detection of neural impulses within the auditory centres of its midbrain. Subsequently, Gavrilov (1980) has proposed that the focused ultrasonic beam may be amplitude modulated and used on profoundly deaf humans to give some form of auditory stimulus. Investigations in the field of infrasound have shown that the difference between that intensity of sound which is just audible and that intensity which is so loud that it damages the hearing mechanism of the ear progressively diminishes as you move further away from the audible range. If this trend also occurs on the high frequency side of the audible region then the margin of safety between that level of ultrasound which is just detectable and that level which will irreversibly destroy the hair cells of the organ of Corti may be extremely narrow.

The inner ear may also be subjected to ultrasound inadvertently as a result of the growing use of ultrasonic descaling devices in dentistry. These devices typically operate at about 20–30 kHz and employ a vibrating hand-held tool to remove calcified plaque from the surface of the teeth. Fortunately, the small surface area of contact together with the large acoustic impedance mismatch between the metal probe and the tooth prevents most of the acoustic energy from entering the tooth, and what does enter should be largely attenuated by scattering and the high absorption coefficient of the tooth and the bones into which they are inserted. Nevertheless it should be noted that Möller *et al.* (1976) report that ultrasonic descaling of maxillary teeth caused tinnitus (an apparent buzzing or ringing sound in the ears) as well as temporary shifts in the audible hearing threshold. An audible sound which is loud enough to cause tinnitus (e.g. gunshots at close range) will almost certainly cause permanent hearing damage if it is repeated. It must therefore be presumed that if the tinnitus and temporary threshold shifts were caused by the conduction of ultrasound through the bones of the skull, the liberal use of dental descaling devices might constitute a potential hearing hazard to the patient.

Another possibility is that the ear could have detected high-frequency airborne sound waves emitted by the ultrasonic descaler. Michael *et al.* (1974) measured the airborne emitted acoustic energy from one descaling device and found approximately 90 dB per 1/3 octave band from 6 kHz to 40 kHz. For comparison, a conventional dental high speed drill emitted approximately 80 dB sound at frequencies ranging from 15 kHz through 100 kHz (Michael *et al.*, 1974).

5.4.2.3.5 The Muscular System. Talbert (1975) has investigated the effect of low (280 kHz) and high (2 MHz) frequency ultrasound on the rate of the spontaneous contractions exhibited by strips of uterine smooth muscle *in vitro*. The 2 MHz ultrasound increased the frequency of the spontaneous contractions and this behaviour could be duplicated by increasing the temperature of the bathing medium which suggests that the ultrasound is merely warming up the tissue.

However, the 280 kHz ultrasound produced variable effects including a decrease or even a cessation of spontaneous contractile activity and then the contradictory response of triggering a strong contraction if the ultrasonic field was suddenly switched on at an intensity of about 3 W/cm^2. It should be noted that the whole of the muscle strip has to be contained within the ultrasonic beam or else the portion of muscle outside the beam acted as a pacemaker for the remainder and these latter effects were not observed.

Hu *et al.* (1978) studied the effects of 5 min exposures of 1·5 W/cm^2 1 MHz ultrasound on rat intestinal muscle. They observed that muscle action potentials were inhibited in a reversible manner after single exposures, but that the extent of the recovery decreased with increasing number of exposures. Mortimer *et al.* (1978) suspended the left anterior papillary muscle from normal rats in Ringers solution under conditions where they could only contract isometrically (i.e. the muscle could not change its length). The magnitude of the tensions developed within the muscle were investigated before, during and after 10 min exposures to 2·4 W/cm^2 of 1 MHz ultrasound. It was found that the ultrasound appeared to decrease the resting tension of the cardiac muscle without causing a significant difference in the active tension developed during muscle contraction. These effects did not appear to have been caused by an increase in temperature.

The degree of contraction of the smooth muscle cells within the walls of blood vessels is largely under the control of the autonomic nervous system. These muscle cells do not have an inbuilt pacemaker to generate a spontaneous rate of contraction as do the smooth muscles of the gut, the ureter, the uterus and the heart. Nevertheless, there are a few reports of transient vasoconstriction or blanching of microvascular networks during ultrasonic irradiation (e.g. Lizzi *et al.*, 1978b). Similarly, unpublished studies in Manchester showed that about 0·5 W/cm^2 of 3 MHz c.w. ultrasound blanched the pinna of a guinea pig's ear, but that the microcirculation returned to normal within seconds of turning off the ultrasonic field. Preliminary experiments in Manchester investigating the effects of MHz ultrasound on denervated (isolated) segments of ear arteries from rabbits and guinea pigs were negative and the experiments on arteries *in situ* were abandoned because of a lack of reproducibility. These observations suggest that any direct effect of ultrasound on the blood vessels themselves is small compared with the normal range of changes in muscular tone resulting from small changes in the neural and hormonal regulatory mechanisms.

5.4.2.3.6 Blood Coagulation System. Section 5.3.5 above outlined some of the experimental evidence showing that therapeutic intensities of MHz ultrasound can damage platelets *in vitro* and consequently activate and accelerate the blood coagulation system. In this case the major damaging mechanism has been shown to be the generation of some form of cavitational activity (Williams *et al.*, 1978). It has also been shown that the generation of acoustic microstreaming within the intact vascular system *in vivo* can result in the production of platelet emboli and the formation of thrombi (Williams, 1977). However, removal of blood from an animal and the addition of an anticoagulant changes its nucleation characteristics and so the effects observed at MHz frequencies *in vitro* may not occur at the same ultrasonic intensities within the intact organism *in vivo*.

As part of a study to identify the "threshold" conditions for platelet damage *in vivo* Williams *et al.* (1981) inserted a siliconized 19 G butterfly needle attached to a plastic cannula into an antecubital vein of healthy adult human volunteers as shown in Fig. 5.14. Sequential 2·5 ml blood samples were collected in plastic tubes containing EDTA and theophylline which anticoagulated the blood and also rendered the platelets functionally inert, i.e. prevented them undergoing the release reaction once they had been removed from the body. Preliminary experiments established that the plasma levels of the platelet specific protein β-TG in the first three samples to be

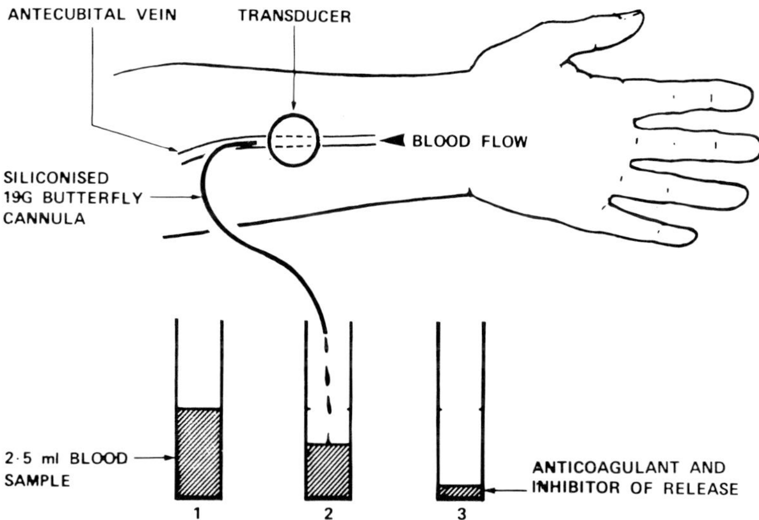

FIG. 5.14 A schematic description of the technique for collecting sequential venous blood samples from adult human volunteers after the blood had been irradiated *in vivo* (from Williams *et al.*, 1981).

collected varied between about 45 and 65 ng/ml. If all of the platelets in any one blood sample had undergone the release reaction before entering the anticoagulant mixture then the plasma levels of β-TG would be of the order of 5000 ng/ml. For the test series, a 0·75 MHz transducer was positioned over the antecubital vein about one inch upstream of the puncture site, but it was not energized while the first sample (a control) was being collected. The transducer was now energized at the highest intensity that the volunteer could tolerate (typically 0·34 to 0·5 W/cm^2, continuous wave) while the second blood sample was being collected. The ultrasound generator was then switched off and about two ml of blood allowed to run to waste before collecting the third sample (another control). There was no significant increase in the levels of β-TG in the blood samples collected while the ultrasound was being administered (Williams *et al.*, 1981).

It was not practicable to increase the ultrasonic intensity delivered to the human volunteers and so a similar technique was developed using the rabbit as an animal model. Rabbits do not have a β-TG-like protein that we could measure, but their platelets are loaded with histamine which is liberated into the plasma when they are disrupted or undergo the release reaction. Thus, cannulae were inserted into either the central ear artery or the abdominal portion of the descending aorta or the vena cava of anaesthetized rabbits and sequential blood samples collected into EDTA/theophylline as described above. A 0·75 MHz transducer was positioned about one inch upstream of the puncture site and only energized for the 25–35 sec that it took to collect the test samples. There was no detectable increase in the plasma histamine levels of the sonicated samples even at the highest intensities generated by our commercial therapeutic apparatus (Rank Sonacel Multiphon; 2·4 W/cm^2 spatial average, continuous wave). As a check that the experimental system was able to detect any platelet damage which might have occurred *in vivo*, the blood was driven to cavitate by means of a 25 kHz ultrasonic disintegrator (Rapidis 50, Ultrasonics Ltd). This apparatus was driven off resonance so that it could be pressed against an intact blood vessel without puncturing it and yet could still produce transient cavitation within freshly-drawn blood. In every case when the 25 kHz probe was applied to the outside of the intact blood vessel, the plasma histamine level was increased and this was also associated with the liberation of free haemoglobin indicating the simultaneous rupture of erythrocytes (Chater and Williams, 1982).

However, Wong and Watmough (1980) found evidence of intravascular haemolysis after they irradiated the beating hearts of anaesthetized mice for 5 min with 0·7–2 W/cm^2 (spatial average, continuous wave) 0·75 MHz ultrasound. This apparently contradictory result almost certainly reflects differences in the nucleation conditions necessary for the induction of cavitation events *in vivo*. Our results indicate that cavitation does not occur readily within orderly flowing or quiescent blood within the intact vascular system *in vivo*. Conversely, the turbulent rheological conditions which must exist within the beating heart and its associated vessels apparently increase both the number and the availability of prospective cavitation nuclei so that cavitational activity may now be generated *in vivo* at ultrasonic intensities which are ineffective in non-turbulent blood. This effect is known as hydrodynamic nucleation and can easily be demonstrated *in vitro*, e.g. when a dilute suspension of erythrocytes in isotonic saline is irradiated with a therapeutic intensity of MHz ultrasound, the rate of haemolysis is critically dependent upon the rate of stirring or agitation of the sample (Chapter 4), or even the frequency at which the sample container is being struck (Fig. 5.15 from Williams, 1982).

Thus, the interactions of ultrasound with platelets *in vivo* are best detected if one irradiates the beating heart or else deliberately introduces minute gas bubbles into the blood stream which may be driven to oscillate by the applied acoustic field. Lunan *et al.* (1979)

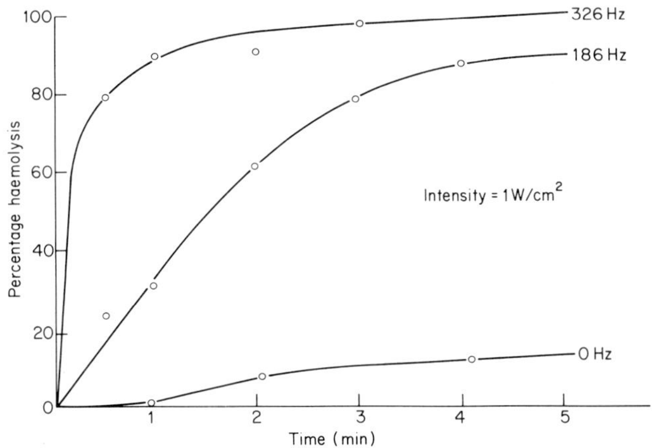

FIG. 5.15 This figure shows the effects of the enhancement of nucleation by a percussive technique. A hexagonal metal bar was driven to rotate while pressed against the outside of a glass sample chamber (having acoustically transparent windows) containing a suspension of human erythrocytes in isotonic saline. The rate of cell lysis increased with the frequency of the percussive impacts even though the acoustic exposure conditions were kept constant (1 W/cm^2, c.w., 0·75 MHz) (from Williams, 1982).

exposed anaesthetized mice to whole body ultrasonic irradiation for 5 min at 2 MHz at a spatial average intensity of 1 W/cm^2 in a 37°C water bath. They found that platelets from animals which had been sonicated 4 hours previously aggregated less efficiently than those from control animals. It is not known if this reflects a direct effect of the ultrasound on the platelets (i.e. the possible induction of a refractory phase following a "sub-threshold" stimulation) or a general systemic effect resulting from the release of substances such as prostaglandins into the bloodstream. Somewhat more controversial results were obtained by Sanada *et al.* (1977) who investigated changes in the morphology of human platelets while patient's hearts were being exposed to ultrasound from a commercially available diagnostic cardiac device (Aloka SSD-110; 2·25 MHz, time averaged intensity 0·5 mW/cm^2). They found that the number of circulating platelets having pseudopodia (one of the earliest detectable changes in "stimulated" platelets) increased with time of exposure and tended to reach a plateau value after about 15 min. The number of circulating platelets having pseudopodia gradually decreased with time after the ultrasound was switched off, but the blood's platelet count was found to have been "significantly" reduced (even though the authors do not provide any statistical evidence or give the raw data on platelet counts) (Sanada *et al.*, 1977).

The implication of the results obtained by Sanada *et al.* (1977) is that diagnostic ultrasound progressively "activates" the circulating platelets, and that these activated platelets leave or are taken out of the circulation. This would explain both the apparent plateau value in the number of cells having pseudopodia after about 15 min sonication (where the rates of production and removal could be equal) and the final reduction in platelet numbers. Their results are apparently supported by similar platelet shape changes produced *in vitro* using the same apparatus (Hattori *et al.*, 1976). However, the magnitude of the observed changes in platelet numbers seems rather large, especially since Lunan *et al.* (1979) found no significant differences between the numbers of circulating platelets in the blood of control mice and those exposed to 1 W/cm^2 of 2 MHz ultrasound. The most probable explanation of the *in vivo* results obtained by Sanada *et al.* (1977) is that they were getting progressive activation of platelets by the needle/catheter system (as observed by Williams *et al.*, 1981) which remained in position for the duration of each experiment (up to 40 or 60 min).

Some early articles report the occasional development of thrombophlebitis as a possible "side effect" of ultrasonic therapy (Kahlert, 1950). In this article 2 out of 250 patients given an unspecified treatment developed intravascular blood clots "a short time after the ultrasonic treatment". No further details were presented. Similarly, Van Went (1954) quotes "a remarkable case—of a physician, whose shoulder joint was irradiated on account of a very painful musculocutaneous neuritis. The patient collapsed after the second session (1 W/cm^2, 5 min), the fingers of the dorsum of the hand of the treated arm swelled markedly, and he suffered violent pains". These symptoms could have been produced by an arterial thrombus. Kadyrova *et al.* (1978) measured various indices relating to the blood coagulation system in 25 patients suffering from chronic diffuse glomerulonephritis subjected to 6–8 treatments each lasting 3–5 min of 0·2 W/cm^2 ultrasound at an unspecified frequency. They observed a small but statistically insignificant increase in the activity of the fibrinolytic system and similar decreases in the circulating level of fibrinogen and in the recalcification time after the patients had been treated with ultrasound.

It can therefore be concluded that diagnostic and low therapeutic intensities of ultrasound are not likely to result in activation of the blood coagulation system *in vivo* provided that areas of turbulence such as the heart, stenoses and fistulae are avoided and microscopic gas bubbles have not been deliberately introduced as contrast agents. Even if it is not possible to avoid these conditions which enhance the development of cavitational activity, the tolerance of the normal coagulation system to a limited amount of activation should protect the subject from large-scale thrombopathy. However, those individuals whose blood is in a hypercoagulable state (Stewart, 1978) will be more susceptible to a thrombogenic stimulus and it would therefore be prudent to exclude such patients from ultrasonic therapy.

5.4.2.3.7 Immunological System. Despite its obvious importance, there have been relatively few investigations concerning the effects of ultrasound on the various components of the immunological system. One notable exception is the article by Anderson and Barrett (1979). They irradiated mice for 5 min with pulsed 2 MHz ultrasound at a time-averaged spatial peak intensity of 8·9 mW/cm^2 over the region of the spleen and then evaluated the ability of the test and control animals to raise antibodies against an intraperitoneal injection of sheep red blood cells. The experimental techniques and terminology described in this article are conventional but may be somewhat confusing to those readers not familiar with immunology. Their results apparently show a small decrease or "desensitization" of the immune response of the irradiated animals. Complementary results have been reported by Koifman *et al.* (1980) who report a small decrease in the amount of antibody formed by mouse lymphocytes irradiated within the intact spleen *in vivo* (0·8 to 6 MHz; 0·05 to 2 W/cm^2) as well as a decrease in the number of plaque-forming cells. However, in both cases the magnitude of this decrease is small compared with the capacity of the immune system to respond to a potentially serious challenge. In addition, the inherent variability of some of these techniques is so large that the statistical evaluations, which are valid for the samples reported in these articles, may not reflect the results one would obtain from a larger study evaluating fewer variables.

Two independent laboratories have been unable to confirm these results. Child *et al.* (1981) and Stratmeyer and his collaborators at the U.S.A. Food and Drug Administration (personal communication) have attempted to duplicate the experimental conditions of Anderson and Barrett (1979) as closely as possible, even using the same strain of mice (Child *et al.*, 1981). In addition, both groups also used much higher time averaged intensities of ultrasound but were still unable to demonstrate a positive effect. However, it must be borne in mind that one (or even two) negative reports do not "cancel out" one positive finding by another author, even though it does suggest that the positive finding may have been due to a peculiarity of the experimental system or to an inadequate control population or even unusual statistical fluctuations.

On the other hand there is a rapidly growing body of evidence which apparently indicates that diagnostic intensities of ultrasound may be able to change the surface characteristics of cells in suspension *in vitro*. The early results of Taylor and Newman (1972) showing that moderately high intensities of MHz ultrasound reduced the electrophoretic mobility of Ehrlich ascites cells have been extended into the diagnostic intensity range by Hrazdira (1981). He found that the electrophoretic mobility of human erythrocytes was transiently changed after the cells had been exposed to pulsed 6 MHz ultrasound at diagnostic intensities and that the irradiated cells had a reduced tendency to aggregate. Similarly, Pinamonti *et al.* (1981) showed that the ultrasonic fields emitted by two commercial diagnostic probes (Kretz-technik 8 MHz ophthalmological; and Aerotech 10 MHz) driven by the same commercial generator/receiver (Panametrics 5052 PR having a PRF of 2·5 kHz) reduced the extent of red cell aggregation produced in response to a low concentration of an incompatible antiserum. They also showed that there was a time-dependent increase in the number of antigens removed from the surface of the erythrocytes during sonication but that these free antigens apparently re-attached themselves to the intact erythrocytes unless the cells were removed from the sonication medium immediately after the ultrasound had been switched off (Pinamonti *et al.*, 1981; and unpublished observations).

The humoral part of the immunological detection mechanism recognizes antigens on the surface of cells and foreign bodies and so anything which changes the surface characteristics of cells is likely to change the efficiency with which these cells are recognized. Two other articles which may be relevant in this context are the report by Liebeskind *et al.* (1981) who used scanning electron microscopy to detect changes in the surface pattern of microvilli on cultured cells, and the report by Siegel *et al.* (1979) that diagnostic intensities of MHz ultrasound reduced the efficiency of attachment of three lines of cultured human cells to a flat plastic surface.

Ford *et al.* (1970) developed a sensitive technique for quantifying an animal's immunological competence by measuring the results of the competition between an animal's lymphocytes and foreign lymphocytes *in vivo*. A known number of foreign lymphocytes are injected into the hind foot pads of mice from which they migrate through the lymphatic system to the popliteal lymph nodes. Here, the host lymphocytes are attracted and proliferate and usually destroy the invaders. The gain in weight of this popliteal lymph node is measured (the node weight may increase by a factor of up to seventyfold). It was found that neither irradiation of the foreign lymphocytes *in vitro* before injection nor irradiation of one of an animal's politeal lymph nodes *in vivo* after both hind limbs had been injected with the same number of foreign cells caused a measurable change in node weights indicating that the ultrasound (3 MHz, 0·8 W/cm^2, c.w.) did not appear to affect either the aggressiveness of the invading lymphocytes or the ability of the host lymphocytes to recognize the invaders and to proliferate (Williams and Ford, to be published).

Another function associated with the immunological system is the mechanism whereby debris and solid particles are cleared from the blood by the cells of the reticuloendothelial system which are found mainly within the liver, spleen, lungs, bone marrow and lymphoid tissue. Colloidal sulphur particles radiolabelled with Technetium 99^{m} which have been injected into an anaesthetized rat via a tail vein are cleared from the blood at an exponential rate having a characteristic half time (Saad and Williams, 1982). Irradiation of the duodenal region of the rats with 0·75 MHz ultrasound for 5 min at an average intensity greater than about 0·8 W/cm^2 increased the half life of removal (i.e. decreased the rate at which the colloid was removed from the blood). This study is not completed but preliminary observations suggest that the ultrasound may be causing the release of materials into the blood stream (possibly the contents of damaged cells) which are competing with the colloidal particles for a finite number of removal sites. Subsequent investigations have implicated blood-borne lymphocytes as the primary target for the ultrasound since isolated lymphocytes irradiated *in vitro* cause a reduction in the half life of colloid removal when re-injected *in vivo*, whereas non-sonicated lymphocytes do not (Saad and Williams, 1982; and article in preparation).

5.4.3.2.8 The Hormonal System. This is another complex field where relatively few quantitative observations have been performed. Most articles on this topic appear in eastern European journals which are frequently difficult to obtain in the West and to have translated into English. In general, the articles describe changes in the blood level of some hormones in patients who are receiving ultrasonic therapy in an attempt to reverse an abnormal condition. For example Muggeo *et al.* (1975) report on the effects of ultrasonic treatment of the pituitary gland on the plasma levels of growth hormone in patients having acromegaly or diabetic retinopathy. Similarly, Stereva and Beleva-Staikova (1976) measured the effects of ultrasonic energy on the level of thyronins within the thyroid gland and Grandesso *et al.* (1967) report the levels of plasma adrenocorticotrophin activity in patients treated with ultrasonic therapy of the pituitary gland by the Arslan method.

On the other hand Cervenka *et al.* (1973) attempted to verify the safety of repeated exposures to diagnostic ultrasound by measuring the neonatal hormonal responses of baby girls who had been irradiated with ultrasound while still *in utero*. The results of this study showed that there was no measurable change which was not too surprising because hormone levels are so important to an individual that they are maintained at an optimum level by complex feedback mechanisms and usually only rise or fall to abnormal levels when something has gone seriously wrong either with their source of production or with part of the homeostatic mechanism. If either of these two systems had been significantly affected by the ultrasonic field the resulting hormonal imbalance would have resulted in easily detectable differences in growth rates or behavioural characteristics in the developing child.

5.4.2.3.9 The Reproductive System. A major use of ultrasound in diagnosis is to replace X-rays in the investigation of the growing foetus. It is therefore of paramount importance that ultrasound does not produce an increased incidence of genetic abnormalities or an increased incidence of growth or developmental (i.e. teratogenic) abnormalities. Consequently, the reproductive system has been investigated extensively with most attention being paid to the effects of ultrasound on the developing foetus.

Several authors have irradiated the testes of adult animals. O'Brien *et al.* (1977) observed histological damage within testes exposed to 1 MHz ultrasound for 30 sec at a spatial peak intensity of 25 W/cm^2. This damage was patchy with damaged areas immediately adjacent to apparently normal areas which suggests that some form of cavitational activity may have occurred. Fry *et al.* (1977) demonstrated that short intense pulses of ultrasound caused a greater temperature increase within mouse testes than the same amount of acoustic energy given in the form of a continuous wave (possibly because of the generation of non-linear effects). Under these pulsed conditions high intensities of ultrasound result in the death of the irradiated animals with LD_{50} values of the order of 200 W/cm^2 at 1·3 MHz (n.b. LD_{50} is the "dose" required to kill 50% of the experimental animals). Fahim *et al.* (1975) advocate the use of ultrasound as a method of causing temporary sterility in the male even though this procedure could be potentially hazardous in that some "damaged" spermatocytes may survive and result in the formation of abnormal infants. Fry *et al.* (1977) demonstrated that there was an increased incidence of foetal abnormalities amongst the offspring of mice whose testes had been irradiated with high ultrasonic intensities prior to mating even though the nature of the observed abnormalities were similar to those found in the controls.

Other authors including Lyon and Simpson (1974) exposed mouse gonads to about 1 W/cm^2 of continuous or pulsed MHz ultrasound and looked for chromosome fragmentaion or translocation in spermatocytes or the induction of dominant lethal mutations in oocytes. However, the relative crudity of the endpoints and the small numbers of animals used in the study meant that only large mutagenic effects would have been detected and so it is not surprising that the results of this study were negative. On the other hand, Bailey *et al.* (1981) showed that 30 sec exposure to 2–5 W/cm^2 of 1 MHz ultrasound *in vivo* resulted in sporadic morphological alterations in the testes of adult mice. Other studies at lower ultrasonic intensities have been uniformly negative. For example, Smyth (1966) exposed either the testes or ovaries of mice to 10 mW/cm^2 of pulsed 2·25 MHz ultrasound for 10 min each day for the 5 days preceding mating and for 10 days during the mating period. No congenital abnormalities were observed in 348 offspring of the irradiated mice or in 391 second generation offspring. Similar negative results at diagnostic exposure levels were obtained by Mannor *et al.* (1971).

The female ovary plays an important role in the maintenance of the state of pregnancy after the fertilized ovum has implanted itself in the uterine wall. Garrison *et al.* (1973) exposed both ovaries of 8-day pregnant rats to various pulsed ultrasonic exposures and found that there was no effect on resorption rates showing that the functioning of the corpus luteum was not affected by those ultrasonic exposures (10 W/cm^2 peak intensity, 10 ms pulse duration, 0·1 duty cycle; and 100 W/cm^2 peak intensity, 0·6 ms pulse duration, 0·006 duty cycle).

These results apparently indicate that the gonads of the reproductive system are relatively resistant to the effects of low intensities of ultrasonic irradiation. At higher intensities the gametes they contain are rendered non-viable instead of being genetically "modified" as occurs with ionizing radiation. However, the fertilized ovum and the developing embryo are exceptionally susceptible to thermal and mechanical damage and consequently operators of diagnostic equipment should aim to subject the pregnant uterus to the least possible ultrasonic exposure compatible with obtaining meaningful clinical information.

REFERENCES

Abdulla, U., Campbell, S., Dewhurst, O. J., Talbert, D., Lucas, M. and Mullarkey, M. (1971). Effect of diagnostic ultrasound on maternal and fetal chromosomes. *Lancet* **2**, 829–831.

Abramovich, A. (1970). Effect of ultrasound on the tibia of the young rat. *J. Dent. Res.* **49**, 1182.

Adler, J. and Hrazdira, I. (1980). The evaluation of ultrasonic action by means of cell electrophoresis. *Scripta Med.* **53**, 329–332.

Anderson, D. W. and Barrett, J. T. (1979). Ultrasound: a new immunosuppressant. *Clin. Immunol. Pathol.* **14**, 18–29.

Anderson, T. P., Wakin, K. G., Herrick, J. F., Bennett, W. A. and Krusen, F. H. (1951). An experimental study of the effects of ultrasonic energy on the lower part of the spinal cord and peripheral nerves. *Arch. Phys. Med.* **32**, 71–83.

Armour, E. P. and Corry, P. M. (1982). Cytotoxic effects of ultrasound *in vitro*. Dependence on gas content, frequency, radical scavengers and attachment. *Rad. Res.* **89**, 369–380.

Bailey, K. I., O'Brien, W. D. Jr and Dunn, F. (1981). Ultrasonically induced *in vivo* morphological damage in mouse testicular tissue. *Arch. Andrology* **6**, 301–306.

Barnet, S. B. (1982). The mutagenic potential of pulsed ultrasound. Proc. 5th World Congress on Ultrasound in Medicine and Biology. *Ultrasound Med. Biol.* **8** (Suppl. 1), 10.

Barnett, S. B. and Kossoff, G. (1977). Negative effect of long duration pulsed ultrasonic irradiation on the mitotic activity in regenerating rat liver. *In* "Ultrasound in Medicine" (Eds D. N. White and R. E. Brown), pp. 2033–2044, Vol. 3. Plenum Press, New York.

Barnett, S. B., Kossoff, G. and Clark, G. M. (1973). Histological changes in the inner ear of sheep following a round window ultrasonic irradiation. *J. Oto-Laryngol. Soc. Aust.* **3**, 508–512.

Barrass, N. ter Haar, G. R. and Casey, G. (1982). The effect of ultrasound and hyperthermia on sister chromatid exchange and division kinetics of BHK 21C13/A3 cells. *Brit. J. Cancer* **45**, 187–191.

Barth, G. and Wachsmann, F. (1949). Biological effects of ultrasonic therapy (article in German), Kongress Bericht, Ter Erlanger Ultraschall-Tagung, 162–205.

Bernstine, R. L. and Dickson, L. G. (1972). Study of effects of ultrasound as determined by electron microscopy. *In* "Interaction of Ultrasound and Biological Tissues (Eds J. M. Reid and M. R. Sikov), pp. 77–81. DHEW Publication (FDA) 73–8008.

Bierman, W. (1954). Ultrasound in the treatment of scars. *Arch. Phys. Med.* **35**, 209–213.

Bleaney, B. I., Blackbourn, P. and Kirkley, J. (1972). Resistance of CHLF hamster cells to ultrasonic irradiation of 1·5 MHz frequency. *Brit. J. Radiol.* **45**, 354–357.

Bobrow, M., Blackwell, N., Unrau, A. E. and Bleaney, B. (1971). Absence of any observed effect of ultrasonic irradiation on human chromosomes. *J. Obst. Gyn. Brit. Commonwlth* **78**, 730–736.

Borovyagin, V. L. and El'piner, I. Ye. (1964). Effect of ultrasonic waves on submicroscopic structures of muscle tissue. *Biofizika* **9**, 312–314.

Borrelli, M. J., Bailey, K. I. and Dunn, F. (1981). Early ultrasonic effects upon mammalian CNS structures (chemical synopses). *J. Acoust. Soc. Amer.* **69**, 1514–1516.

Boyd, E., Abdulla, U., Donald, I., Fleming, J. E. E., Hall, A. J. and Ferguson-Smith, M. A. (1971). Chromosome breakage and ultrasound. *Brit. Med. J.* **2**, 501–502.

Braeman, J., Coakley, W. T. and Gould, R. K. (1974). Human lymphocyte chromosomes and ultrasonic cavitation. *Brit. J. Radiol.* **47**, 158–161.

Brock, R. D., Peacock, W. J., Geard, C. R., Kossoff, G. and Robinson, D. E. (1973). Ultrasound and chromosome aberrations. *Med. J. Aust.* **2**, 533–536.

Brug, E., Braunsteiner, E. and von Gemmern, C. (1976). Ultrasonic welding of bones. Preliminary results. *Chirung* **47**, 555–558.

Bucher, N. L. R. (1963). Regeneration of mammalian liver. *Int. Rev. Cytol.* **15**, 245–300.

Buchtala, V. (1948). Ultraschalltherapie-Anwendungsbereich und Kontraindikationen. Ärztl. Wschr. **3**, 321.

Buckton, K. E. and Baker, N. V. (1972). An investigation into possible chromosome damaging effects of ultrasound on human blood cells. *Brit. J. Radiol.* **45**, 340–342.

Bugnon, C., Cottin, Y., Kraehenbuhl, J. and Weill, F. (1972). Aberrations chromosomiques provoquées par des ultra-sons diagnostiqués sur des lymphocytes humains en culture. *J. Radiol. Electrol. Med. Nucl.* **53**, 750–755.

Bundy, M. L., Lerner, J., Messier, D. L. and Rooney, J. A. (1978). Effects of ultrasound on transport in avian erythrocytes. *Ultrasound Med. Biol.* **4**, 259–262.

Campbell, P. N. and Kernot, B. A. (1962). The incorporation of [^{14}C]Leucine into serum albumin by the isolated microsome fraction from rat liver. *Biochem. J.* **82**, 262–266.

Carney, S. A., Lawrence, J. C. and Ricketts, C. R. (1972). Some effects of ultrasound on guinea-pig ear skin. *Brit. J. Ind. Med.* **29**, 214–219.

Cervenka, J., Jezek, J. and Suk, V. (1973). Neonatal hormonal responses of baby girls from pregnancies followed up by ultrasound. Investigations to verify the safety of repeated ultrasonic examinations. *Cesk. Gynekol.* **38**, 725–727.

Chapman, I. V. (1974). The effect of ultrasound on the potassium content of rat thymocytes *in vitro*. *Brit. J. Radiol.* **47**, 411–415.

Chater, B. V. and Williams, A. R. (1977). Platelet aggregation induced *in vitro* by therapeutic ultrasound. *Thrombosis and Haemostatis* **3**, 640–651.

Chater, B. V. and Williams, A. R. (1982). Absence of platelet damage following the exposure of non-turbulent blood to therapeutic ultrasound. *Ultrasound Med. Biol.* **8**, 85–87.

Child, S. Z., Hare, J. D., Carstensen, E. L., Vives, B., Davis, J., Adler, A. and Davis, H. T. (1981). Test for the effects of diagnostic levels of ultrasound on the immune response of mice. *Clin. Immunol. Immunopathol.* **18**, 299–302.

Ciaravino, V. and Miller, M. W. (1978). The effect of 1 MHz ultrasound on synchronized mammalian cell proliferation. *Rad. Res.* **74**, 583–584.

Clarke, P. R. and Hill, C. R. (1969). Biological action of ultrasound in relation to the cell cycle. *Exp. Cell Res.* **58**, 443–444.

Coakley, W. T. and Dunn, F. (1972). Interaction of megahertz ultrasound and biological polymers. *In* "Interaction of Ultrasound and Biological Tissues—Workshop Proceedings" (Eds J. R. Reid and M. R. Sikov), pp. 43–45. DHEW Publication (FDA) 73–8008 BRH/DBE 73–1.

Coakley, W. T. and Dunn, F. (1975). Ultrasound and DNA (Letter). *Lancet* **2**, 1037.

Coakley, W. T., Hughes, D. E., Slade, J. S. and Laurence, K. M. (1971). Chromosome aberrations after exposure to ultrasound. *Brit. Med. J.* **1**, 109–110.

Coakley, W. T., Slade, J. S., Braeman, J. M. and Moore, J. L. (1972). Examination of lymphocytes for chromosome aberrations after ultrasonic irradiation. *Brit. J. Radiol.* **45**, 328–332.

Coble, A. J., and Dunn, F. (1976). Ultrasonic production of reversible changes in the electrical parameters of isolated frog skin. *J. Acoust. Soc. Amer.* **60**, 225–229.

Combes, R. D. (1975). Absence of mutation following ultrasonic treatment of *Bacillus subtilis* cells and transforming deoxyribonucleic acid. *Brit. J. Radiol.* **48**, 306–311.

Crowell, J. A., Kusserow, B. K. and Nyborg, W. L. (1977). Functional changes in white blood cells after microsonation. *Ultrasound Med. Biol.* **3**, 185–190.

Crum, L. A. and Eller, A. (1970). Motion of bubbles in a stationary sound field. *J. Acoust. Soc. Amer.* **48**, 181–189.

David, H., Weaver, J. B. and Pearson, J. F. (1975). Doppler ultrasound and fetal activity. *Brit. Med. J.* **2**, 62–64.

Dimitrieva, N. P. (1964). Electron microscopic study of changes in the structure of mitochondria in tumour cells after exposure to ultrasound. *Biofizika* **9**, 571–579.

Drastichova, V., Samohyl, J. and Slavetinska, A. (1973). Strengthening of sutured skin wounds with ultrasound in experiments on animals. *Acta Chir. Plast. (Praha)*. **15**, 114–119.

Duarte, L. R. (1976). Ultrasonic stimulation of fracture healing. *In* "Digest of 11th International Conference, Med. Bio. Eng. and 6th Can. Med. Biol. Eng. Conference", pp. 248–249. Ottawa, Canada.

Dunn, F. (1971). *In* "Proceedings 1st World Congress on Ultrasonic Diagnostics in Medicine" (Eds J. Bock and K. Ossoinig), pp. 451–455. Verlag der Weiner Medizinischen Akademie, Vienna.

Dunn, F. and Fry, F. J. (1971). Ultrasonic threshold dosages for the mammalian central nervous system. *IEEE Trans. Bio-Med. Eng.* July, 253–256.

Dunn, F. and MacLeod, R. M. (1968). Effects of intense non-cavitating ultrasound on selected enzymes. *J. Acoust. Soc. Amer.* **44**, 932–940.

Dvorak, M. and Hrazdira, I. (1966). Changes in the ultrastructure of bone marrow cells in rats following exposure to ultrasound. *Z. Mikr. Anat. Forsch.* **75**, 751–760.

Dyer, H. J. (1972). Structural effects of ultrasound on the cell. *In* "Interaction of Ultrasound and Biological Tissues" (Eds J. M. Reid and M. R. Sikov), pp. 73–75. DHEW Publication (FDA) 73–8008.

Dyson, M. and Brookes, M. (1982). Stimulation of bone repair by ultrasound. *In* Proc. Fifth World Congress of Ultrasound in Medicine and Biology. *Ultrasound Med. Biol.* **8** (Suppl. 1), 50.

Dyson, M., Pond, J. B., Joseph, J. and Warwick, R. (1968). The stimulation of tissue regeneration by means of ultrasound. *J. Clin. Sci.* **35**, 273–285.

Dyson, M., Pond, J. B., Woodward, B. and Broadbent, J. (1974). The production of blood cell stasis and endothelial damage in the blood vessels of chick embryos treated with ultrasound in a stationary wave field. *Ultrasound Med. Biol.* **1**, 133–148.

Dyson, M., Franks, C. and Suckling, J. (1976). Stimulation of healing of varicose ulcers by ultrasound. *Ultrasonics* **15**, 232–236.

Edmonds, P. D. (1980). Effects of ultrasound on biological structures. *In* "Ultrashalldiagnostik in der Medizin" (Eds M. Hinselmann, M. Anliker and R. Meudt), pp. 2–18. Georg Thieme Verlag, Stuttgart, New York.

El'piner, I. Ye., Faikin, I. and Basurmanova, O. (1966). Intracellular microcurrents caused by ultrasound waves. *Fed. Proc. (Trans. suppl.)* **25**, 716–720.

Engel, D. (1971). The influence of ultrasonic waves on the permeability of the synovial membrane. *Z. Ges. Exp. Med.* **155**, 1–12.

Esmat, N. (1975). Investigations of the effects of different doses of ultrasonic waves on the human nerve conduction velocity. *J. Egypt. Med. Assoc.* **58**, 395–402.

Fischman, H. (1973). Ultrasound and marrow-cell chromosomes. *Lancet* **2**, 920–921.

Ford, W. L., Burr, W. and Simonsen, M. (1970). A lymph node weight assay for the graft-versus-host activity of rat lymphoid cells. *Transplantation* **10**, 258–266.

Frenkel, J. (1944). Orientation and rupture of linear macromolecules in dilute solution under the influence of viscous flow. *Acta Physicochimica U.R.S.S.* **19**, 51–59.

Fridd, C. W., Linke, C. A., Barbaric, Z., Elbadawi, A. and Carstensen, E. L. (1977). Unfocused ultrasound for localized tissue destruction in rabbit kidneys. *Investigative Urol.* **15**, 19–22.

Fry, F. J. (1979). Biological effects of ultrasound—a review. *Proc. I.E.E.E.* **67**, 604–619.

Fry, F. J., Ades, H. W. and Fry, W. J. (1958). Production of reversible changes in the central nervous system by ultrasound. *Science* **127**, 83–84.

Fry, F. J., Kossoff, G., Eggleton, R. C. and Dunn, F. (1970). Threshold ultrasonic dosages for structural changes in the mammalian brain. *J. Acoust. Soc. Amer.* **48**, 1413–1417.

Fry, F. J., Johnson, L. K., Erdmann, W. A. and Baird, A. I. (1977). Ultrasonic toxicity in the mouse. *In* "Symposium on Biological Effects and Characterizations of Ultrasound Sources" (Eds D-W. G. Hazzard and M. L. Litz), pp. 153–161. HEW Publication (FDA) 78–8048.

Fung, H., Cheung, K., Lyons, E. A. and Kay, N. E. (1978). The effect of low dose ultrasound on human peripheral lymphocyte function *in vitro*. *In* "Ultrasound in Medicine" (Eds D. N. White and E. A. Lyons), pp. 583–586. Vol. 4. Plenum Press, New York.

Galitsky, A. B. and Levina, S. I. (1964). Vascular origin of trophic ulcers and application of ultrasound and preoperative treatment to plastic surgery. *Acta Chir. Plast. (Praha)* **6**, 271–278.

Galperin-Lemaitre, H., Gustot, P. and Levi, S. (1973). Ultrasound and marrow cell chromosomes. *Lancet* **2**, 505–506.

Galperin-Lemaitre, H., Kirsch-Volders, M. and Levi, S. (1975). Ultrasound and Mammalian DNA. *Lancet* **2**, 662.

Garrison, B. M., Bo, W. J., Kreuger, W. A., Kremkau, F. W. and McKinney, W. M. (1973). The influence of ovarian sonication on fetal development in the rat. *J. Clin. Ultrasound* **1**, 316–319.

Gavrilov, L. R. (1980). Using of focussed ultrasound to introduce the information through different sensor channels. Paper C7 presented at the Ultrasound Interaction in Biological and Medical Symposium, Reinhardsbrunn, G. D. R., Nov. 10–14.

Gavrilov, L. R., Tsirulnikov, E. M. and Shchekanov, E. E. (1975). Responses of the auditory centers of the frog midbrain to labyrinth stimulation by focussed ultrasound. *Fiziol. Zh. S.S.S.R.* **61**, 213–221.

Glick, D., Adamovics, A., Edmonds, P. D. and Taenzer, J. C. (1979). Search for biochemical effects in cells and tissues of ultrasonic irradiation of mice and of the *in vitro* irradiation of mouse peritoneal and human amniotic cells. *Ultrasound Med. Biol.* **5**, 23–33.

Goldblat, V. I. (1969). Processes of bone tissue regeneration under the effect of ultrasound. *Ortopedia Travmotologii Protezirovanie* **30**, 57–61. (Transl. in *Phys. Med.* **34**, 362–369.)

Goldman, O. E. and Lepeschkin, W. W. (1952). Injury to living cells in standing sound waves. *J. Cell Comp. Physiol.* **40**, 255–268.

Gornia, F. I., Shepeleva, I. S., Semenova, V. A. and Ianovskaia, E. M. (1974). Certain problems of reparative regeneration of open fractures after ultrasonic osteosynthesis (Russ). *Ortop. Traumatol. Protez.* **10**, 24–27.

Graham, E., Hedges, M., Leeman, S. and Vaughan, P. (1980). Cavitational bioeffects at 1·5 MHz. *Ultrasonics* **18**, 224–228.

Grandesso, R., Ambrosio, G., Babighian, G., Bianchini, R., Binda, F., Crepaldi, G., Muggeo, M. and Spandri, P. (1967). Plasmatic adrenocorticotrophin activity in patients treated with ultrasonic therapy of the pituitary gland according to the Arslan method. *Folia Endocrinol (Roma)* **20**, 453–463.

Greguss, P. (1975). Ultrasonic irradiation and pupillary response. *Ultrasonics* **13**, 63–65.

Greguss, P. and Bertenyi, A. (1976). A critical analysis of ultrasonic therapy. *Ultrasonics* **14**, 81–82.

Hara, K. (1980). Effects of ultrasonic irradiation on chromosomes, cell division and developing embryos. *Acta Obst. Gynaecol. Japan* **32**, 61–68.

Hattori, A., Sanada, M., Ohno, M., Watanabe, T., Shu, T., Kasahara, T., Tamura, K. and Matsuoka, M. (1976). *In vitro* effect of ultrasound upon (human) blood platelets. *Nikon Choompa Igakukai, Koen Rombunshu* **Nov**. 117–118.

Hawley, S. A. and Dunn, F. (1964). Some effects of megacycle ultrasound upon rotifers. *Naturwissenschaften* **51**, 555–558.

Herrick, J. F. (1953). Temperatures produced in tissues by ultrasound: Experimental study using various techniques. *J. Acoust. Soc. Amer.* **25**, 12–16.

Hill, C. R., Clarke, P. R., Crowe, M. R. and Hammick, J. W. (1970). Biophysical effects of cavitation in a 1 MHz ultrasonic beam. *In* "Ultrasonics for Industry Conference Papers", pp. 26–30. Iliffe Press, London.

Hill, C. R., Joshi, G. P. and Revell, S. H. (1972). A search for chromosome damage following exposure of Chinese Hamster cells to high intensity, pulsed ultrasound. *Brit. J. Radiol.* **45**, 333–334.

Hrazdira, I. (1979). Ultrasonically-induced stimulation of haemopoietic tissue. Proc. IV Ultrasound. *Bio. Med. Symp.* **1**, 112–116.

Hrazdira, I. (1981). Cellular effects of ultrasound. Proc. 4th European Conference Ultrasound in Medicine and Biology, Dubrovnik, Yugoslavia, May 18–22.

Hrazdira, I., Raková, A., Vacek, A. and Horký, D. (1974). Effect of ultrasound on the formation of haemopoietic tissue colonies in spleen. *Folia Biologica (Praha)* **20**, 430–432.

Hu, J. H. and Ulrich, W. D. (1976). Effects of low-intensity ultrasound on the central nervous system of primates. *Aviat. Space Environ. Med.* **47**, 640–643.

Hu, J. H., Taylor, J. D., Press, H. C. and White, J. E. (1978). Ultrasonic effects on mammalian interstitial muscle membrane. *Aviat. Space Environ. Med.* **49**, 607–609.

Hughes, D. E. (1972). The interaction of ultrasound with cells. *In* "Interaction of Ultrasound and Biological Tissues" (Eds M. Reid and M. R. Sikov), pp. 61–63. HEW Publications (FDA) 73–8008.

Hughes, D. E., Chou, J. T. Y., Warwick, R. and Pond, J. (1963). The effect of focussed ultrasound on the permeability of frog muscle. *Biochim. Biophys. Acta* **75**, 137–139.

Hussey, M. (1975). "Diagnostic Ultrasound, an Introduction to the Interaction Between Ultrasound and Biological Tissues". Blackie, Glasgow.

Hustler, J. E., Zarod, A. P. and Williams, A. R. (1978). Ultrasonic modification of experimental bruising in the guinea-pig pinna. *Ultrasonics* **Sept**. 223–228.

Ikeuchi, T., Sasaki, M., Oshimura, M., Azumi, J. and Tsuji, K. (1973). Ultrasound and embryonic chromosomes. *Brit. Med. J.* **1**, 112.

Istomina, O. and Ostrovskij, E. (1936). The effects of ultrasonic vibration on plant development. *Dokl. Akad. Nauk. S.S.S.R.* **11**, 155–160.

James, J. A. and Halliwell, M. (1970). Thermal effects of ultrasound on the temporal bone. *Med. Biol. Eng.* **8**, 477–481.

James, J. A., Dalton, G. A., Freundlich, H. F., Bullen, M. A., Wells, P. N. T., Hughes, D. E. and Chou, J. T. Y. (1963). Histological, thermal and biochemical effects of ultrasound on the labyrinth and temporal bone. *Acta Oto-Laryngol.* **57**, 306–312.

Johnston, R. L. and Dunn, F. (1976). Ultrasonic absorbed dose, dose rate and produced lesion volume. *Ultrasonics* **July**, 153–155.

Joshi, G. P., Hill, C. R. and Forrester, J. A. (1973). Mode of action of ultrasound in the surface charge of mammalian cells *in vitro*. *Ultrasound Med. Biol.* **1**, 45–48.

Kadyrova, R. Kh., Sumanova, Sh. B., Shairo, A. K. and Shigirt, G. P. (1978). Effect of ultrasonic irradiation on the blood circulation system, the permeability of the vascular walls and the state of the cardiovascular system in patients with chronic diffuse glomerulonephritis. *Zoravookhr. Kaz.* **11**, 34–36.

Kahlert, K. von (1950). Successes, failures and dangers of the clinical use of ultrasound. *Deutsche Med. Wchnschr.* **75**, 1–2.

Kashkooli, L. A., Rooney, J. A. and Roxby, R. (1980). Effects of ultrasound on catalase and malate dehydrogenase. *J. Acoust. Soc. Amer.* **67**, 1798–1801.

Kaufman, G. E. and Miller, M. W. (1978). Growth retardation in chinese hamster V-79 cells exposed to 1 MHz ultrasound. *Ultrasound Med. Biol.* **4**, 139–144.

Kaufman, G. E., Miller, M. W., Griffiths, T. D., Ciaravino, V. and Carstensen, E. L. (1977). Lysis and viability of cultured mammalian cells exposed to 1 MHz ultrasound. *Ultrasound Med. Biol.* **3**, 21–25.

Kaufmann, J. S. and Kremkau, F. W. (1978). Influence of ultrasound on mouse leukemia cell DNA synthesis, membrane integrity, and uptake of anticancer drugs *in vitro. In* "Ultrasound in Medicine" (Eds D. White and E. A. Lyons), pp. 586–590, Vol. 4. Plenum Press, New York.

Keller, M. and Tanka, D. (1977). Effects of ultrasonic treatment on the organs of experimental animals. III. Enzyme-histochemical examination of the prolonged effect. *Acta Morph. Acad. Sci. Hung.* **25**, 107–119.

Knokh, G. and Knaut, K. (1975). Ultrasonic therapy of bone fractures (Russ). *Sov. Med.* **6**, 115–118.

Koifman, M. M., Vacilieva, T. N., Maslov, K. I., Maev, R. G. and Levin, V. M. (1980). Antibody secretion changes induced by ultrasound in lymphoid cells. Paper C16 at the Ultrasound Interaction in Biol. and Med. Symposium, Reinhardsbrunn, East Germany, Nov. 10–14.

Kremkau, F. W., Kaufmann, J. S., Burch, P. G. *et al.* (1977). Ultrasonic enhancement of anticancer agents. *In* "Ultrasound in Medicine" (Eds D. N. White and R. E. Brown), p. 2109, Vol. 36. Plenum Press, New York.

Kremkau, F. W. and Witcofski, R. L. (1974). Mitotic reduction in rat liver exposed to ultrasound. *J. Clin. Ultrasound* **2**, 123–126.

Krenztlin, C. and Cattaneo, V. (1968). Ultrasound in dentistry. First histochemical studies on dental pulp. *Arch. Stomatol (Napoli)* **9**, 381–392.

Kunze-Mühl, E. and Golob, E. (1972). Chromosomenanalysen nach Ultraschalleinwirkung. *Humangenetik* **14**, 237–246.

Latt, S. A. and Schreck, R. R. (1980). Sister chromtid exchange analysis. *Amer. J. Hum. Genet.* **32**, 297.

Lakshminarayanaiah, N. and Siddiqi, F. (1972). Studies with composite membranes. III. Measurements of water permeability. *Biophys. J.* **12**, 540.

Lehmann, J. F. and Biegler, R. (1954). Changes of potentials and temperature gradients in membranes caused by ultrasound. *Arch. Phys. Med.* **35**, 287–295.

Lehmann, J. F. and Krusen, F. H. (1954). Effect of pulsed and continuous application of ultrasound on transport of ions through biologic membranes. *Arch. Phys. Med. Rehabil.* **35**, 20–23.

Lele, P. P. (1979). Safety and potential hazards in the current applications of ultrasound in obstetrics and gynaecology. *Ultrasound Med. Biol.* **5**, 307–320.

Li, G., Hahn, G. M. and Tolmach, L. (1977). Cellular inactivation by ultrasound. *Nature* **267**, 163–165.

Liebeskind, D., Bases, R., Elequin, F., Neubort, S., Leifer, R., Goldberg, R. and Koenigsberg, M. (1979a). Diagnostic ultrasound: effects on the DNA and growth patterns of animal cells. *Radiology* **131**, 177–184.

Liebeskind, D., Bases, R., Mendez, F., Elequin, F. and Koenigsberg, M. (1979b). Sister chromatid exchanges in lymphocytes after exposure to diagnostic ultrasound. *Science* **205**, 1273–1275.

Liebeskind, D., Bases, R., Koenigsberg, M., Koss, L. and Raventos, C. (1981). Morphological changes in the surface characteristics of cultured cells after exposure to diagnostic ultrasound. *Radiology* **138**, 419–423.

Liebeskind, D., Padawer, J., Wolley, R. and Bases, R. (1982). Diagnostic ultrasound: time-lapse and transmission electron microscopic studies of cells insonated *in vitro. Brit. J. Cancer* **45**, 176–186.

Lizzi, F. L., Packer, A. J. and Coleman, D. J. (1978a). Experimental cataract production by high frequency ultrasound. *Ann. Ophthalmol.* **10**, 934–942.

Lizzi, F. L., Coleman, D. J., Driller, J., Franzen, L. A. and Jakobiec, F. A. (1978b). Experimental, ultrasonically induced lesions in the retina, choroid and sclera. *Invest. Ophthalmol. Vis. Sci.* **17**, 350–360.

Loch, E. G., Fischer, A. B. and Kuwert, E. (1971). Effect of diagnostic and therapeutic intensities of ultrasonics on normal and malignant human cells *in vitro*. *Amer. J. Obstet. Gynecol.* **110**, 457–460.

Lota, M. J. and Darling, R. C. (1955). Changes in permeability of the red blood cell membrane in a homogeneous ultrasonic field. *Arch. Phys. Med.* **36**, 282–287.

Love, L. A. and Kremkau, F. W. (1980). Intracellular temperature distribution produced by ultrasound. *J. Acoust. Soc. Amer.* **67**, 1045–1050.

Ludlam, C. A., Moore, S., Bolton, A. E., Pepper, D. S. and Cash, J. D. (1975). The release of a human platelet-specific protein measured by a radioimmunoassay. *Thrombosis Res.* **6**, 543–548.

Lunan, K. D., Wen, A. C., Barfod, E. T., Edmonds, P. D. and Pratt, D. E. (1979). Decreased aggregation of mouse platelets after *in vivo* exposure to ultrasound. *Thrombosis and Haemostasis* **40**, 568–570.

Lyon, M. F. and Simpson, G. M. (1974). An investigation into the possible genetic hazards of ultrasound. *Brit. J. Radiol.* **47**, 712–722.

Macintosh, I. J. C. (1971). Chromosome breakage and ultrasound. *Brit. Med. J.* **5**, 703.

Macintosh, I. J. C. and Davey, D. A. (1970). Chromosome aberrations induced by an ultrasonic foetal pulse detector. *Brit. Med. J.* **4**, 92–93.

Macintosh, I. J. C. and Davey, D. A. (1972). Relationship between intensity of ultrasound and induction of chromosome aberrations. *Brit. J. Radiol.* **45**, 320–327.

Macintosh, I. J. C., Brown, R. C. and Coakley, W. T. (1975). Ultrasound and *in vitro* chromosome aberrations. *Brit. J. Radiol.* **48**, 230–232.

Madsen, P. W. and Gersten, J. W. (1961). The effects of ultrasound on conduction velocity of peripheral nerves. *Arch. Phys. Med. Rehabil.* **42**, 645–649.

Majewski, C. and Jankowiak, J. (1967). Electron microscopic studies of the reaction activity of acid phosphatase in kidneys of rats treated with ultrasonic rays. *Arch. Phys. Ther.* **19**, 69–72.

Majewski, C., Kalinowski, M. and Jankowiak, J. (1966). Electron microscopic studies of acid phosphatase activity in the liver of rats subjected to ultrasound. *Amer. J. Phys. Med.* **45**, 234–237.

Mannor, S. M., Serr, D. M., Tamari, I., Meshorer, A. and Frei, E. H. (1971). The safety of ultrasound in foetal monitoring. *Amer. J. Obstet. Gynecol.* **113**, 653–661.

Marmur, R. K. and Plevinskis, V. P. (1978). Ultrasonic effects of various intensities on manifestations and duration of poststimulatory cytochemical changes in the retina. *Oftalmol. Zh.* **33**, 287–290.

Markham, D. E. and Wood, M. R. (1980). Ultrasound for Dupuytren's contracture. *Physiotherapy* **66**, 55–58.

Martin, C. J. and Gregory, D. W. (1979). A microscopic investigation of changes in mouse liver produced by ultrasound. Paper presented at the British Bio-Acoustics Group Meeting, British Institute of Radiology, London, July 6th.

McClain, R. M., Hoar, R. M. and Saltzman, M. B. (1972). Teratologic study of rats exposed to ultrasound. *Amer. J. Obstet. Gynecol.* **114**, 39–42.

McNally, N. (1977). Sub-lethal effects of ultrasound induced in rat thymocytes *in vitro*. M.Sc. Thesis, University of Dundee, Dundee, Scotland.

Mendez, J., Franklin, B. and Kollias, J. (1976). Effect of ultrasound on the permeability of D_2O through cellulose membranes. *Biomedicine* **25**, 121–122.

Mermut, S., Katayama, K. P., Castillo, R. del. and Jones, H. W. (1973). The effect of ultrasound on human chromosomes *in vitro*. *Obstet. Gynecol.* N.Y. **41**, 4–6.

Michael, P. L., Kerlin, R. L., Bienvenue, G. R. and Prout, J. H. (1974). An evaluation of industrial acoustic radiation above 10 kHz. Pennsylvania State University Final Report on US-DHEW Contract 99G72–125.

Miller, D. L., Nyborg, W. L. and Whitcomb, C. C. (1979). Platelet aggregation induced by ultrasound under specialised conditions *in vitro*. *Science* **205**, 505–507.

Miller, M. W., Kaufman, G. E., Cataldo, F. L. and Carstensen, E. L. (1976). Absence of mitotic reduction in regenerating rat livers exposed to ultrasound. *J. Clin. Ultrasound* **4**, 169–172.

Miller, M. W., Ciaravino, V. and Kaufman, G. E. (1977). Colony size and giant cell formation from mammalian cells exposed to 1 MHz ultrasound. *Rad. Res.* **71**, 628–634.

Molinari, G. A. (1968). Low intensity ultrasonic irradiation of the cochlea through the round window. II. Changes in the action potentials. *Bull. Soc. Ital. Biol. Sper.* **44**, 406–408.

Möller, P., Grevstad, A. O. and Kristoffersen, T. (1976). Ultrasonic scaling of maxillary teeth causing tinnitus and temporary hearing shifts. *J. Clin. Periodontol.* **3**, 123–127.

Moore, J. L. and Coakley, W. T. (1977). Ultrasonic treatment of Chinese Hamster cells at high intensities and long exposure times. *Brit. J. Radiol.* **50**, 46–50.

Morris, S. M., Palmer, C. G., Fry, F. J. and Johnson, L. K. (1978). Effect of ultrasound on human leucocytes: Sister chromatid exchange analysis. *Ultrasound Med. Biol.* **4**, 253–258.

Mortimer, A. J., Roy, O. Z., Taichman, G. C., Keon, W. J. and Trollope, B. J. (1978). The effects of ultrasound on the mechanical properties of rat cardiac muscle. *Ultrasonics* 179–182.

Morton, K. I., ter Haar, G. R., Stratford, I. J. and Hill, C. R. (1982). The role of cavitation in the interaction of ultrasound with V79 chinese hamster cells *in vitro*. *Brit. J. Cancer* **45**, 147–150.

Mueller, P. and Rudin, D. O. (1968). Action potentials induced in bimolecular lipid membranes. *Nature (Lond.)* **217**, 713–719.

Muggeo, M., Molinari, M., Fedele, D., Tiengo, A. and Crepaldi, G. (1975). Effect of pituitary ultrasonic treatment on plasma human growth hormone levels in diabetic retinopathy and acromegaly. *Acta Diabetol. Lat.* **12**, 137–149.

Mummery, C. L. (1978). The effect of ultrasound on fibroblasts *in vitro*. Ph.D. Thesis, University of London, England.

Murai, N., Hoshi, K., Kang, C-H. and Suzuki, M. (1975). Effects of diagnostic ultrasound irradiated during foetal stage on emotional and cognitive behaviour in rats. *Tohoku J. Exp. Med.* **117**, 225–235.

Neppiras, E. A. and Hughes, D. E. (1964). Some experiments on the disintegration of yeast by high intensity ultrasound. *Biotech. Bioeng.* **6**, 247–270.

Neumann, A., Müller, Th. and Wehner, W. (1980). Physical and technical problems of bone welding. Paper E20 at the Ultrasound Interaction in Biology and Medicine Symposium, Reinhardsbrunn, G.D.R.

Nyborg, W. L. (1977). Physical mechanisms for Biological Effects of Ultrasound. HEW Publication (FDA) 78–8062.

Oakley, E. M. (1982). Evidence for effectiveness of ultrasound treatment in physical medicine. *Brit. J. Cancer* **45**, 233–237.

O'Brien, W. D. Jr (1976). Ultrasonically-induced fetal weight reduction in mice. *In* "Ultrasound in Medicine" (Eds D. N. White and R. Barnes), pp. 531–532, Vol. 2. Plenum Press, New York.

O'Brien, W. D. Jr, Brady, J. K., Graves, C. N. and Dunn, F. (1977). Preliminary report on morphological changes to mouse testicular tissue from *in vivo* ultrasonic irradiation. *In* "Symposium on Biological Effects and Characterizations of Ultrasound Sources", pp. 182–191. HEW Publication (FDA) 78–8048.

Ochs, A. L. and Burton, R. M. (1974). Electrical response to vibration of a lipid bilayer membrane. *Biophys. J.* **14**, 473–489.

Pasechnik, V. I. and Sokolov, V. S. (1973). Changes in the permeability of modified bimolecular phospholipid membranes on periodic stretching. *Biofizika* **18**, 655–660.

Paul, B. J., La Fratta, C. W., Dawson, A. R., Baab, E. and Bullock, F. (1960). Use of ultrasound in the treatment of pressure zones in patients with spinal cord injury. *Arch. Phys. Med. Rehabil.* **41**, 438–440.

Petrov, V. I., Bazhanov, N. N., Loschilov, V. I., Ter-Asaturov, G. P. and Kuspangaliev, M. V. (1975). Osteosynthesis of fragments of the mandible by homologous grafting of sections and ultrasonic coagulation (Russ). *Stomatologiia (Mosk).* **54**, 29–34.

Pinamonti, S., Mazzeo, V., Pedrielli, F., Bredael, I., Rosetti, S., Merli, F. and Jehenson, P. (1981). Functional changes in human erythrocytes caused by pulsed ultrasound *in vitro*. Proc. 4th European Conference Ultrasound in Medicine and Biology, Dubrovnik, Yugoslavia, May 18–22.

Pinchuk, V. G., Heklman, B. S. and Lazaretnyk, A. Sh. (1971). Ultrastructural changes in the kidney under the effect of ultrasound. *Fiziol. Zh.* **17**, 109–113.

Pizzarello, D. J., Wolsky, A., Becker, M. H. and Keegan, A. F. (1975). A new approach to testing the effect of ultrasound on tissue growth and differentiation. *Oncology* **31**, 226–232.

Pizzarello, D. J., Vivino, A., Madden, B., Wolsky, A., Keegan, A. F. and Becker, M. (1978). Effects of pulsed low-power ultrasound on growing tissues. *Exp. Cell Biol.* **46**, 179–191.

Poliakov, V. A. (1970). Repair of bone defects and regeneration of bone tissue by ultrasonic welding. *Voen. Med. Zh.* **7**, 8–11.

Pospisilova, J. and Rottova, A. (1977). Ultrasonic effect on collagen synthesis and deposition in differently localized experimental granuloma. *Acta Chir. Plasticae* **19**, 148–157.

Pospisilova, J., Pospisil, M. and Rottova, A. (1971). The influence of ultrasound on the early developmental stage of experimental granuloma in mice. *Scripta Medica* **44**, 441–449.

Prasad, N., Prasad, R., Bushong, S. C., North, L. B. and Rhea, E. (1976). Ultrasound and mammalian DNA. *Lancet* **1**, 1181.

Repacholi, M. H. (1981). Ultrasound: Characteristics and biological action. Nat. Res. Council of Canada. Publ. No. 19244, Code ISSN 0316–0114.

Robinson, H. P., Sharp, F., Dinald, I., Young, H. and Hall, A. J. (1972). The effect of pulsed and continuous wave ultrasound on the enzyme histochemistry of placental tissue *in vitro*. *J. Obst. Gyn. Brit. Commonwlth* **79**, 821–827.

Rohr, K. R. and Rooney, J. A. (1978). Effect of ultrasound on a bilayer lipid membrane. *Biophys. J.* **23**, 33–40.

Rott, H-D., Huber, H. J., Soldner, R. and Schwanitz, G. (1972). Examinations of chromosomes after *in vitro* exposure of human lymphocytes to ultrasound. *Electromedica* **40**, 14–16.

Ruban, E. L. and Dolgopolov, N. N. (1952). The effect of ultrasonic waves on the early developmental phases of plants. *Dokl. Akad. Nauk. S.S.S.R.* **84**, 623–626.

Ryabchenko, N. I., Braginskaya, F. I., El'piner, I. E. and Tseitlin, P. I. (1964). *Biofizika* **9**, 31–40.

Saad, A. H. and Williams, A. R. (1982). Effects of therapeutic ultrasound on clearance rate of blood borne colloidal particles *in vivo*. *Brit. J. Cancer* **45**, 202–205.

Saggio, M. A. and Sommer, C. V. (1980). Morphological and functional studies of macrophages and lymphocytes exposed to ultrasound. *Ultrasound Med. Biol.* **00**, 000–000.

Samosudova, N. V. and El'piner, I. Ye. (1966). Ultrastructure of myofibrils exposed to ultrasonic waves. *Biofizika* **11**, 713–715.

Sanada, M., Hattori, A., Watanabe, T., Shu, T., Kasahara, T., Ohn, M. and Tamura, K. (1977). The *in vivo* effect of ultrasound upon human blood platelets. Nikon Choompa Igakukai, Koen Rombunshu, Nov. 149–150.

Sarvazyan, A. P. and Pashovkin, T. N. (1979). Effect of ultrasound on active and passive transport of ions through biological membranes. *Proc. IV Ultrasound Med. Biol. Symp.* **1**, 103–105.

Serr, D. M., Padeh, B., Zakut, H., Shaki, R., Mannor, S. M. and Kalner, B. (1971). Studies on the effects of ultrasonic waves on the fetus. *In* "Proc. 2nd Europ. Congr. Perinatal Med." (Ed. P. J. Huntingford), pp. 302–307. Karger, Basel.

Shoji, R., Momma, E., Shimizu, T. and Matsuda, S. (1971). An experimental study on the effect of low-intensity ultrasound on developing mouse embryos. *J. Fac. Sci. Hakkaido Univ. (VI), Zool.* **18**, 51–56.

Siegel, E., Siegel, E. P., Goddard, J., Fleischer, A. C. and Everette James, A. (1978a). Effects of diagnostic and therapeutic ultrasound on cultured human cells. 26th Annual Meeting of the Association of Univ. Radiologists, Univ. of Texas Health Science Center, San Antonio, April 30–May 4.

Siegel, E., Siegel, E. P., Goddard, J. and Everette James, A. (1978b). Effects of diagnostic ultrasound on T-1 human kidney cells. 25th Annual Meeting of the Society of Nuclear Medicine.

Siegel, E., Goddard, J., Everette James, A. and Siegel, E. P. (1978c). Attachment: a sensitive indicator of diagnostic ultrasound exposure for cultured human cells. 64th Scientific Assembly and Annual Meeting of the Radiological Society of North America, Chicago, Illinois, Nov. 28th.

Siegel, E., Goddard, J., James, A. E. Jr and Siegel, E. P. (1979). Cellular attachment as a sensitive indicator of the effects of diagnostic ultrasound exposure on cultured human cells. *Radiology* **133**, 175–179.

Sikov, M. R. and Hildebrand, B. P. (1976). Effects of ultrasound on the prenatal development of the rat. Part I; 3·2 MHz continuous wave at nine days of gestation. *J. Clin. Ultrasound* **5**, 357–363.

Sikov, M. R., Hildebrand, B. P. and Stearns, J. D. (1977). Postnatal sequelae of ultrasound exposure at 15 days of gestation in the rat (work in progress). *In* "Ultrasound in Medicine" (Ed. D. N. White), pp. 2017–2023, Vol. 3. Plenum Press, New York.

Smyth, M. G. (1966). Animal toxicity study at diagnostic power levels. *In* "Diagnostic Ultrasound" (Eds C. C. Grossman, J. H. Holmes, C. Joyner and E. W. Purnell), pp. 296–299. Plenum Press, New York.

Spencer, J. L. (1952). Effects of intense ultrasonic vibrations on Pisum. II. Effects on growth and their inheritance. *Growth* **16**, 255–277.

Stefanovic, V. Djukanovic, A., Velasevic, K. and Zivanovic, D. (1960). *Experientia* **14**, 486–487.

Stefanovic, V., Kostic, I. L., Bresjanac, M. and Zivanovic, D. (1959). *Bull. Soc. Chim. Belgrade* **24**, 175–178.

Stephens, R. J., Torbit, C. A., Groth, D. G., Taenzer, J. C. and Edmonds, P. D.

(1978). Mitochondrial changes resulting from ultrasound irradiation. *In* "Ultrasound in Medicine" (Eds D. N. White and E. A. Lyons), pp. 591–594, Vol. 4. Plenum Press, New York.

Stereva, S. and Beleva-Staikova, R. (1976). Influence of ultrasonic energy on the level of the thyronins in the thyroid gland. *Folia Med. (Plovdiv)* **18**, 155–159.

Stewart, G. J. (1978). The role of hypercoagulability in thrombosis. *Brit. J. Haematol.* **40**, 359–362.

Stolzenberg, S. J., Torbit, C. A., Edmonds, P. D. and Taenzer, J. C. (1980a). Effects of ultrasound on the mouse exposed at different stages of gestation: Acute studies. *Rad. Environ. Biophys.* **17**, 245–270.

Stolzenberg, S. J., Edmonds, P. D., Torbit, C. A. and Sadmore, D. P. (1980b). Toxic effects of ultrasound in mice: Damage to central and autonomic nervous systems. *Toxicol. Appl. Pharmacol.* **53**, 432–438.

Straburzynski, G., Jendykiewicz, Z. and Szulc, S. (1965). Effect of ultrasonics on glutathione and ascorbic acid contents in blood and tissues. *Acta Physiol. Polon.* **16**, 612–619.

Stratmeyer, M. E. (1977). Research directions in ultrasound bioeffects—a public health view. *In* "Proceedings Symposium on Biological Effects and Characterization of Ultrasound Sources", pp. 240–245. HEW Publication (FDA) 78–8048.

Stratmeyer, M. E., Simmons, L. R., Pinkavitch, F. Z., Jessup, G. L. and O'Brien, W. D. Jr (1977). Growth and development of mice exposed *in utero* to ultrasound. *In* "Symposium on Biological Effects and Characterizations of Ultrasound Sources" (Eds D-W. G. Hazzard and M. L. Litz), pp. 140–143. HEW Publication (FDA) 78–8048.

Stuhlfauth, K. (1952). Neural effects of ultrasonic waves. *Brit. J. Phys. Med.* **15**, 10–14.

Summer, W. and Patrick, M. K. (1964). "Ultrasonic Therapy—A Textbook for Physiotherapists". Elsevier, London.

Takagi, S. F., Higashino, S., Shihuya, A. and Osawa, N. (1960). The action of ultrasound on the myelenated nerve, the spinal cord and the brain. *Jap. J. Physiol.* **10**, 183–193.

Talbert, D. G. (1975). Spontaneous smooth muscle activity as a means of detecting biological effects of ultrasound. *In* Proceedings of "Ultrasonics International 1975", pp. 279–284. IPC Science and Technology Press.

Taylor, K. J. W. and Newman, D. L. (1972). Electrophoretic mobility of Ehrilich cell suspensions exposed to ultrasound of varying parameters. *Phys. Med. Biol.* **17**, 270–276.

ter Haar, G. R. and Wyard, S. J. (1978). Blood cell banding in ultrasonic standing wave fields: a physical analysis. *Ultrasound Med. Biol.* **4**, 111–123.

ter Haar, G. R., Dyson, M. and Smith, S. P. (1979). Ultrasound changes in the mouse uterus brought about by ultrasonic irradiation at therapeutic intensities in standing wave fields. *Ultrasound Med. Biol.* **5**, 167–179.

Thacker, J. (1973). The possibility of genetic hazard from ultrasonic radiation. *Curr. Topics. Rad. Res. Q.* **8**, 235–258.

Thacker, J. (1974). An assessment of ultrasonic radiation hazard using yeast genetic systems. *Brit. J. Radiol.* **47**, 130–138.

Thacker, J. (1975). Ultrasound and mammalian DNA. *Lancet* **2**, 770.

Tien, H. T. (1974). "Bilayer Lipid Membranes", pp. 134–145. Marcel Dekker, New York.

Tippe, A. (1979). Sound field effects on the electrophysiological functions of myelinated nerves. *Proc. IV. Ultrasound Med. Biol. Symp.* **1**, 106–111.

Tirrell, M. and Middleman, S. (1978). Shear deformation effects in enzyme calalysis. *Biophys. J.* **23**, 121–128.

Treton, J. A., Courtois, Y. and Lang, J. (1977). Action of ultrasonic irradiation on the DNA of cultivated bovine epithelial lens cell. *Biomedicine* **27**, 303–307.

Tsokh, R. M. (1974). Effect of ultrasound on the course of chemical burns of the eyes. *Vestn. Oftalmol.* **4**, 56–58.

Tsutsumi, Y., Sano, K., Kuwabara, T., Takakura, K., Hayakawa, I. Suzuki, T. and Katanuma, M. (1964). A new portable echo-encephalograph, using ultrasonic transducers; and its clinical applications. *Med. Electron. Biol. Eng.* **2**, 21–29.

Tuchman, L. S. (1956). Role of ultrasound in scleroderma. *Amer. J. Phys. Med.* **35**, 118–122.

Tucker, S. H. (1976). The effects of ultrasound on rat thymocytes *in vitro*. M.Sc. Thesis, University of Dundee, Dundee, Scotland.

Valtonen, E. (1967). Influences of ultrasonic radiation in the medical therapeutic range on the fine structure of the liver parenchymal cell. *Virchows Arch. (Path. Anat.)* **343**, 26–33.

Van der Decken, A. and Campbell, P. N. (1964). The effect of ultrasonic vibrations on the proteinsynthesizing activity of microsome preparations from rat liver. *Biochem. J.* **91**, 195–201.

Van Went, J. M. (1954). "Ultrasonic and Ultrashort Waves in Medicine". Elsevier, Amsterdam.

Warwick, R., Pond, J. B., Woodward, B. and Connolly, C. C. (1970). Hazards of diagnostic ultrasonography—a study with mice. *IEEE Trans. Sonics Ultrasonics* **SU-17**, pp. 158–164.

Watmough, D. J., Dendy, P. P., Eastwood, L. M., Gregory, D. W., Gordon, F. C. A. and Wheatley, D. N. (1977). The biophysical effects of therapeutic ultrasound on HeLa cells. *Ultrasound Med. Biol.* **3**, 205–219.

Watson, J. D. (1976). "Molecular Biology of the Gene", 3rd Edn. W. A. Benjamin, Menlo Park, California.

Watson, J. D. and Crick, F. H. C. (1953). A structure for Deoxyribose Nucleic Acid. *Nature* **171**, 737.

Watts, P. L. and Stewart, C. R. (1972). The effect of fetal heart monitoring by ultrasound on maternal and fetal chromosomes. *J. Obst. Gyn. Brit. Commonwlth* **79**, 715–716.

Watts, P. L., Hall, A. J. and Fleming, J. E. E. (1972). Ultrasound and chromosome damage. *Brit. J. Radiol.* **45**, 335–339.

Webster, D. F. (1980). The effect of Ultrasound on Wound Healing. Ph.D. Thesis, University of London.

Webster, D. F., Pond, J. B., Dyson, M. and Harvey, W. (1978). The role of cavitation in the *in vitro* stimulation of protein synthesis in human fibroblasts by ultrasound. *Ultrasound Med. Biol.* **4**, 343–351.

Webster, D. F., Harvey, W., Dyson, M. and Pond, J. B. (1980). The role of ultrasound-induced cavitation in the *in vitro* stimulation of collagen synthesis in human fibroblasts. *Ultrasonics* **January**, 33–37.

Wegner, R-D. and Meyenburg, M. (1982). Investigations into possible genetic effects of diagnostic ultrasound. Proc. 5th World Congress on Ultrasound in Medicine and Biology. *Ultrasound Med. Biol.* **8** (Suppl. 1), 206.

Wells, P. N. T. (1974). The possibility of harmful biological effects in ultrasonic diagnosis. *In* "Cardiovascular Applications of Ultrasound" (Ed. R. S. Reneman), pp. 1–77. North-Holland, Amsterdam.

Wells, P. N. T. (1977). "Biomedical Ultrasonics". Academic Press, London, New York, San Francisco.

Williams, A. R. (1973). A possible alteration in the permeability of ascites cell membranes after exposure to acoustic microstreaming. *J. Cell Sci.* **12**, 875–885.

Williams, A. R. (1977). Intravascular mural thrombi produced by acoustic microstreaming. *Ultrasound Med. Biol.* **3**, 191–203.

Williams, A. R. (1982). Absence of meaningful thresholds for bioeffect studies on cell suspensions *in vitro*. *Brit. J. Cancer* **45**, 192–195.

Williams, A. R., Hughes, D. E. and Nyborg, W. L. (1970). Hemolysis near a transversely oscillating wire. *Science* **169**, 871–873.

Williams, A. R., O'Brien, W. D. Jr and Coller, B. S. (1976b). Exposure to ultrasound decreases the recalcification time of platelet rich plasma. *Ultrasound Med. Biol.* **2**, 113–118.

Williams, A. R., Sykes, S. M. and O'Brien, W. D. Jr. (1976a). Ultrasonic exposure modifies platelet morphology and function *in vitro*. *Ultrasound Med. Biol.* **2**, 311–317.

Williams, A. R., Chater, B. V., Allen, K. A., Sherwood, M. R. and Sanderson, J. H. (1978). Release of β-thromboglobulin from human platelets by therapeutic intensities of ultrasound. *Brit. J. Haematol.* **40**, 133–142.

Williams, A. R., Chater, B. V., Allen, K. A. and Sanderson, J. H. (1981). The use of β-thromboglobulin to detect platelet damage by therapeutic ultrasound *in vivo*. *J. Clin. Ultrasound* **9**, 145–151.

Wong, Y. S. and Watmough, D. J. (1980). Haemolysis of red blood cells *in vitro* and *in vivo* caused by therapeutic ultrasound at 0·75 MHz. Paper C14 at the Ultrasound Interaction in Biology and Medicine Symposium, Reinhardsbrunn, East Germany, Nov. 10–14.

Wood, R. W. and Loomis, A. L. (1927). The physical and biological effects of high-frequency sound waves of great intensity. *Phil. Mag.* **4**, 417–436.

Yamamoto, Y. and Baba, M. (1963). A new supersonic applicator head of opthalmic use. *J. Clin. Ophthalmol.* **17**, 295.

Young, R. R. and Henneman, E. (1961). Reversible block of nerve conduction by ultrasound. *Arch. Neurol.* **4**, 83–89.

6. ASSESSMENT OF RISK FROM ULTRASONIC EXPOSURES

As with any other physical or chemical agent, too large a "dose" of ultrasound is known to cause deleterious changes to occur within living tissues, i.e. a large "dose" of ultrasonic radiation is hazardous. At the other extreme, a minute "dose" is almost certainly completely safe in that the living tissues are apparently unchanged by the ultrasonic exposure or else are transiently perturbed by the ultrasonic wave but rapidly repair or recover their original state without having undergone any structural or functional change which would cause them to deviate from their normal pattern of behaviour. The problems to be addressed in this chapter are therefore whether or not we can define the boundary zone between these two extreme situations and to attempt to estimate how crucial observance of the lower limit of this boundary zone is to the safety of the irradiated individual, i.e. what are the nature and long-term implications of the bioeffects resulting from exceeding this lower limit. On the basis of this information it is presumed that a meaningful assessment can then be made of the relative risk of a given ultrasonic exposure which can be offset against the real or imagined benefits to be gained by applying that procedure. It is further presumed that if the potential risks outweigh the potential benefits then that ultrasonic exposure may not be performed.

However, the whole concept of what constitutes a "potential risk" depends upon one's point of view. For example, a clinician who exposes the human foetus to ultrasound so as to obtain diagnostic information such as its orientation *in utero*, its growth rate or its heart rate could regard the exposure conditions which are known to produce *any* biological change *in vivo* as being potentially hazardous on the grounds that these same exposure conditions could also produce other (as yet unknown) biological changes *in vivo*, some of which may be potentially hazardous to that foetus. Conversely, a physiotherapist is deliberately attempting to change or modify the irradiated tissues so as to inaugurate, accelerate or complement the existing repair processes. Thus, the ultrasonic exposure parameters in the "presumably safe" zone as applied to the diagnostician is interpreted by the physiotherapist as exposures which are unlikely to produce any beneficial results in the treatment of patients. If their treatments are to be effective, physiotherapists must therefore use ultrasonic exposure levels which diagnosticians regard as being "potentially hazardous". Physiotherapists must therefore be concerned with the implications of any observed bioeffect on the future well-being of the irradiated tissues as well as on the acoustic exposure conditions under which that bioeffect occurred.

The confusion which currently exists within this field is exemplified by the fact that the average ultrasonic intensities recommended for some therapeutic applications of ultrasound (i.e. where the tissues are meant to be affected by the ultrasound) are less than those emitted by many diagnostic devices used for the investigation of peripheral vascular blood flow (where it is presumed that the ultrasound is not affecting any biological structures).

There are three main approaches to the problem of attempting to assess the safety of a given ultrasonic exposure. These may be briefly summarized as:

(1) The mechanistic approach. This approach attempts to identify the various mechanisms by which the energy of an ultrasonic wave may be transformed within the irradiated tissues so as to result in a modification of that tissue. The physical and biochemical parameters which affect the transduction process may then be modified to accentuate the occurrence of any given mechanism in an endeavour to determine under which combination of exposure conditions could that mechanism occur *in vivo*. The potential biological consequences resulting from the occurrence of that mechanism *in vivo* could serve as guidelines to pinpoint specific potential bioeffects which could be investigated using the other two approaches outlined below.

(2) The bioeffects approach. This relies upon a large number of investigators subjecting a wide variety of biological systems to ultrasound and attempting to identify the "threshold" acoustic parameters which cause a detectable change in the morphology, function, behaviour, reproductive capacity or survival of that biological system. Ideally, the range of exposure parameters used in these investigations should converge towards those currently employed in diagnosis or therapy so as to yield information which may

be extrapolated directly into the clinical situation. Those bioeffects which might be expected to occur under the exposure conditions used in diagnosis or therapy, and which would constitute a potential risk to the well-being of the patients or their offspring, give the epidemiologist some idea of what to look for in the comparison of ultrasound exposed and control populations.

(3) The epidemiological approach. These studies involve the statistical comparison of a population of patients who have been exposed to ultrasound with a control population which has been chosen to be as similar as possible in terms of age, sex distribution, general state of health etc., except that they have not been exposed to ultrasound. Providing that the parameter or parameters which might have been changed by the ultrasonic exposure can be and have been measured, then this statistical comparison is the only technique which can provide a meaningful estimate of the magnitude of any potential hazard resulting from the use of ultrasound. The disadvantages of this approach are that if the biological parameters which might have been changed by the ultrasonic exposure have not been measured, or these measurements not included in the statistical analysis, then this procedure must yield a false negative result. Conversely, if the control population had been chosen incorrectly or the numbers in either group were inadequate, then either a false negative or a false positive result might be obtained.

Each of the three different but complementary approaches to the assessment of the potential safety of ultrasound outlined above will now be discussed in more detail.

6.1 THE MECHANISTIC APPROACH

It is generally accepted that ultrasound causes changes within biological tissues either via the deposition of heat (i.e. thermal effects) or via some form of cavitational activity or else via some "other mechanisms" which include radiation pressure effects and any other effects which cannot be conveniently pigeonholed under "thermal" or "cavitational". In practice several or perhaps even all possible mechanisms will be coexisting together within any irradiated tissue and even though one may apparently be dominant under some particular combination of exposure conditions, it may not require a large change within that tissue or in the exposure conditions to change this order of dominance. Each mechanism also does not act in isolation, but interacts with the other mechanisms to enhance their potential destructive effects; thus, an elevated temperature increases the susceptibility of tissue to damage by acoustic microstreaming forces generated by "stable" cavitation bubbles, while a strong standing wave component increases the magnitude of the radiation pressure effects leading to blood cell stasis as well as providing a greatly improved environment for the growth of cavitation nuclei.

6.1.1 Thermal Effects

The thermal effects of ultrasound have been described in Chapter 3 and are apparently a function of the maximum temperature attained within that tissue, the time for which that elevated temperature was maintained and the inherent thermal susceptibility of that tissue. Lele (1975) has shown that the mammalian foetus is exceptionally sensitive to elevated temperatures and has proposed that a temperature rise of only a few degrees centigrade at certain critical stages of its development will result in teratogenic malformations. On the other hand, adult non-dividing tissues appear to be remarkably tolerant of moderate temperature elevations, for example, small portions of mammalian brain tissue *in vivo* appear to be capable of withstanding temperatures of up to about 44°C for prolonged periods without the constituent cells being killed (Lele, 1977). However, it must be presumed that functional changes would have occurred below this temperature which would have resulted in the death of the animal had those portions of the central nervous system regulating the activity of the respiratory or cardiovascular systems been irradiated.

The biological effects resulting from moderate temperature elevations are not fully understood. It is known that the blood flow through human skin (and to a lesser extent muscle) is increased due to a reflex relaxation of the muscular tone to the arterioles (i.e. the hyperaemic response) once the tissue temperature exceeds a certain threshold value. This threshold value is not constant and is influenced by a number of physical and hormonal factors as well as by many pharmacologically active agents such as ethanol or nicotinic acid. Experiments performed on certain reptilian species have shown that infected animals re-adjusted their body temperature to a higher value throughout the course of their infection and that their symptoms were less severe and their infections of shorter duration than similar infected animals whose bodies were deliberately cooled to their pre-infection normal value. Experiments on mammals have not yielded such clear-cut results. Nevertheless, there is an appreciable body of evidence which suggests that moderate temperature elevations do have a beneficial effect upon the regenerative processes occurring within mammalian tissues (ter Haar, in press).

The solubility coefficient of most common gases decreases with increasing temperature and so a temperature rise of a few degrees centigrade would tend to supersaturate a tissue which was already saturated with that gas at 37°C (e.g. carbon dioxide or nitrogen). This increased partial pressure of gas would tend to accelerate the rate of growth of any micronuclei within the irradiated tissue volume.

Cells in suspension in a medium having a low absorption coefficient are not usually affected by elevated temperatures because the heat deposited within any part of the cell rapidly diffuses into the surrounding large volume of colder suspending medium (Love and Kremkau, 1980). Conversely, cells in suspension are more liable to be damaged by a cavitational mechanism because of the enhancement of nucleation inevitably associated with the use of foreign (especially hydrophobic) surfaces and the use of well-mixed aerated solutions which have not been filtered through innumerable biological membranes.

6.1.2 Cavitational Effects

Cavitational effects are described at length in Chapter 4 and would be expected to be the dominant damaging mechanism affecting cells in suspension *in vitro* or when the irradiated medium *in vivo* already contains stable gas bubbles as is frequently found in plant tissues and insects at various stages of their life cycle. Even when stable gas bubbles are already present, the biological effects produced by a given ultrasonic intensity can be highly variable because the amplitude of oscillation of the gas bubble interface is crucially dependent upon the dimensions of that bubble, i.e. how close it is to being of a resonant size for that particular driving frequency, and whether or not the geometrical constraints acting on that gas bubble permit the development of surface wave activity. This situation is further complicated by the fact that the dimensions of the stable gas bubbles within an irradiated body are continually changing; for example, the rise in ambient temperature is causing the bubbles to expand and is also increasing the partial pressure of the dissolved gas within the surrounding medium. At the same time the bubble oscillation and surface wave activity (if present) induce microstreaming at the bubble interface which is continually replenishing the liquid at that site which tends to accelerate the rate of shrinking of that bubble if its average internal gas partial pressure is higher than that of the surrounding medium, or to accelerate its rate of growth if its average partial pressure is lower than that of the surrounding medium. The intensity-dependent effects of rectified diffusion are superimposed on top of these other effects which are also greatly influenced by the mechanical properties of any membrane or film of denatured protein which may be present at the gas/air interface.

In the absence of stable pre-formed gas bubbles the rate limiting factors which determine whether or not cavitational activity is going to occur are the number and availability of the minute pockets of undissolved gas molecules commonly called micronuclei, the available gas content of the medium and its composition, and the acoustic exposure conditions which can drive that micronucleus to grow by rectified diffusion until it approaches its resonant dimensions. The rate of microbubble growth is increased by the use of high ultrasonic intensities, long exposure times and the use of lower ultrasonic frequencies. It apparently occurs more rapidly in continuous wave exposures than in pulsed mode exposures and may apparently be suppressed altogether by the use of extremely short exposures separated by long intervals. Microbubble growth may also be suppressed if the ambient pressure is increased. Geometric factors which enhance microbubble growth include the use of reflectors to produce standing wave fields, the use of unfocused acoustic beams and the presence of hydrophobic or artificial interfaces which may have micronuclei adsorbed onto their surface.

The acoustic factors which determine the rate of microbubble growth have received a great deal of attention in many laboratories, but the non-acoustic factors have been largely ignored. Cells in suspension are usually more dense than their suspending medium and consequently tend to settle during the ultrasonic exposure period. Various investigators have adopted a wide range of agitation systems to maintain their cells in homogeneous suspension; these include stirring the sample or rotating or rocking or oscillating the entire sample chamber. However, all of these techniques could enhance the nucleation characteristics of the enclosed cell suspensions to a greater or lesser extent so that it might be possible to obtain a wide range of reproducible cavitation-induced biological effects under the same acoustic exposure conditions simply by altering the magnitude of the agitation system used to maintain the cells in suspension. Williams (1982) has shown that the amount of haemolysis obtained when a suspension of human erythrocytes in isotonic saline was exposed for 1 min to c.w. 0·75 MHz ultrasound at a space averaged intensity of 0·6 W/cm^2 increased with the increasing speed of rotation of the stirring bar used to keep the cells in suspension. The percentage of lysed cells increased from about 5% at a stirring rate of 500 r/min to about 80% at 1800 r/min even though the ultrasonic exposure conditions and the sample geometry had been kept constant (these measurements are described in more detail in Chapter 4 and in Figs 4.17 and 4.18. It was also found that a similar enhancement of the rate of haemolysis was obtained when the stirring bar was removed and the entire sample chamber vibrated at various frequencies or even when the sample chamber was kept stationary but repeatedly struck by the corners of a hexagonal rod driven to rotate while pressed against the outside of the sample chamber (see Fig. 5.15). The mechanism of the enhancement of cavitational activity observed using the latter percussive technique is presumed to be analagous to that observed during the degassing of a liquid by reduced atmospheric pressure where an apparently bubble-free liquid will abruptly form a number of visible bubbles when the containing vessel is struck or agitated.

It should be noted that the reproducibilities of the erythrocyte haemolysis measurements outlined above were greatest at the high-speed stirring rates and progressively decreased as the stirring rate was decreased. At zero or low-speed stirring rate it was possible to distinguish between cell suspensions which had been vigorously agitated immediately before sonication (these gave more haemolysis) and similar samples which had been mixed by gentle rotation or inversion (these gave lower and more variable haemolysis values). At high stirring speeds it was not possible to distinguish between these same two samples, presumably because of the "swamping" effect of the large number of nucleation sites produced by the high-speed stirring.

Complementary results have been obtained by Church *et al.* (1981) who found that ultrasonically-induced lysis of cell suspensions at low therapeutic intensities only occurred when the sample chamber was rotated (i.e. duplicating the exposure arrangement developed by Hill, 1972). These authors concluded that active bubbles were being maintained at pressure antinodes in a partial standing wave field and that the rate of cell lysis became detectable only when the tube was rotated which swept the cell suspension through the regions occupied by these active bubbles. It is not known how the rotation of the sample chamber may have contributed to the growth of these active bubbles.

The magnitude of the increases in erythrocyte haemolysis produced under constant acoustic exposure conditions merely by changing the rate of stirring of the sample brings into question the relative merit of the various commonly measured dosimetric parameters. The observed changes were so large that it is a pointless exercise to attempt to ascribe a meaningful value to the "threshold" acoustic parameters needed to change some property of cells in suspension *in vitro* without considering the effects of the pre-treatment of that sample and the time-scale over which it was performed as well as the effects of any stirring or agitation system on the nucleation properties of that suspension.

Similar non-acoustic considerations may also apply to the generation of cavitational activity within soft tissues *in vivo*. For example, early experiments designed to investigate the factors determining the rate of growth of gas bubbles in mammalian tissues during decompression showed that the number of bubbles formed within limbs which were being actively stimulated to contract greatly exceeded those found in adjacent (passive) limbs (Fulton, 1951). If this enhancement of nucleation sites by muscular activity also occurs during ultrasonic irradiation at ambient atmospheric pressure, then one would expect to obtain more cavitational activity during insonation of actively contacting muscle than passive muscle (under the same acoustic exposure conditions). This may be of some relevance in physiotherapy where a patient may be asked to exercise a limb while it is being irradiated with ultrasound. Some of the ultrasonic generators currently being sold to physiotherapists also incorporate a "Faradic Stimulator", i.e. the ability to use the metal face of the transducer housing as a live electrode so that the irradiated tissues may be simultaneously subjected to an ultrasonic beam and to pulses of electrical current. The effects of the various electrical pulsing parameters on the nucleation characteristics of the tissue are not known. Until evidence is forthcoming that there is a proven therapeutic advantage in using these mixed modality techniques, it is advisable that they should be used with extreme caution.

An additional complicating factor is that the rate of growth of a micronucleus to form a cavitational "event" is dependent upon the maximum value of the acoustic pressure amplitude and its temporal characteristics and not necessarily the time averaged intensity at that site. Preliminary investigations in Manchester using the stirred suspensions of human erythrocytes in isotonic saline described above have shown that the amount of haemolysis increased with increasing peak pulse intensity even though the time averaged intensity was kept constant at 0·1 or 0·5 W/cm^2. These results indicate that the 2 ms pulse lengths emitted by a common British physiotherapy device (the Rank Sonacel Multiphon) are long enough to enable cavitational activity to occur *in vitro* providing the micronuclei are "activated" by stirring. Thus, the probability of generating cavitational activity *in vivo* with a therapeutic device may be increased if the ultrasound is delivered in the form of 2 ms pulses having peak pulse intensities of the order of 0·5 to 5 W/cm^2 separated by long intervals to give a low time averaged intensity. It is probable that the use of shorter pulse lengths would reduce or eliminate this effect.

6.1.3 Radiation Pressure Effects

The radiation pressure force in a c.w. plane progressive travelling wave beam is typically small and is unlikely to damage well-anchored soft tissues at diagnostic or even therapeutic intensities. However, this same force is greatly enhanced if the acoustic wave retraces its path through the tissue after being reflected from a strong reflector (i.e. in a strong standing wave field). In this case the force acts over a distance equal to one quarter of a wavelength (about 0·04 cm in soft tissue at 1 MHz) so that particles which are more dense than their suspending medium and are free to move are driven to regions of maximum acoustic pressure amplitude. This effect is particularly noticeable within small blood vessels where the blood cells may gather to form clumps separated by zones of clear plasma (i.e. blood cell stasis). The biological hazards one would envisage arising from these radiation pressure forces in a standing wave field are mainly those of tissue anoxia resulting from the interruption of the normal flow of blood. The extent of the damage caused by this interruption of the blood supply depends upon the time for which the stasis is maintained, the nature and size of the affected tissue volume, the number of unoccluded collateral blood vessels and the inherent susceptibility of that tissue (e.g. nervous tissue is very sensitive to anoxia whereas muscular tissue is more tolerant).

However, the experimental conditions responsible for the production of blood cell stasis can also change the magnitude of both the thermal and cavitational effects occurring within that sample. For example, the strong reflector causes the reflected wave to return through the tissue so that more of its energy is absorbed and converted into heat. Also, the impairment of the microvascular blood flow caused by blood cell stasis means that the heat is not removed as efficiently and so the resulting temperature rise within the irradiated tissues may be substantially greater than it would have been under free field conditions. Finally, cavitation is enhanced because micro-nuclei and gas bubbles smaller than resonant size are driven to the pressure antinodes where they may fuse together and are maintained under the ideal conditions necessary to enhance their rate of growth to become active cavitation bubbles.

6.1.4 Summary of Mechanistic effects

The above sections have attempted to outline some of the complexities inherent in the interpretation of which mechanism or mechanisms might have been responsible for any given biological effect. Some of the vagaries of cavitational activity have been emphasized so that the readers may appreciate the difficulties of extrapolating from a bioeffect observation obtained using a stirred cell suspension *in vitro* to a determination of a probability as to whether or not that same effect is likely to occur at similar intensities *in vivo*. Until more evidence is forthcoming on the behaviour of cavitational bubbles within soft tissues *in vivo* and the geometrical, mechanical and acoustic exposure conditions necessary for them to be formed in sufficient numbers to produce detectable biological effects, it is reasonable to presume that those bioeffects which can definitely be ascribed to a cavitational mechanism in cell suspension *in vitro* are unlikely to occur to the same extent at comparable acoustic exposures within intact soft tissues *in vivo*.

However, fluid tissues such as blood may be an exception to this general statement in that there is some indirect evidence that under certain circumstances blood can be driven to cavitate *in vivo* at acoustic exposures comparable with those producing similar bioeffects *in vitro*. Williams *et al*. (1981) showed that low therapeutic intensities of 0·75 MHz ultrasound did not cause detectable damage to the cellular elements of blood flowing in a non-turbulent manner in both humans and rabbits *in vivo* (see Chapter 5, Section 5.3.6). Conversely, Wong and Watmough (1980) showed that similar intensities lysed erythrocytes (presumably by a cavitational mechanism) when the beating hearts of anaesthetized rats were irradiated with 0·75 MHz ultrasound. This apparently contradictory result almost certainly reflects an enhancement of the nucleation conditions existing within the heart's turbulent blood flow analagous to that observed when cell suspensions are stirred or agitated *in vitro*.

Until these observations have been investigated in more detail, and the potential hazards resulting from a small amount of intravascular haemolysis (with the concomitant activation of some platelets) have been quantified, it would appear to be prudent to avoid exposing the beating heart to unnecessary acoustic exposure. Neither adult nor foetal hearts should be exposed to therapeutic intensities of ultrasound under any circumstances. Diagnostic intensities of 2–5 MHz ultrasound (especially if pulsed) are less likely to be hazardous but long-term cardiac exposures as used in foetal monitoring ought not to be used routinely but only when their use can be justified on medical grounds.

Other routine clinical procedures which should be viewed with caution include the deliberate introduction of small gas bubbles to enhance the contrast of blood to aid in the ultrasonic visualization of the chambers of the adult heart (Meltzer and Roelandt, 1982; Ziskin *et al.*, 1972). There is a high probability that some of these intravascular gas bubbles will be of resonant size while they are being irradiated. *In vitro* investigations by Miller *et al.* (1979) have shown that gas bubbles of resonant size can be driven to oscillate at amplitudes which result in the disruption of platelets (and erythrocytes) by the acoustic field emitted by a commercial doppler diagnostic apparatus. These investigations have not yet been attempted using commercial pulse-echo diagnostic equipment, but erythrocyte lysis has been detected *in vitro* at c.w. intensities as low as 4 mW/cm^2 at 2 MHz (Miller, Williams and Nyborg, unpublished observations). It is therefore probable that the use of gas microbubbles in ultrasonic cardiographic investigations will result in a limited amount of cellular disruption with the consequent activation of some platelets. However, the normal mammalian coagulation system is able to tolerate a limited amount of "activation" without forming large numbers of platelet emboli or thrombi. The magnitude of the potential hazard resulting from any intravascular cavitational activity therefore depends upon the magnitude of the resulting "activation" of the coagulation system and the inherent "coagulability" of that individual's blood. It would therefore be prudent to exclude patients whose blood is in a hypercoagulable state (Stewart, 1978), or who have a history of thromboembolic complications, from ultrasonic cardiac investigations involving the deliberate introduction of gas bubbles until further work has been performed to quantify the magnitude of the cavitation-induced blood cell damage and its effects upon both the normal and abnormal coagulation systems *in vivo*.

One proposed technique for the generation of intracardiac gas bubbles to enhance ultrasonic contrast which should be actively discouraged is that suggested by Higashiizumi *et al.* (1979). These authors proposed that the intensity of an ultrasound beam passing through the heart be increased until it was "just" below the "threshold" necessary for the production of "cavitational activity". Under these conditions it is presumed that gas bubbles would grow from blood-borne nuclei by rectified diffusion. However, Higashiizumi *et al.* (1979) had not considered the fact that it requires much more energy to cause a micronucleus to grow into a bubble than is needed to drive that bubble so that it can destroy biological tissues. Thus, the high acoustic intensities necessary to enable this technique to be used will almost certainly result in intracardiac haemolysis with potentially serious consequences.

6.2 THE BIOEFFECTS APPROACH

Some of the vast literature describing the plethora of changes produced by ultrasound in a wide variety of biological systems have been briefly summarized in Chapter 5. Much of the early work in this field has been omitted either because the dosimetry is considered to be inadequate or because the acoustic intensities used were so high that it was felt that their results are not applicable at the lower exposure levels encountered in diagnostic and therapeutic applications. This review is also unbalanced in that positive bioeffect articles greatly outnumber those reporting negative effects, although this largely reflects the selection procedures adopted by the various scientific journals and the motivations of the authors who tend not to publish negative results even though these can (under certain circumstances) be as important as positive effects. In addition, most of the reported effects have not been confirmed by independent investigators in other laboratories. Some reports of positive bioeffects have had such profound biological implications and/or have been reputed to occur at such low acoustic power densities that other investigators have been stimulated to duplicate their experimental systems. In general, it has not been possible to confirm many of these startling effects in other laboratories.

The objective of this section is to briefly but critically review some of these "low power" positive bioeffects obtained *in vivo* and to attempt to assess both the probability of these effects occurring in humans during diagnostic and/or therapeutic applications and the possible biological consequences which could result from their occurrence. This will be followed by a review of the various so-called "boundary curves" which have attempted to define the "threshold" exposure conditions above which positive bioeffects have been reported.

6.2.1 Chromosome Damage Reported at Diagnostic Intensities

The first startling report of a potentially serious biological effect induced by diagnostic intensities of ultrasound was the report by Macintosh and Davey in 1970 that they had observed an increased incidence of chromosome aberrations in irradiated human lymphocytes. The significance of this report lay in the fact that this technique is used to assess the mutagenic potential of different ionizing radiations. Therefore, if ultrasound is also able to increase the number of chromosome aberrations (especially at diagnostic intensities) then it is probably also mutagenic, i.e. may be capable of inducing tumours. The majority of the attempts to replicate this work yielded negative results (see Chapter 5, Section 5.3.8.2). However, two or more negative reports do not cancel out one positive report, and so detailed investigations were undertaken to try to determine how the positive results could have been obtained in the first place. In this case the cardinal error committed by Macintosh was that he did not conduct his assessments of the number of chromosome aberrations in a double-blind manner, i.e. he knew which samples had been exposed to ultrasound which unconsciously biased his decision as to whether or not any given dubious agglomeration of chromosomes was "normal—but tangled" or "abnormal". When this work was repeated using exactly the same ultrasonic Doppler instrument and exposure apparatus, but this time scoring the slides under double-blind conditions, no increase in chromosomal abnormalities was detected (Macintosh *et al.*, 1975).

The mechanisms of energy transfer from an ultrasonic wave into the medium through which it is propagating are so different from those of ionizing radiation that it is unlikely that they would both produce similar biological effects by the same mechanism. Unlike ionizing radiation, ultrasound is more likely to kill or disrupt the cell by a thermal or cavitational mechanism instead of leaving an intact cell whose intracellular components have been chemically modified. Extensive investigations on plants by Khokhar and Oliver (1975) and Miller (1978) have shown that even high intensities of ultrasound do not produce the "classical" chromosome aberrations commonly seen after X-ray treatment. However, Cataldo *et al.* (1973) developed a new technique for scoring chromosomal abnormalities in plants and described many new types of aberrations produced by high intensities of ultrasound called "bridge prophases", "bridged metaphases" and bizarre "agglomerated mitotics". Later, Cataldo *et al.* (1976) found that these "agglomerated mitotics" had not passed through the phase of DNA synthesis and were therefore interphase cells whose nuclei had been damaged by a thermal and/or cavitational mechanism. The nature and magnitude of these unusual abnormalities was such that even if they could be produced within intact mammalian cells (where the relative contributions of the various interactive mechanisms would be expected to be different), then the damaged cells would almost certainly be unable to reproduce themselves.

6.2.2 Effects on Adult and Developing Nervous Tissue

The unique role of the nervous system in the control and regulation of all vital functions in mammals has stimulated many series of investigations to determine whether or not diagnostic intensities of ultrasound may have an adverse effect upon its function or development. Hu and Ulrich (1976) applied pulsed 2·25–5 MHz ultrasound (SATA intensity 3 mW/cm^2; SATP intensity 1·5 W/cm^2) to the brains of anaesthetized squirrel monkeys and were able to detect the presence of additional electronic discharges (evoked potentials). These signals appeared immediately after the ultrasound was switched on and disappeared after 2 or 3 min even though the ultrasound remained on. The authors attempted to eliminate sources of artifact resulting from direct electrical pickup, but all electroencephalographic measurements are so extremely sensitive to electrical interference that artifactual effects have not been ruled out. These observations should be repeated in an independent laboratory.

It is known that higher intensities of pulsed ultrasound can stimulate the frog cochlea to generate trains of impulses at the pulse repetition rate of the acoustic source (Gavrilov *et al.*, 1975). It has also been reported that about 10 mW/cm^2 of 6–150 kHz ultrasound provoked modifications of the bioelectric potentials recorded from *in vitro* preparations of frog neuromuscular preparations (Nagy and Mihalas, 1974). It is therefore possible that some signals may have been generated at the diagnostic intensities used by Hu and Ulrich (1976) even though they were not deliberately irradiating the internal ear (a structure which is designed for the conversion of acoustic waves into nervous impulses). Providing that the acoustic intensities employed are not high enough to damage the sensory or neural tissues, the reversible elicitation of trains of impulses should not constitute a potential hazard to the wellbeing of adult animals.

In 1975 David *et al.* reported the novel observation that pregnant women were able to detect an increased number of foetal movements during the period that their foeti were being subjected to ultrasound from either of two common doppler foetal heart monitors. Each patient was studied for 30 min with the ultrasonic transducer taped to her abdomen, but it was only energized (without the patient's knowledge) for the second half of the observation period. A mean increase of 90% in the number of foetal movements was recorded. Attempts to duplicate these observations in other laboratories using a more stringent experimental protocol have yielded one positive (Sheldon, 1978) and three negative results (Docker and Petoussi, 1982; Hertz *et al.*, 1979; Powell-Phillips and Towell, 1979).

Murai *et al.* (1975a) exposed the shaved abdomen of pregnant rats to the acoustic field emitted by a commercial doppler device (2·3 MHz, SATA intensity 20 mW/cm^2) for 5 hours. Seven types of reflex neuromotor responses were evaluated in the subsequent offspring during the first 21 days after birth; these were the righting reflex, cliff-top aversion, negative geotaxis, grasp reflex, vibrissa placing response and acceleration righting reflex. Only the grasp reflex showed a statistically significant (though not large) difference between the irradiated and the sham-irradiated animals. A complication of these experiments is that the pregnant animals had to be immobilized for the duration of the 5 hour treatment and there were even more significant differences between the offspring of the restrained but mock-irradiated animals and those of the unrestrained controls. Another criticism of these investigations is that they were not performed double-blind. This is important because many of the reflexes measured involved an element of subjective assessment and were thus open to unconscious bias in the case of an ambiguous response.

In a subsequent paper, Murai *et al.* (1975b) reported on the results of four "emotional reactions" in the three groups of animals who had been irradiated *in utero* as described above (i.e. their mothers had either been restrained and irradiated, restrained but not irradiated, or unrestrained and unirradiated). The animals were evaluated at 120 days of age and the parameters evaluated were ambulation in an open field, defecation and urination responses in an open field, vocalization response in reaction to being handled by humans and their escape response from an electric shock. The vocalization as a result of handling and response to electric shock both showed significant differences between the irradiated and mock-irradiated groups, even though these results are subject to the same criticisms as those of their first article. Two tests for cognitive (learning) response made at the age of 150 days showed negative results.

Brown *et al.* (1981) have performed an extensive series of well-devised studies to measure the effects of low levels of continuous wave ultrasound on reflex development and the aquisition of conditioned reflexes in the offspring of ICR mice which had been irradiated *in utero*. The pregnant animals were shaved and immersed in a water bath and exposed to 1 MHz c.w. ultrasound for 3 min at intensities of 0, 50 or 500 mW/cm^2. Their offspring were evaluated for the number of days required to develop a conditioned food cup response to a compound stimulus (light plus a 2800 Hz tone) as well as the time taken to attain five developmental reflexes (pivoting, walking, forelimb grip, accelerated righting, and the ability to stay on a rotating rod). Neither of the two groups of offspring exposed to ultrasound differed significantly from their mock-exposed controls for any of the measured parameters.

The positive bioeffect observations reported by Murai *et al.* (1975a,b) give cause for concern but, even if proved to be reproducible, do not necessarily indicate that diagnostic intensities of ultrasound are hazardous. The experimental problems inherent in restraining conscious animals throughout long irradiation periods cause many psychological and physiological changes which may mask or accentuate any effects caused by ultrasound (Brodie and Hanson, 1960). This is illustrated in the article by Tsutsumi *et al.* (1964) who found elevated levels of some enzymes in the cerebrospinal fluid of dogs after they had been restrained in gypsum for 6–10 hours (see Chapter 5, Section 5.4.2.2). These animals had also been subjected to diagnostic intensities of MHz ultrasound from a new portable device, but it is reasonably certain that similar elevated enzyme levels would have been detected in the absence of ultrasound. It has also been demonstrated experimentally that a psychological stress during pregnancy produces some effects upon the behavioural development of the offspring (Thompson, 1957).

6.2.3 Effects on Embryonic Growth and Differentiation

In view of the widespread exposure of the human foetus to diagnostic intensities of ultrasound, a wide variety of developing embryonic systems have been investigated in an attempt to ascertain whether or not ultrasound could be exerting an adverse effect upon their growth and differentiation. Some early reports gave cause for concern; for example, Shoji *et al.* (1971) reported a significant increase in the number of skeletal abnormalities in the offspring of both DHS and A/He mice which had been exposed for 5 hours *in utero* to 2·25 MHz ultrasound at an intensity of 40 mW/cm^2 on the 9th day of gestation. Similarly, Mclain *et al.* (1972) reported a slight increase in skeletal variations in the offspring of Charles Rivers rats exposed for 30 min to c.w. 2·5 MHz ultrasound at an intensity of 10 mW/cm^2 on the 8th, 9th, 10th, 11th and 13th days of gestation. However, 2-hour exposures under the same experimental conditions yielded negative results and so these observations should be viewed with some scepticism. Complementary studies by Shimizu and Tanaka (1980) in Chinese hamsters and Takeuchi *et al.* (1966) in rats showed that diagnostic intensities of pulsed ultrasound caused no detectable increase in the number of foetal anomalies or any change in foetal weight. Similarly, Takabayashi *et al.* (1979) were able to demonstrate an increase in the number of foetal anomalies in mice only after the time-averaged power density had been increased to levels approaching those employed in therapeutic applications, i.e. indicating that the damaging mechanism may be thermal. Sekiba *et al.* (1980) also found an increased incidence of foetal anomalies in rats exposed to therapeutic intensities of 2 MHz c.w. ultrasound *in utero*.

Akamatsu and Sekiba (1977) exposed the pre-implantation embryos of mice and rats to diagnostic levels of pulsed ultrasound and therapeutic intensities of continuous wave ultrasound *in vitro*. Diagnostic exposures (2·25 MHz, pulse width 3 μs, repetition rate 1 kHz, SATA intensity 2·2 mW/cm^2) for up to 12 hours had no effect on the rate of growth of the expanding blastocyst after transplantation into a pseudopregnant recipient uterus. However, therapeutic intensities (SATA values of 2–3 W/cm^2) for 60 min resulted in a marked inhibition of growth even though this inhibition was less marked if the morulae and early blastocytes were cooled during the irradiation period (this again suggests a thermal mechanism of damage).

The Japanese Ministry of Health and Welfare sponsored an intensive series of investigations from 1977 until 1979 to study the potential safety of pulsed ultrasound. All participants used the same ultrasonic generator (the USP-1) which emitted 3–10 μs wide pulses of 2 MHz ultrasound at repetition rates between 150 and 1000 Hz with a maximum SATA intensity of 0·5 W/cm^2 and a maximum SATP intensity of 50–60 W/cm^2. Shimizu and Tanaka (1981) found no increase in the number of foetal abnormalities under any exposure conditions when chinese hamster foeti were irradiated *in utero* on days 8, 9 or 10 of gestation. However, Takabayashi *et al.* (1981) and Sakamoto *et al.* (1981) did find an increased incidence of foetal anomalies after 2 different strains of pregnant mice were irradiated on the 8th day of gestation. The minimum exposure levels reported to produce morphological abnormalities were an SATA intensity of 600 mW/cm^2 (pulse width 180 μs, repetition frequency 150 Hz for 5 min) with ICR mice (Sakamoto *et al.*, 1981), and an SPTA intensity of 1·24/cm^2 with C3H/He mice (Takabayashi *et al.*, 1981). This latter article is of special interest in that the authors claim that foetal abnormalities were not obtained either if the peak intensity within the pulse was reduced below about 50 W/cm^2 or if the pulse width was reduced to less than 3 μs even at the highest peak intensity.

The eggs and larvae of the fruit fly *Drosophila melanogaster* have been used extensively to detect and quantify the damaging effects of ionizing radiation. Several groups of workers have used this system to look for potential mutagenic effects arising from the use of ultrasonic irradiation. Pizzarello *et al.* (1978a) reported that a high proportion of *Drosophila* larvae exposed for 2·5 min to pulsed 2·25 MHz ultrasound (SATA intensity 1·5 mW/cm^2; pulse repetition rate 500 Hz) were either killed or modified in that when they eventually hatched they were of normal proportions but very small. These miniature flies were not seen in control populations. Child *et al.* (1981) were unable to confirm the production of dwarf flies under any circumstances or to obtain any lethal effects at similar SATA intensities. However, lethal effects were obtained at SATA intensities greater than about 3 mW/cm^2 if the peak pulse intensity was increased. Child *et al.* (1981) found that the spatial average temporal peak intensity (SATP) was a more meaningful index of biological damage to the flies than the time averaged intensity; an apparent "threshold" being evident in that the larvae were killed at temporal peak (i.e. pulse) intensities above about 10 W/cm^2. The eggs and larvae of *Drosophila* contain stable gas bodies which will be driven to oscillate by the applied ultrasonic field. Mammalian cells which do not normally contain gas bodies would therefore not be damaged in the same way at these low average intensities.

Pizzarello *et al.* (1978a) also exposed one horn of the pregnant rat uterus for 5 min to the same 2·25 MHz pulsed ultrasonic beam (SATA intensity 1·5 mW/cm^2). They reported that insonation on days 3, 5 or 6 of gestation resulted in smaller foeti which also weighed less after drying, and that there was an increase in the number of resorption sites (spontaneous abortions). This result is at variance with the many studies performed at higher exposure levels in other laboratories (see Chapter 5, Section 5.4.2.3). Another unusual finding reported by Pizzarello *et al.* (1978b) was that a transplantable mouse lymphosarcoma irradiated *in vivo* for 5 min using the same diagnostic apparatus described above was "less infective" when subsequently innoculated into donor mice, i.e. the transplantability of the tumour was reduced. The authors were unable to detect any change in the mitotic index of the irradiated tumours. These investigations have not been repeated in an independent laboratory, but other investigators have found that the growth rate of transplantable tumours is not apparently affected by ultrasound at intensities below those which result in an appreciable temperature elevation within that tumour. Pizzarello *et al.* (1975) also claim that the same diagnostic apparatus (Picker, Ultrasonoscopy Model 102) slowed down the rate of regeneration of the amputated forelimb of the adult newt *Notophthalmus viridescens*. The regenerating amphibian limb would appear to offer many advantages as an experimental model, but has apparently not been adopted by other investigators.

Regenerating mammalian liver is another convenient non-embryological model of rapidly proliferating tissue. Kremkau and Witcofski (1974) exposed the livers of adult Charles River CD rats to c.w. 1·9 MHz ultrasound (spatial average intensity 60 mW/cm^2) before surgically removing about 70% of the liver mass. They reported a significant reduction in the number of cells undergoing mitosis in the irradiated livers compared with their mock-irradiated controls. Miller *et al.* (1976) and Barnett and Kossoff (1977) attempted to duplicate these observations but concluded that there was no difference in the number of mitotic figures in the control and the irradiated livers.

6.2.4 Effects on Immunological Tissues

The immunological system may be viewed in a simple manner as having two discrete functions; one is that of the humoural system (whose role is to produce antibodies to "foreign" materials) and the other is the reticuloendothelial system (whose role is to engulf "foreign" particles or materials to which antibodies have attached themselves). The effects of ultrasound on both of these functions are currently being actively investigated and some surprising results have been reported.

Anderson and Barrett (1979) irradiated the spleens of anaesthetized mice *in vivo* with ultrasound from a diagnostic apparatus (2 MHz; SATA intensity 8·9 mW/cm^2, SPTP 28 W/cm^2; 691 pulses/sec) for 1·6, 3·3 or 5 min. The animals were subsequently challenged with sheep erythrocytes and the authors report a small decrease in the IgM compartment of the immunoglobulin response which was also reflected in a decreased number of direct antibody plaque-forming cells present in the spleen compared to sham treated controls. Stratmeyer and his collaborators (personal communication) and Child *et al.* (1981) have attempted to duplicate these observations at similar and higher acoustic intensities, but were unable to demonstrate a positive effect. It should be noted that the reactive capacity of the mammalian immunological system is so large that even if the small "desensitization" reported by Anderson and Barrett (1979) is subsequently shown to be a repeatable effect, this would not constitute a significant functional impairment of the system in its ability to combat an invading organism.

Fung *et al.* (1978) exposed human peripheral lymphocytes to continuous wave ultrasound from a Sonicaid Foetal doppler device for zero to 30 min. They measured the uptake of tritiated thymidine over an 18 hour period one day after the ultrasound exposure and found that their results fell into two groups. In one group of individuals the lymphocytes showed a significant stimulation of uptake after short exposures (3–12 min) returning to control values after longer exposures (15–30 min). In the other group of individuals, their lymphocytes exhibited no stimulatory effects at short exposure times, but a significant reduction in uptake after 12–30 min exposures. Stimulation of the lymphocytes with phytohemagglutinin following the ultrasound exposure resulted in qualitatively similar results.

Anderson and Barrett (1981) have also investigated the effects of diagnostic intensities of ultrasound (SATA intensity 8·9 mW/cm^2 at 2 MHz, emitted by their same commercial apparatus—a Sperry echoencephalogram, type UM), on the ability of the reticuloendothelial system to phagocytose inert particles and bacteria. In one series of experiments they irradiated the livers of anaesthetized mice for either 1·6, 3·3 or 5 min and investigated the rate at which particles of colloidal carbon were being cleared from the animal's bloodstream. They reported that the half life of the exponential clearance curve was reduced when assayed immediately after the ultrasonic exposure, and that the magnitude of this reduction increased as the time interval between sonication and assay was increased. These observations are also perplexing in that there is an apparent negative "dose" correlation, i.e. longer exposure times have less effect even though the animals have received more ultrasonic power. Their experimental protocol can be seriously criticized on the grounds that the carbon particles were injected as a suspension in 0·25 ml of water containing 0·5% gelatine. The entire blood volume of a mouse is only of the order of about 3 ml and so the injection of 0·25 ml of water will result in the immediate osmotic lysis of a significant number of erythrocytes and ultimately result in a significant reduction in the osmolarity of the animal's blood plasma. The measured clearance rates therefore reflect the rate at which carbon particles coated with gelatine (i.e. thermally denatured collagen from another animal) are removed in a hypotonic environment in competition with free haemoglobin and cellular debris. These criticisms could be countered by repeating the experiments using carbon particles suspended in isotonic saline. These experiments have not been repeated in an independent laboratory, but Saad and Williams (1981) have found that continuous wave 0·75 MHz ultrasound began to affect the rate of clearance of blood-borne colloidal sulphur particles (injected in isotonic saline) in rats only at intensities greater than about 0·8 W/cm^2.

In another series of experiments, Anderson and Barrett (1981) irradiated the shaved abdomen of anaesthetized mice for 10 min using the same diagnostic apparatus described above and then isolated the free peritoneal macrophages by lavage. These macrophages were subsequently incubated with cultures of *Staphyloccoccus aureus*. It was found that a significantly larger fraction of the macrophages from the irradiated animals did not engulf bacteria which resulted in a lower net bacteriocidal rate. However, those sonicated macrophages which did engulf bacteria ingested approximately the same number of cells as their controls even though they were apparently less able to kill them once they were inside the body of the macrophage. These investigations merit being repeated in an independent laboratory.

6.2.5 Effects on the Cell Surface and its Properties

Siegel *et al.* (1979) exposed three human cell lines (kidney, embryonic kidney and amniotic fluid cells) growing in tissue culture on plastic Petrie dishes to the output of a diagnostic apparatus (SATA intensity 1·76 ±0·18 mW/cm^2) for 0·25–60 min. The authors concluded that there was a significant reduction in the attachment of the sonicated cells to the plastic surface even after only 0·5 min exposure. This effect was reported to be more marked in amniotic cells than in the kidney cell lines. Although the effects described are adequately controlled and the measurement criteria and analysis are convincing, some of the underlying assumptions are not necessarily valid. The experimental arrangement would have been greatly improved if the cells had been attached to an acoustically transparent membrane such as Cling-film or Saran Wrap® instead of to a rigid plastic because of the uncertainties involved with the nature of the acoustic interaction at the liquid/plastic interface. Local heating due to the formation of shear waves at the interface or the ultrasound-induced release of minute quantities of plasticizer or partially polymerized monomer from the plastic cannot be ruled out. These experiments ought to be repeated under similar but more rigorous experimental conditions.

Another possible explanation for the results obtained by Siegel *et al.* (1979) comes from the experiments reported by Liebeskind *et al.* (1981a). Balb/c 3T3 cells were grown in tissue culture and suspensions of the cells were exposed to the acoustic field of an Ekoline-20 clinical diagnostic ultrasonoscope (SATA intensity 15 mW/cm^2) for 30 min before being seeded into dishes and being left to grow. It was noted that sonicated cells developed abundant membraneous folds or projections (microvilli) from their surface as early as three days post sonication, and that they were still present up to 37 days later. When the cells grew to confluence (i.e. the cells were all touching their neighbours) the cells stopped growing and regained the normal smooth appearance characteristic of the controls. In a related series of experiments 3T3 mouse fibroblast cells were exposed to the same acoustic field and a time-lapse motion picture made of the subsequent cellular behaviour (Liebeskind *et al.*, 1981b). This motion picture film shows that the surface of the sonicated cells is thrown into a large number of folds or projections which apparently undergo frenetic bubbling activity when compared with their controls filmed at the same frame rate. In general, the sonicated cells exhibited less contact inhibition and did not spread well over surfaces. It is possible that this increased surface activity could be responsible for the decreased cellular adhesiveness reported by Siegel *et al.* (1979). It is surprising that the dramatic increase in cell surface activity observed by Liebeskind *et al.* (1981b) had not been noticed by earlier investigators making time-lapse motion pictures of the behaviour of sonicated fibroblasts (e.g. Mummery, 1978).

Other unconfirmed reports that diagnostic intensities of MHz ultrasound can change the physico-chemical nature of the cell surface are those of Hrazdira (1981) and Pinamonti *et al.* (1981). Hrazdira (1981) found that the electrophoretic mobility of human erythrocytes was transiently changed and the cells had a reduced tendency to aggregate after they had been exposed to diagnostic intensities of pulsed 6 MHz ultrasound. Similarly, Pinamonti *et al.* (1981) showed that pulsed beams of 8 or 10 MHz ultrasound as used in ophthalmology apparently removed some antigens from the surface of human erythrocytes. Even if these effects are proved to be reproducible and shown to occur *in vivo*, they should not constitute a significant hazard to the well-being of the irradiated animal.

6.2.6 Effects on the Cell Nucleus

The original quantitative test for the assessment of mutagenicity was to count the number of chromosomal abnormalities in preparations of treated cells. Recently, more sensitive tests have been developed such as sister chromatid exchange (SCE; i.e. the exchange of discrete portions of homologous genetic material between the sister chromatids) where there is apparently a linear relationship between the number of SCEs and the mutation frequency caused by a chemical mutagen or an ionizing radiation (Carrano *et al.*, 1978). There is still some disagreement concerning the precise mechanism of occurrence of SCE and their interpretation but it is generally accepted that an increased number of SCEs reflects an increased amount of repair activity within the nucleus, even though it is not adequate to replace the classical methods of analysis of structural chromosome damage (Gebhart, 1981).

Liebeskind *et al.* (1979a) exposed human lymphocytes to diagnostic intensities of pulsed ultrasound (SATA intensity 6·6 mW/cm^2; SATP intensity approximately 35 W/cm^2) and reported a small but statistically significant increase in SCE rates in all experiments. In contrast, Morris *et al.* (1978) exposed suspensions of human lymphocytes in Go-phase for 10 min to 15–36 W/cm^2) of continuous wave 1 MHz ultrasound. A slight but significant increase in SCE rates was found in the mock-irradiated cells relative to the controls which were not put into the sample chamber. However, the ultrasonic treatment had no detectable additional effect. The results of Morris *et al.* (1978) indicate that the manipulation of the cells during the various experimental procedures increases the SCE rates and so the results obtained by Liebeskind *et al.* (1979a) should be viewed with caution until confirmed.

Wegner *et al.* (1980) found no increase in the rate of SCE in Chinese hamster ovary cells growing in culture following ultrasound exposure. They also found no significant increases in chromosome breakage or DNA single strand breaks. Similar negative effects were obtained with fibroblast cells in culture by Barrass *et al.* (1982). Conversely, a positive increase in the frequency of SCE rates has been reported by Haupt *et al.* (1981). They exposed human peripheral lymphocytes *in vitro* to the 3·5 MHz pulsed acoustic field generated by a commercially available real time scanner (ADR Model 2130, delivering an SPTA intensity of 0·021 mW/cm^2 in the form of 0·89 μs long pulses at a rate of 2420 per sec). They observed a statistically significant increase in the number of cells having higher frequencies of occurrence of SCEs amongst the sonicated cells. The magnitude of this increase was small (an average of 1–2 SCE/s per cell) and there was considerable overlap between the control and test populations. Nevertheless, a positive dose response was obtained in that the number of SCEs apparently increased in a non-linear manner with increasing time of exposure, tending to an asymptotic value after exposures longer than about 10 min. The authors interpret their results in terms of a small sub-population of the lymphocytes which is apparently abnormally sensitive to ultrasound and undergoes many SCEs (Haupt *et al.*, 1981). While the dosimetry appears to be adequate, the exposure system leaves much to be desired in that the cell suspensions were irradiated inside closed plastic culture vessels while resting on a highly reflecting surface. It is not known how much of the incident acoustic energy entered the exposure chamber, or what effects the large surface area of hydrophobic plastic had on the acoustic nucleation conditions within that vessel.

Au *et al.* (1982) have extended ultrasonic SCE investigations into the *in vivo* situation. They exposed 11 and 15-day-old mice embryos *in utero* to 0·3–0·5 W/cm^2 of 2·0 MHz ultrasound for 15 min while measuring the maternal rectal temperature to ensure that it did not exceed 38°C. The foeti were removed, and minced tissues were scored for SCEs. They did not observe any significant increase in SCE frequency either following exposure to ultrasound or following a similar rectal temperature rise (from 35°C) caused by raising the temperature of the water bath in which the animals were partially immersed. Similar negative results were obtained with both the 11 and 15-day-old embryos. Lundberg *et al.* (1982) also observed no significant change in the SCE rates of human amniotic fluid cells irradiated *in vivo* for various times during amniocentesis compared with similar cells from patients who had not been exposed to any ultrasonic diagnostic investigations and whose amniocenteses were performed without the benefit of ultrasonic visualization.

Kurachi *et al.* (1981) exposed human lymphocytes *in vitro* to pulsed 2 MHz ultrasound (SATA intensity 0·5 W/cm^2, SATP intensity 50 W/cm^2, pulse width 10 s and repetition frequency 1000 Hz) and were unable to detect an increase in chromosomal abnormalities.

Another startling series of observations reported by Liebeskind *et al.* (1979b) were that diagnostic intensities of pulsed ultrasound (2·5 MHz; SATA intensity 6·6 mW/cm^2, pulse repetition rate 200 Hz, SATP intensity 35·4 W/cm^2) caused a variety of changes to occur in cultured HeLa cells. These changes included a small but transient increased immunoreactivity to antinucleoside antibodies for the cells in G_1 phase (i.e. the possible presence of single stranded or denatured DNA even though the authors were unable to demonstrate DNA strand damage by the moderately sensitive technique of alkaline-sucrose ultracentrifugation) and low levels of DNA repair synthesis. Other reported effects included a positive increase in the SCE rates of human lymphocytes and changes in the growth pattern of the same C3H mouse cell line which was used for the Liebeskind *et al.* (1981b) article. The C3H cells were seeded into Petri dishes and allowed to grow for 6 weeks when it was found that the sonicated cells had formed more abnormal focal colonies than their controls. Normal cells cease growing when they touch their neighbours (i.e. contact inhibition) and so the presence of focal colonies indicates that some of the cells have apparently lost this normal check on their growth and had continued to divide until they piled up on top of each other. When large numbers of cells from these abnormal foci were injected into mice some rapidly growing sarcomas were formed. While this article apparently demonstrates several potentially hazardous effects of diagnostic intensities of ultrasound on mammalian cells, it should be interpreted with caution. The work appears to have been well designed and performed with a thoroughness which makes it a more substantial and significant article than many of the other reports of possible deleterious biological effects caused by diagnostic intensities of ultrasound. However, the experimental procedures are so complex that interpretation of the results is extremely difficult. A major criticism of the experimental arrangement is that the cell suspensions were irradiated in a polypropylene tube through a window of Parafilm®. Preliminary experiments have shown that Parafilm® (a plastic film which may be stretched over the open end of a tube to seal it) is a very poor choice of material for what is supposed to be an acoustically-transparent window, and it is not known what chemical species may be leached from it during the irradiation procedure. These investigations ought to be repeated on a larger scale using an inert cylindrical sample chamber (e.g. glass) having acoustically transparent windows of an inert material such as cling film at either end.

6.2.7 Effects on Blood

Sanada *et al.* (1977) collected blood samples via an indwelling catheter from 10 patients whose hearts were being exposed to 0·5 mW/cm^2 (SATA intensity) of pulsed 2·25 MHz ultrasound as part of a routine clinical echocardiographic investigation. They found that some of the circulating platelets had undergone pseudopod formation (one of the earliest signs of platelet activation) and that the number of these morphologically altered platelets increased with time of exposure up to about 15 min. If these observations are shown to be repeatable, this may be an example of the increase in the number of "available micronuclei" resulting from the turbulent rheological conditions existing within the beating heart. However, it is more probable that the observed effect is an artifact caused by the progressive build up of thrombogenic materials around and within the indwelling catheter. Several studies have subjected non-turbulent blood *in vivo* to therapeutic intensities of continuous wave MHz ultrasound and have been unable to detect significant amounts of damage to any of the cellular elements of the blood (Chater and Williams, 1982; Howkins and Weinstock, 1970; Williams *et al.*, 1981).

Some perplexing *in vivo* haematological observations have been reported by Dauksha *et al.* (1975) and Yaroniene (1978). They investigated the long-term effects of exposure of rabbit hearts to diagnostic intensities of both continuous wave and pulsed 2 MHz ultrasound. They devised a lightweight portable ultrasonic generator which was attached to the rabbits so that the animals were irradiated for either 15 days or one month. The spatial average intensity in the continuous wave mode was 10 mW/cm^2 and the pulsed mode had a pulse duration of 4 μsec and a pulse repetition rate of 1 kHz giving a space and time averaged intensity of 0·36 mW/cm^2. Despite the marked differences in the acoustic parameters of the two exposures, the observed effects were remarkably similar and reputedly consisted of nonspecific changes such as capillary expansion, microthrombus formation, erythrocyte aggregation, vacuolization of hepatocytes and the emergence of binuclear cells in liver, etc.

Bause *et al.* (in press) investigated the osmotic fragilities of the erythrocytes of women whose foeti had been monitored during labour with a doppler diagnostic device. They found that eight control women who had not received ultrasound did not show any significant change between their individual pre- or postpartum median corpuscular fragilities. However, eight women monitored for 7 hours or more had a significant decrease in their mean corpuscular fragility. These observations are remarkable in that they imply either that the plasma membranes of the maternal erythrocytes had been rendered more fragile or that a significant fraction of these erythrocytes had irreversibly lost some ions (and water) so that their surface area/volume ratio had been increased making them more resistant to osmotic lysis. Both of these alternatives are extremely unlikely. It is more probable that the women who received ultrasound were more at risk from possible complications and therefore received additional attention and possibly higher levels of analgesic or anaesthetic agents which would dissolve in the erythrocyte membrane and consequently increase its surface area, resulting in a decreased osmotic fragility. Chiba *et al.* (1980) examined blood cells exposed to diagnostic intensities of pulsed ultrasound and were unable to detect adverse changes.

Takemura and Moriyaki (1977) reported that significant numbers of erythrocytes were lysed after 6–8 hours exposure *in vitro* to the acoustic field emitted by a commercially available doppler device (estimated c.w. intensity 10–20 mW/cm^2). They also report that the amount of haemolysis obtained using another commercially available diagnostic instrument emitting a pulsed ultrasonic beam was higher than that of the controls but much lower than that of the samples exposed to the c.w. doppler device. Experiments performed over time scales as long as these are fraught with experimental pitfalls, not least being the growth of bacteria and fungi and the death of the mammalian cells if the suspending medium had not contained adequate levels of glucose. It is therefore highly probable that these observations will not be repeatable if performed under more stringent experimental conditions. For example, Koh (1981) found no haemolysis after 2–12 hours exposure of blood *in vitro* to a doppler diagnostic device (intensity = 20 mW/cm^2).

6.2.8 Boundary or "Threshold" Curves

The rapid growth in the number of bioeffect publications appearing throughout the 1950s, 60s and 70s has prompted several authors to attempt to define the boundary zone between those acoustic exposure conditions which are known (or presumed) to result in positive biological effects and those acoustic exposure conditions which are believed to result only in negative effects. The most common approach has been simply to catalogue all the available bioeffect information and to display the data in graphical form using different symbols for positive and negative effects. One axis of the graph is usually a measure of the acoustic intensity to which that biological system has been exposed while the other axis is usually a measure of the duration of that exposure. A line or series of lines are then drawn between those portions of the intensity/time domain where positive bioeffects dominate and those portions where no confirmed biological effects have been demonstrated. It is then presumed that the acoustic exposures below this line represent a "safe" region where biological tissue may be exposed to ultrasound without the possibility of hazard. It is further presumed that the position of this boundary zone will be progressively lowered to smaller intensity/time combinations as more sensitive biological endpoints are investigated and effects are obtained at smaller acoustic "doses".

Figure 6.1 presents four of the most commonly quoted attempts to define this boundary zone. For the curves by Ulrich (1974), Nyborg (1977) and that based on the data compiled by Edmonds (1972) the intensity refers to the spatial peak time averaged value (i.e. the spatial peak intensity multiplied by the duty cycle if the sound was pulsed) and the time axis refers to the total time of the acoustic exposure (i.e. the "on" time plus the "off" time if the transducer was pulsed). However, for the curve by Wells (1974) the intensity is the "on" intensity and the time is the total "on" time of the transducer. Another difference is that the curve by Ulrich (1974) incorporates a "safety factor" of two, whereas the other curves were drawn immediately below the positive bioeffect results available at that time. All four curves are in reasonable agreement with the area above and to the right of each curve containing the zone of positive bioeffect reports.

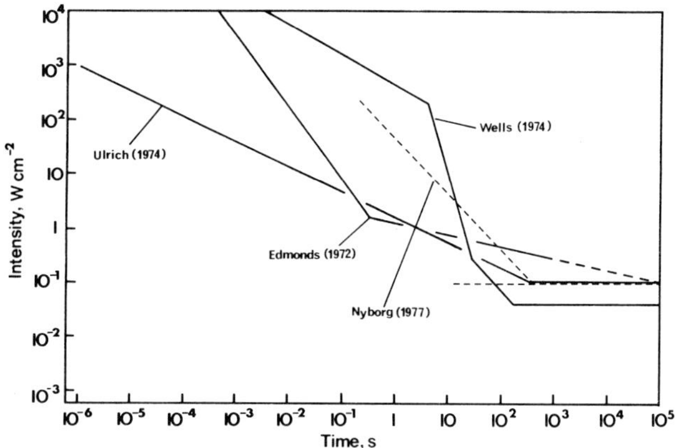

FIG. 6.1 A compilation of the most commonly quoted "threshold" or "boundary" curves. For details see text.

It should be noted that most of the positive bioeffect data points used to define the position of each curve represent the intensity/time combinations which happened to be used by those authors for their experiments and generally do not represent the minimum or "threshold" conditions below which that observed bioeffect did not occur. In addition, most of the data points have not been confirmed in an independent laboratory using a different experimental arrangement. Thus, these curves do not represent the "threshold" exposure conditions below which ultrasound has no biological effects. Instead, they indicate the "transition zone" where bioeffects may or may not be found to occur at those acoustic exposure levels depending upon the experimental arrangement and the sensitivity of the detection system.

The main advantage of these so-called boundary curves is that they enable clinical users of diagnostic equipment to subject their patients to ultrasound in the secure knowledge that their time-averaged exposure levels are well within the "safe" zone and are therefore not likely to be hazardous. However, these curves are also restricting in that any article reporting a positive bioeffect which apparently occurs at exposure conditions far below the existing boundary curves tends to be automatically discredited even though it may have been designed and executed to a much higher standard than some of the accepted positive bioeffect observations which were used to compile the original boundary curves. This is particularly important in view of the growing number of recent positive bioeffect reports obtained using commercially available diagnostic equipment as the source of the ultrasound. Most of these articles (some of which are briefly reviewed above) were not available when the various boundary curves were being compiled. While it is not possible to single out any one of these articles as proof that diagnostic intensities of ultrasound may be hazardous, when viewed as a whole they suggest that a number of subtle biological effects may occur at exposure levels significantly below the present boundary curves, i.e. at diagnostic exposure levels and/or using conventional diagnostic apparati.

Another interpretation is that the existing boundary curves are invalid in that they attempt to combine the biological effects produced by continuous wave and pulsed ultrasound on a common axis. This approach is only valid if the common measured parameters (i.e. "average" intensity and exposure time) are the major physical parameters determining the magnitude of the dominant damaging mechanism. This would be true if all of the observed biological effects were caused by a thermal mechanism. However, this would not necessarily be true if the damaging mechanism involved some form of cavitational activity or non-linear effect. In the latter case the peak pressure amplitude and its duration as well as the pulse repetition rate would be the crucial physical parameters to be measured and not the space and time averaged intensity. Here, one would expect to obtain different biological effects from continuous or pulsed mode applications even though the time and spaced averaged intensities may have been identical.

The report by Sarvazyan *et al.* (1980) is one of the few sources which indicate that different pulsing regimes can exert different biological effects even though the time averaged intensity and duty factor have been kept constant. They exposed entire frog embryos and explants of ectomesodermal tissue for 5–15 min to 0·88 MHz ultrasound at pulse repetition frequencies between 1 and 1000 Hz with a duty factor of 0·5, all at a space and time averaged intensity of 25 mW/cm^2. They observed that the tissue explants subjected to pulse repetition frequencies between 10 and 20 Hz were destroyed whereas those irradiated with continuous wave ultrasound or pulsed ultrasound at a PRF of 30–50 Hz continued to divide and roll as if imitating the gastrulation of entire embryos. These experiments ought to be repeated in an independent laboratory. If it can be confirmed that the pulse repetition rate and/or the peak pressure amplitude of an acoustic pulse are major determinants of the magnitude of a biological effect (and independent of the time averaged intensity) then it must result in a re-evaluation of our attitude towards the dosimetry of pulsed diagnostic equipment and raise new questions as to their potential hazard.

6.2.9 Synergistic Effects

Positive bioeffects may also be obtained at extremely low acoustic power densities if the ultrasound has a synergistic interaction with another modality, i.e. it potentiates or accentuates the tissue damaging effects of other potentially toxic agents such as ionizing radiation or chemicals. Thus, under certain circumstances it is possible that a diagnostic ultrasonic exposure together with a low concentration of certain drugs will result in an observed biological effect whereas neither agent would have had any detectable effects in isolation. However, most of the investigations searching for possible synergistic effects have been directed towards improving therapeutic efficiency, i.e. attempting to use therapeutic intensities of ultrasound to enhance the lethal effects of cytotoxic drugs and X-rays (Kremkau, 1982).

Kremkau (1979) has reviewed the literature concerned with the effects of ultrasound on the growth of tumour cells both alone and in combination with drugs and/or ionizing radiation. Studies reporting that ultrasound enhanced the toxicity of cytotoxic drugs *in vivo* include Armour *et al.* (1977), Kavetsky *et al* (1977) and Fry *et al.* (1978). These observations are supported by the *in vitro* results of Kaufmann and Kremkau (1978) who showed that mouse leukaemia cells irradiated with "sub-lethal" intensities of ultrasound in the presence of some cytotoxic drugs (e.g. cyclophosphamide's active metabolite and melphalan) were less infective when re-injected into host animals than similar cells treated with drugs alone. This effect was only observed with 5 out of the 10 cytotoxic drugs which were investigated and no synergistic effects were detected in the pioneering investigations of Hill (1967). Negative results were also obtained by Hering and Shepstone (1977) who were unable to demonstrate any synergistic effects between the cytotoxic agents vincristine or hydroxyurea and either pulsed or continuous 1 MHz ultrasound on the growth rates of the roots of *Zea mays*. However, it should be noted that the diffusion characteristics and hence the rate of penetration of these drugs through the plant cell wall and plasmalemmal membrane is different to that through the mammalian cell membrane and so these two experimental systems may not be directly comparable.

There are several applications in physiotherapy which suggest a possible synergistic interaction between ultrasound and various drugs. For example, Goralchuk and Koshik (1976) reported that rabbits having suppurative corneal ulcers induced with *Staphylococcus aureus* responded better and exhibited more tissue regeneration when they were treated with penicillin plus a 5 min treatment of $0{\cdot}4\ W/cm^2$ of 1·6 MHz ultrasound than with penicillin alone. Other reports are less well documented but apparently claim that ultrasonic irradiation markedly improved the therapeutic effectiveness of viricidal ointments (Fahim, 1978). All of these apparent synergistic effects imply or assume that the role of the ultrasound is to increase the penetration of the active chemical species into the cells or tissues to be affected, i.e. the phonophoretic effect. Some of the claims for the effectiveness of phonophoresis exceed the bounds of credulity, for example, Fahim (1978) is reputed to have radioactively labelled an ointment and applied it to the backs of rats. Treatment with an unspecified therapeutic level of ultrasound is said to have caused the ointment to have "penetrated at least $1\frac{1}{2}$ inches" into the animal's body. Other more believable reports include that of Dohnalek *et al.* (1965) who showed that a 10 min treatment using 800 kHz ultrasound caused a 3·6-fold increase in the rate of penetration of radioactive iodine through the skin of human volunteers as well as dogs and rabbits. Surprisingly, the phonophoretic properties of ultrasound have received relatively little attention despite their obvious importance and potential applications in therapy.

Other forms of synergistic effects may also exist. For example, Taylor and Pond (1972) demonstrated that hypoxia and ultrasound exhibited synergism in that the exposure time required to produce a haemorrhagic lesion in the rat spinal cord was decreased at low oxygen tensions.

There are several reports which indicate that therapeutic intensities of ultrasound may have a synergistic interaction with X-rays in reducing the growth rate of mammalian tumours *in vivo* (Gavrilov *et al.*, 1975; Kremkau, 1979; Spring, 1972; Woeber, 1965). Other investigators have observed a synergistic interaction between ultrasound and X-rays in one tumour type but not in another (Witcofski and Kremkau, 1978). Negative reports following *in vivo* irradiation by ultrasound and X-rays include Harkanyi *et al.* (1978) who found no increased damage to the chromosomes of mouse bone marrow cells. Unfortunately, many of these *in vivo* experimental series can be criticized on the basis of inadequate dosimetry or of having too subjective an end point giving rise to the possibility of observer bias if the experiments were not conducted in a double blind manner.

In vitro investigations where the living cells are irradiated in suspension and then inoculated into an animal or spread on a nutrient medium should yield less ambiguous results. Kunze-Mühl (1975) irradiated venous blood with X-rays and therapeutic intensities (3 W/cm^2) of 0·8 MHz ultrasound and observed a significant increase in the number of chromosome aberrations if the ultrasound was administered after the X-rays, but not if given before the X-rays. However, these same authors obtained exactly the opposite effect when they used low intensities (0·02 W/cm^2) of 2 MHz ultrasound plus X-rays. Todd and Schray (1974) reported that 0·14 W/cm^2 of 920 kHz ultrasound decreased the dose of 50 kV$_p$ X-rays required to prevent 99% of cultured Chinese hamster cells from forming colonies by a factor of 1·3 providing that the ultrasound exposure took place within 10 min of the X-ray treatment. However, Clarke and Hill (1970) found that exposure of L5 178y mouse lymphoma cells in suspension to ultrasound and/or X-rays did not result in an enhancement of the X-ray induced damage. Both X-rays and ultrasound have been reported to reduce the electrophoretic mobility of Erlich ascites tumour cells (i.e. their rate of migration in a potential gradient). Repacholi *et al.* (1971) reported that exposure of the cells to both modalities resulted in an additive (but not synergistic) reduction in their electrophoretic mobilities implying that both modalities exert their effects via independent mechanisms.

In a preliminary communication, Burr *et al.* (1978) demonstrated a highly significant (P 0·00001) relative increase in the number of chromosome aberrations observed in human lymphocytes *in vitro* when ultrasound was administered at the same time as, or immediately after 200 Rads of gamma irradiation. This synergistic effect was not observed if the ultrasound (1 MHz, c.w., 2 W/cm^2 for 30 min) was given either before the gamma rays or more than 2 hours afterwards. This observation may be the result of enhanced ultrasound-induced cavitational activity brought about by the production of nuclei within the suspending medium by the ionizing radiation.

It is not possible to derive any meaningful conclusions as to whether or not low levels of ultrasound may have a synergistic interaction with ionizing radiation from the experimental evidence which is currently available. Some of the mechanisms by which ultrasound may enhance the radiosensitivity of tumours and other cells have been discussed by Fleischer (1975). In general, these interaction mechanisms are so different that any additional effects resulting from exposure to therapeutic levels of ultrasound are most probably a reflection of the additional strain placed upon the regenerative capacity of the damaged cells. It is therefore unlikely that diagnostic intensities of ultrasound will have a synergistic interaction with ionizing radiation.

However, therapeutic intensities of ultrasound are likely to have an additive or synergistic interaction with ionizing radiation. In this case the additional damage probably results from the temperature rise induced within the irradiated tissues. Diagnostic intensities of ultrasound do not produce significant temperature elevations and so are unlikely to result in similar effects. However, synergistic interactions with cytotoxic or other chemical species appear to be due to a change in the rate at which these species are able to enter the cell (i.e. phonophoresis). It is not known which of the parameters of an acoustic wave are most important in the acceleration of transmembrane transport, and so it is still an open question as to whether or not diagnostic intensities of pulsed ultrasound may enable phonophoresis to occur.

6.3 THE EPIDEMIOLOGICAL APPROACH

In the final analysis, it is only the results of epidemiological surveys which can provide the information needed to assess the possible "harmfulness" of any given ultrasonic exposure. In essence, this approach is straightforward in that one merely has to compare a population which has been exposed to ultrasound with another population which is identical in every relevant respect except that it has not been exposed to ultrasound. However, this is extremely difficult to achieve in practice for several reasons. One is that the foetus is subjected to a diagnostic investigation whenever the clinician suspects that that pregnancy may be abnormal (i.e. there is a past history of abnormal pregnancies or there are some abnormal clinical signs). Thus, the uncritical comparison of the outcome of pregnancies which received ultrasound with those which did not receive ultrasound (because they appeared to be normal and uncomplicated) must yield the spurious result that ultrasound exposure is associated with an increased incidence of foetal abnormalities or obstetric complications. It is therefore apparent that those mothers-to-be who were going to deliver abnormal children or who were going to have obstetric complications even if they had not been subjected to ultrasound must be eliminated from the test group. Unfortunately, these women are best chosen with hindsight, i.e. after their children have been born and the abnormalities or complications have been verified. However, if this is done it is possible that some of these abnormalities may have been caused by the ultrasonic exposure and so their elimination from the test group would give a false negative result.

One of the largest and most frequently quoted epidemiological studies is that conducted by Hellman *et al.* (1970). They followed up 3297 women from the United States, Sweden and Scotland whose foeti had been exposed to diagnostic ultrasound on at least one occasion. After the "abnormal" pregnancies had been eliminated only 1114 patients were considered to have been completely normal and 35 of these had spontaneous abortions and were also omitted from the data analysis or were analysed separately. They concluded that neither the frequency of the ultrasound examination nor the time of the first examination seemed to have an adverse effect on the incidence of foetal abnormality or abortion, i.e. the incidence of abnormalities in the test group were no more common than those in the general population. This result is reassuring, but the study has been criticized on several points in addition to the possible bias introduced by the over enthusiastic elimination of abnormal pregnancies as discussed above. The nature of the ultrasonic exposure, its duration, the number of investigations performed and the output characteristics of the various diagnostic machines employed varied from country to country. The measured end points were crude (i.e. the visual detection of relatively gross abnormalities) and these would be expected to result from damage to the foetus at an early stage in its development whereas most of the ultrasonic investigations were performed in the third trimester of pregnancy when organogenesis is complete. Also, the follow-up period was limited to a brief period after birth so that no information is available about the possibility of subtle developmental or delayed effects on the growing children.

The major weakness in any epidemiological survey is the choice of the controls since they have to be as similar as possible to the test group in every relevant respect except that they have not been exposed to ultrasound. If the study is being devised before data collection has begun (i.e. it is a prospective study) then suitable controls can usually be found. However, it is more common to have a study which analyses data which has already been obtained (i.e. a retrospective survey) and in this case it is virtually impossible to find an adequate control group. Most investigators choose to compare the results obtained from their test group (after elimination of "abnormal" individuals, with demographic statistics derived from unrelated large-scale studies (where incidentally "abnormal" individuals have usually not been excluded).

Other epidemiological studies which are commonly quoted in support of the hypothesis that foetal diagnostic exposures are harmless include Bernstine (1969), Falus *et al.* (1972), Koranyi *et al.* (1972), Timmermans (1973) and Scheidt *et al.* (1978). Bernstine (1969) studied the effects of continuous wave ultrasound emitted by a commercial Doppler device on the progeny of 720 women. Falus *et al.* (1972) selected 400 apparently normal patients who had received a diagnostic ultrasonic investigation for follow-up, but only 171 responded to a questionnaire, and these were predominantly from the higher socio-economic groups. It is to be expected that this group would be better nourished and have a lower rate of obstetric complications than the population at large and so it is not surprising that no increase in gross developmental abnormalities was found. Scheidt *et al.* (1978) compared 297 infants whose mothers had received both amniocentesis and ultrasound with a similar group of 661 infants whose mothers had had amniocentesis only. A third group of 949 infants had not been exposed to either amniocentesis or ultrasound, but there was no control group that had been exposed to ultrasound alone. The authors collected data on 123 outcome variables obtained at approximately 1 year of age including a history of certain health problems, a physical and neurological examination and the Denver Developmental Screening Test. Patients whose pregnancies were terminated by abortion or infants who had been diagnosed as having genetic or metabolic defects via the amniocentesis were excluded from the analysis. The authors concluded that there were more abnormalities in grasp and tonic neck reflexes in the amniocentesis with ultrasound group than in the group with neither. However, there were no statistically significant differences between the amniocentesis with ultrasound group and the group with amniocentesis only.

The other major problem which bedevils all epidemiological surveys is the inability to obtain a sufficient number of individuals to be able to detect anything less than large-scale effects. The natural variability of any given biological parameter is so large that this parameter has to be measured in a relatively large number of individuals before a meaningful estimate of its mean or median magnitude and its standard deviation can be obtained. If ultrasonic exposure has caused a change in the magnitude or rate of occurrence of this parameter then the test population will have a different mean or median value but will probably still have a similar standard deviation. If the magnitude of the change in the mean of the test population is larger than both standard deviations, then it is relatively easy to prove that the control and test populations are significantly different. However, if the change in the magnitude of the mean value is smaller than the standard deviations then the ability of a statistical test to be able to discriminate between the test and control populations becomes critically dependent upon the number of individuals or measurements used to obtain each distribution. A discussion of the relevant statistics is beyond the scope of this book, but as a rough guide it should be appreciated that the number of individuals in any trial has to be increased 100-fold to achieve the same level of statistical significance if the rate of occurrence of any measurable event is decreased by a factor of ten. Thus, any small-scale epidemiological survey utilizing coarse endpoints which have a relatively high incidence of spontaneous variability is only able to detect large-scale effects. Some epidemiological surveys openly admit their statistical limitations; for example Scheidt *et al.* (1978) indicated that they only had a 75% chance of detecting a 3-fold increase in the rate of occurrence of an event which normally occurred with a 1% frequency, or put another way, only a 37% chance of detecting a 2-fold increase in adverse outcomes.

However, some larger studies are currently under way; for example Lyons and his collaborators intend comparing the offspring of 10 000 pregnancies exposed to ultrasound with a matched control group of 5000. In a preliminary report of this study Lyons and Coggraves (1979) found that more of the 2428 children irradiated with ultrasound *in utero* had developed neoplasms, but that there was apparently no increased incidence of congenital malformations, chromosomal abnormalities or developmental problems.

The results of some large-scale epidemiological surveys conducted in Japan have recently become available in the West. Koh (1981) examined the offspring of 6788 pregnant women who had been exposed to continuous wave Doppler ultrasound at a space averaged intensity of 20 mW/cm^2. There were no significant differences in the rate of spontaneous abortion or premature delivery, the incidence of foetal malformations or the foetal size for that gestational age between the exposed and control groups, and this result was apparently not influenced by the number of ultrasonic investigations performed or the timing of the first examination. Similar negative findings were reported by Mukubo *et al.* (1981) in a novel statistical study which involved comparing the incidence of neonatal defects in all of the births at the Tokyo University Hospital in the period before the introduction of ultrasound (1966–1968) with those obtained during the period of the enthusiastic use of Doppler ultrasound (1969–1974) and the period of use of real-time scanning (1978–1980). It is implied that every pregnancy admitted after 1968 was investigated using the ultrasonic technique currently in vogue. Before the introduction of ultrasound 3036 deliveries had 77 neonatal abnormalities (2·53%); the 5298 deliveries subjected to Doppler ultrasound had 134 birth defects (2·53%); while the 2429 deliveries investigated with a real time scanner had 53 abnormalities (2·13%). There was no statistical difference between either of these three groups (Mukubo *et al.*, 1981). The largest epidemiological study currently available is that by Morohashi and Iizuka (1977) who compared the offspring of 1950 pregnancies examined using Doppler ultrasound (SATA intensity less than 10 mW/cm^2) in the early stages of pregnancy when organogenesis had not been completed, with 10 133 pregnancies which had been irradiated during the later stages of pregnancy. It was found that there were no differences in the incidence of neonatal abnormalities between these two groups and that there was no increased rate of abortion in the group irradiated after organogenesis was complete. This is in agreement with the findings of Koh (1981) who found no increase in the rate of spontaneous abortions in 6788 pregnancies exposed to Doppler ultrasound. The lack of experimental details and descriptions of the criteria used to choose the control groups (if any) in these Japanese articles mean that it is virtually impossible to do anything other than take them at face value.

Another set of variables which further complicate the interpretation of any epidemiological survey is that the emphasis has had to have been placed on the physical characteristics of the ultrasonic procedure (e.g. whether the exposure was continuous or pulsed) rather than on the "dose" received by the foetus. This "dose" is dependent not only upon the output characteristics of the various machines used for the diagnostic investigations, but also upon the number of investigations, the stage of development of the foetus at the time of each irradiation and other less determinate variables such as the thickness and composition of the tissues between the transducer and the foetus, the duration of each exposure and the "dwell time" (i.e. the length of time that the acoustic beam is passing through any given tissue segment) of the transducer. Ideally, each epidemiological study should be subdivided so that each different exposure variable is studied independently. However, this would further decrease the sample numbers in each group with the consequent reduction in the discriminatory sensitivity of the statistical procedure.

Finally, it is especially difficult to decide which end points are to be evaluated in studies on human beings. Gross morphological abnormalities evident at the time of birth are the easiest parameter to measure, but they are unlikely to be produced by ultrasonic irradiation in the third trimester of pregnancy (which is the time when most diagnostic investigations are performed). Subtle effects such as behavioural changes may not manifest themselves until adulthood. It is therefore obvious that even the most sophisticated epidemiological trial cannot yield a meaningful estimate of the potential hazard resulting from ultrasonic exposure unless the "correct" endpoints have been measured and included in the trial. Unfortunately, we do not yet know which are the "correct" endpoints to be evaluated. If one also bears in mind the added limitations imposed by the choice of an inadequate control population and the selection of the individuals to be included in the test group, it is not surprising that nearly all the existing epidemiological surveys have reported negative effects. In fact, the practical and financial limitations inherent in the construction of the well-defined large-scale prospective epidemiological study necessary to detect any subtle small-scale effects of diagnostic ultrasound make it unlikely that such a survey will ever be devised.

6.4 SAFETY LEVELS FOR DIAGNOSTIC EXPOSURES

The bioeffects committee of the American Institute of Ultrasound in Medicine issued a statement in both 1976 and 1978 which attempted to set the stage for a more simple and rational approach to the assessment of a "safe level" for diagnostic exposures. The statement was based upon the bioeffect data which was used to compile the "Boundary Curves" presented in Fig. 6.1 and is specifically a verbal version of the curve proposed by Nyborg (1977):

Statement on Mammalian *in vivo* Ultrasonic Biological Effects (1976, 1978)

In the low megahertz frequency range there have been (as of this date) no demonstrated significant biological effects in mammalian tissues exposed to intensities* below 100 mW/cm^2. Furthermore, for ultrasonic exposure times † less than 500 seconds and greater than one second, such effects have not been demonstrated even at higher intensities when the product of intensity and exposure time† is less than 50 Joules/cm^2.

---

* Spatial Peak, Temporal Average intensity as measured in a free field in water (using a miniature hydrophone whose sensitive area is smaller than the distance over which the local value of the ultrasonic field intensity shows a significant variation).

† Total exposure time: this includes off-time as well as on time for a repeated-pulse regime.

It should be noted that most of the data used to derive this statement was obtained before 1975 and applies to mammals other than man. It is not clear how to relate this data to the human situation. The authors also state that "Data available at present on intensity levels at which bioeffects would occur are, in general, not minimum levels (if indeed, definite minima exist). Further research is urgently needed to determine whether significant biological changes occur at levels lower than those corresponding to the statement. As more results become available, it is reasonable to expect at least some lowering of the observed "threshold" levels for some biological systems, especially as more sensitive biological tests are discovered, and as more critical physical conditions are identified".

Some of the positive bioeffect reports obtained at diagnostic intensities of ultrasound are presented in Section 6.2 above. All of these reports occur at intensity levels below those referred to in the A.I.U.M. statement. While it cannot be said that any one of these reports constitutes a "demonstrated significant biological effect" so as to invalidate the statement, it does beg the question as to whether or not the statement would have used a lower "cut off" intensity (or would even have been written at all) if these reports had all been available at the time of its formulation.

The statement was intended to provide a starting point which may be helpful in arriving at recommendations for the confident use of ultrasound in medicine. Its authors acknowledge that "the statement does not, in itself, imply specific advice on 'safe levels' which might be universally valid. Determination of recommended maximum levels will require consideration of such difficult topics as: adequacy of present knowledge of bioeffects; expected reliability of equipment specifications; assessment of patient benefits; and others. So far these matters have not been treated systematically".

Unfortunately, despite the clear instructions of its authors, the statement has been used out of context as "proof" that diagnostic intensities of ultrasound are "safe". In some ways this could be a good thing in that it gives clinicians the confidence to treat their patients secure in the knowledge that they are not going to harm them. However, complacency usually leads to abuse in that unnecessary investigations may be undertaken for social, legal or financial reasons.

Evidence is now accumulating that the A.I.U.M. statement (which was derived mainly from experiments using continuous wave ultrasonic exposures) appears to be valid for continuous wave exposures similar to those used in many Doppler applications. However, it may not be valid for many pulsed mode applications, particularly those employing extremely short pulses of high spatial peak intensities. Many more experimental bioeffect investigations employing pulsed ultrasound are therefore needed so that a tentative "boundary" curve similar to those depicted in Fig. 6.1 can be constructed to determine whether or not a different, and perhaps lower, "threshold" may exist.

The major disadvantage of the tentative boundary curves presented in Fig. 6.1, and bioeffect statements similar to that presented above, is that they tend to be adopted by the regulatory agencies of various national governments and incorporated into legislation as the "scientific basis" for a maximum upper limit of emission from diagnostic devices. If the scientific community is certain that the boundary curves do represent the "lower limit of intensity" at which bioeffects occur, then the application of an upper limit will protect the foetal population from potentially hazardous acoustic intensities. However, there is insufficient evidence to support the belief that our present boundary curves do represent a "threshold" or lower limit, or even that "intensity" is the crucial physical parameter to be measured (especially in the case of pulsed ultrasonic exposures).

Another disturbing feature of having any legal maximum permitted value of acoustic emission for a diagnostic device is that there is the tendency for the manufacturers of an existing device which operates at an emitted intensity far below the legal maximum to increase his acoustic output so as to improve the signal to noise ratio of his equipment. Thus, the average ultrasonic "dose" received by the patients will gradually increase as the outputs of the various devices increase and converge towards the legal maximum. This could be particularly hazardous in the case of pulsed mode devices because their "threshold" intensity for the production of adverse bioeffects could be much less than that of continuous mode devices (especially if the mechanism of interaction is different and the dominant physical parameter is peak particle pressure and not intensity).

6.4.1 Summary

It is not possible at the present time (1982) to define any single acoustic intensity below which ultrasonic exposures will be "safe". It is unlikely that continuous wave exposures at megahertz frequencies *in vivo* at spatial peak intensities less than about 100 mW/cm^2 will be associated with detectable hazard (providing that areas of blood turbulence such as the heart are avoided and gas bubbles are not present). Pulsed mode diagnostic equipment emitting space and time averaged intensities significantly less than about 10 mW/cm^2 do not appear to be causing any detectable adverse changes *in vivo*. However, the scientific evidence underlying and complementing this belief is woefully inadequate and it is therefore imperative that more bioeffect investigations are conducted using pulsed ultrasound.

6.5 RECOMMENDATIONS

6.5.1 General

Despite our present inability to precisely quantify any potential risk arising from the diagnostic uses of ultrasound, its magnitude is almost certainly very much smaller than the risk inherent in obtaining the same diagnostic information using ionizing radiation. It is therefore recommended that ultrasound be used instead of X-rays wherever practicable, especially in those investigations involving the foetus or other rapidly growing tissues.

Until the potential risk associated with the diagnostic uses of ultrasound have been quantified it would be prudent to minimize the patient exposure by: (1) avoiding unnecessary investigations; (2) using the apparatus which emits the lowest acoustic intensity without sacrificing the quality of the clinical information; and (3) by minimizing the time for which the transducer is held in a stationary position (i.e. the dwell time).

The manufacturers of all diagnostic and therapeutic equipment should be encouraged or coerced into displaying the acoustic emission characteristics of their equipment in a prominent position on the outer casing of the apparatus. This information must include: (1) the emitted frequency; (2) the waveform if not sinusoidal; (3) a beam profile or its verbal description indicating any focal regions; (4) the maximum value of the spatial peak and time averaged intensity which could be attained within living tissues using that device (in the absence of reflections); and (5) whether the acoustic beam is continuous or pulsed. If the beam is pulsed, this description must also include an estimate of the temporal peak intensity during the pulse, the duration of each pulse, the interpulse interval (or the mark/space ratio or duty factor of the device), and a pictorial representation of the shape of the envelope enclosing the acoustic pulse.

In general, it would be more satisfactory if the scientific community could formulate a non-regulatory code of practice (in collaboration with the manufacturers of ultrasonic equipment) for the "safe use" of ultrasound in medicine. The Ultrasound Imaging

Section of the American National Electrical Manufacturers Association (N.E.M.A.) is currently developing such a voluntary code of practice or "safe-use" guideline in collaboration with the Bioeffects and Standards Committees of the American Institute of Ultrasound in Medicine (A.I.U.M.). This joint A.I.U.M./N.E.M.A. guideline sets out to define those parameters which relate to acoustic output levels and to specify the labeling requirements for their equipment. The guideline also sets forth the electrical and mechanical safety guidelines as well as attempting to summarize the known or suspected biological effects resulting from exposure to ultrasound.

A "safe-use" guideline has a number of advantages over a regulation or legal standard issued by a national government in that it can be introduced more rapidly and modified quickly in response to changes in technology or the discovery of adverse bioeffects. On the other hand, these guidelines have severe limitations in that they are a voluntary code of practice and therefore there is no compulsion for the individual manufacturers to comply with their recommendations. These guidelines are also less useful to the manufacturers in that they do not provide the detailed requirements on design, construction and functioning which would be embodied within a legal standard. Nevertheless, manufacturers would generally prefer a "safe-use" guideline to a legal standard because it permits greater freedom and flexibility and has less stringent recommendations. Given the present inadequate data base for the assessment of the potential hazard resulting from exposure to ultrasound, the adoption of "safe-use" guidelines provides a useful interim measure which anticipates the eventual promulgation of a legal standard.

6.5.2 Diagnostic Exposures

As a general rule humans should be exposed to diagnostic ultrasound only when there is a valid medical reason for its use, i.e. individuals should not be used to produce multiple test images when servicing equipment or for the commercial demonstration of new devices. Despite repeated statements issued by the various bioeffect committees, many manufacturers still persist in using live models (usually female and occasionally heavily pregnant) to assist in the selling of their diagnostic equipment by permitting clinicians to reproduce their routine clinical investigations using the new instrument on display. This is an extremely effective sales technique which satisfies the main objectives of the prospective purchasers (i.e. does this new machine yield new, better or more interpretable information than their existing equipment?). However, at a recent (1981) European meeting the foetus of one model in her third trimester of pregnancy received a total of about three hours ultrasonic exposure on only one day of the five day long meeting. It is recommended that the organizers of major clinical meetings forbid the use of pregnant women as aids to the demonstration of new equipment and encourage the use of inanimate phantoms (which can be a more rigorous and objective evaluation of the performance of any instrument) and pre-recorded audio-visual displays.

It would be prudent to avoid the use of diagnostic ultrasound to examine pregnant women during their first trimester of gestation since this is the period when the foetus will be especially sensitive to agents producing teratologic or more subtle developmental effects.

Each instrument should clearly display its output characteristics (as discussed above) in a standardized format so that the operators may elect to choose the device which emits the least acoustic power. Ideally, each instrument should be equipped with an adjustable power output so that the clinician can use the minimum acoustic exposure necessary to obtain the relevant diagnostic information. If it is found that a certain diagnostic procedure requires a relatively high acoustic exposure, then all the preliminary site detection and alignment procedures should be performed at a lower exposure level and the higher value used only for the data gathering portion of the investigation.

In the interest of patient safety, the duration of the ultrasonic investigation should be kept as low as is practicably possible. All of these recommendations are directed towards the goal of obtaining the least patient exposure compatible with retaining the optimum diagnostic information.

Apart from foetal exposures, the only other diagnostic investigations which presently give cause for concern are those involving the irradiation of the beating heart. The turbulent rheological conditions within the heart favour hydrodynamic nucleation as discussed above. This, coupled with the use of deliberately introduced gas bubbles (to increase the ultrasonic contrast) greatly enhances the probability of obtaining cavitational activity with the consequent "activation" of the blood coagulation system. It is therefore recommended that further investigations be performed to evaluate the potential hazard incurred by this procedure (especially in patients whose blood may be in a "hypercoagulable" state). Similarly, the long-term irradiation of the foetal heart as a technique for monitoring foetal distress during protracted labour ought to be used less frequently and supervised more rigorously.

6.5.3 Therapeutic Exposures

These devices pose particular problems for the various regulatory agencies because they emit acoustic intensities up to about 3–4 W/cm^2, i.e. well within the exposure range where numerous undesirable biological effects have been reported. According to Lehmann and de Lateur (1982) the fibrous connective tissues have to be heated to temperatures within the range 42–45°C before there is evidence of a beneficial therapeutic effect. If these devices are restricted to the same maximum intensity output as diagnostic devices (i.e. 100 mW/cm^2) then they are unlikely to be of any therapeutic value. There must therefore be an element of risk associated with any effective therapeutic ultrasound exposure. This can only be offset if the real or presumed benefits of the treatment are significantly greater than any potential risk. Unfortunately physiotherapists have generally been reluctant to carry out well-controlled trials to evaluate the effectiveness of their treatments. As a consequence, it is not possible to quantify the beneficial effects of therapeutic ultrasound despite the fact that many experienced therapists frequently quote spectacular results which have convinced them that their ultrasonic treatments are extremely effective. On the other hand there is as yet no evidence of adverse sequelae resulting from therapeutic exposures. This stalemate situation where one is trying to balance or equate two unquantified parameters is unsatisfactory in that any study which clearly indicates that there is a significant hazard associated with therapeutic treatments will probably lead to the banning of therapeutic ultrasound. It is therefore recommended that physiotherapists devise well-controlled clinical trials to evaluate the effectiveness of their ultrasonic treatments while they still have the opportunity to do so.

It is beyond the scope of this book to describe in detail the various experimental protocols leading to the design of a satisfactory clinical trial, but it might be worthwhile to indicate some of the inherent disadvantages of using patients attending a typical physiotherapy clinic to generate the required data. If it is decided that the initial hypothesis is that one wishes to determine whether or not ultrasound has a beneficial effect upon a given "condition", the first requisite is to assemble enough patients having that condition to ensure a reasonable probability of obtaining a statistically significant result. Unfortunately, there is usually a wide variability in the severity of any given condition in different individuals or even at different sites on that same individual. Also, each condition to be treated is at a different stage of development in that it is either getting worse, getting better or in a quasi-steady state as the body attempts unsuccessfully to combat it. Bearing in mind that each individual may respond to a different extent to the same treatment, it is apparent that a small scale trial can only be expected to yield a positive result if the treatment is dramatically effective. One way to circumvent these statistical limitations is to use each patient as his or her own control and to treat only one of a pair of symmetrical lesions or to treat only half of a large visible structure such as a bruise, a burn or a scar. If the treated half improves faster than its control then the treatment has been shown to be effective, and the control half may now be treated in the same way.

An alternative approach to the problem of the design of a clinical trial is not to treat the "condition" directly but to develop an experimental model which duplicates many (if not all) of the characteristics of that original condition, and to treat the model under rigorous experimental conditions (Williams, 1981). For example, if you wish to investigate the effect of ultrasound on bruises or burns, then you can produce reproducible pairs of these lesions in a painless manner (by suction in the case of a bruise or by heating anaesthetized skin in the case of a burn) on an acceptable site on adult human volunteers or experimental animals. One of each pair of lesions is now treated while the other is left as a control. The choice of a reproducible model system (where applicable) obviates most of the statistical limitations inherent in the use of patients and permits the evaluation of a broad range of treatment variables to determine which treatment regime (if any) is most likely to be effective in the clinic.

Different physiotherapists use widely different ultrasonic "doses" to treat similar conditions. This is partly a reflection of the different advice given by their teachers (which reflects different empirically-derived exposure regimes which different pioneer therapists believed to be effective) and partly a reflection of the acoustic outputs of the different machines which are currently available. Recent surveys where ultrasonic therapy machines in the clinic have been re-calibrated have shown that the majority of the instruments did not deliver the amount of acoustic energy indicated on their meter or output control (Stewart *et al.*, 1973; Repacholi and Benwell, 1979). Both the American and Canadian governments are consequently devising performance and accuracy specifications to ensure that this latter source of variability is reduced. In addition, the Canadian regulation (RED Act, 1981) stipulates that the maximum attainable spatial average intensity (in the absence of reflections) emitted by any device should not exceed 3 W/cm^2.

In practice, patients generally receive space and time averaged intensities significantly less than 3 W/cm^2. Providing that the patient has normal sensation and is fully conscious, the ultrasonic intensity is gradually raised until the patient experiences a sensation of mild warmth. If a tingling sensation or pain is experienced in the area of treatment then the power level must be reduced. This approach is preferable to the one adopted by many physiotherapists whereby the same extremely low "dose" is recommended for all patients having a given "condition". This has the twin disadvantages that the patient may be exposed to any hazards associated with the use of ultrasound whilst at the same time receiving an ultrasonic "dose" which may be too low to be effective.

The major recommendation which can be made regarding the therapeutic uses of ultrasound is that physiotherapists should conduct clinical trials to evaluate the effectiveness of their treatments. By this means, ineffective treatments can be identified and either eliminated or modified so as to be effective.

Patients who are anaesthetized or who have an impaired ability to detect or express the sensation of pain ought not to be exposed to ultrasound. Liberal amounts of coupling medium should be used to ensure efficient acoustic transmission and the transducer ought to be kept in motion throughout the duration of the treatment. The use of a stationary transducer ought to be strongly discouraged as it may result in the formation of standing waves and the development of cavitational activity as well as possibly overheating the tissues. Another disadvantage of using a stationary transducer mounted in a clamp or rack is that the therapist may be distracted or even leave the treatment room so that the patient who is experiencing discomfort (caused by too high an intensity) has no one to complain to and bears the pain until the therapist returns, with consequent thermal damage.

The major contraindications for the use of therapeutic ultrasound are: (1) the foetus should not be exposed under any circumstances and so ultrasound should never be used to alleviate back pain of muscular origin during pregnancy; (2) the beating heart should never be irradiated; (3) if possible, the irradiation of blood pools and major blood vessels should be avoided in patients with a history of thrombophlebitis or thromboembolic disorders; (4) the irradiation of tissues which have an ineffectual blood supply with high intensities of ultrasound should be avoided; (5) the subjection of the epiphyses of children's bones to excessive intensities should be avoided as this may restrict their rate of growth.

Therapists ought to ensure that they do not expose themselves unnecessarily to ultrasound by refraining from immersing their hands in the water bath during a treatment period. Similarly, therapists ought not to use any portion of their anatomy as a portable "meter" to check that their transducer is emitting acoustic energy. They should also report any device which is behaving abnormally (i.e. it elicits a feeling of warmth at an unusually low output intensity or else is unable to produce a feeling of warmth even at its maximum output). A therapist complained that one European device was unpleasant to operate in that her fingers tingled and became stiff after treating two or more patients. On examination it was found that due to a faulty design of the transducer mounting, the acoustic intensity at a portion of the therapist's fingertips was greater than that received by the patient (Oosterbaan, 1981).

6.5.4 Dental Exposures

Despite the large number of patients each year whose teeth are descaled using an ultrasonic device, there has been relatively little research on the effects of the treatment on the tooth, and most of this has been concerned with the efficiency of the cleaning process. In general, ultrasonic descalers can result in a smoother "burnished" appearance of the tooth surface (as seen by scanning electron microscopy) than manual descaling techniques (Pameijer *et al.*, 1972; Jones *et al.*, 1972), even though this depends largely upon the skill and experience of the operator. Forrest (1967) showed that dental hygienists at the beginning of their training course produced considerable scratching and gouging of the surface of artificial teeth whether they used an ultrasonic or hand-held probe. However, towards the end of their apparently one-year long course, the descaled teeth showed little or no evidence of macroscopic surface defects.

Ultrasonic descalers are potentially hazardous in that they operate at low ultrasonic frequencies (*ca.* 15–40 kHz) where the threshold for the generation of cavitational activity is much lower than that at megahertz frequencies. In fact, these devices utilize the cavitational activity produced within their cooling spray to assist in the removal of calcified plaque and other surface debris. It has been shown that dental descaling devices are efficient initiators of thrombus formation within mammalian blood *in vitro* (Williams and Chater, 1980), but it is not known if this ultrasonic energy may be conducted through the tooth and possibly initiate thrombus formation within the blood vessels of the pulp cavity. This is currently under investigation. It is nevertheless reassuring to know that despite the inherently destructive nature of this ultrasonic technique, there have been no reports of adverse sequelae (Goldman, 1961; Zach *et al.*, 1959) provided that the temperature within the pulp cavity is minimized by cooling (Frost, 1977).

The other potential hazards associated with the use of ultrasonic dental descaling devices include: (1) thermal damage resulting from a suboptimal flow of cooling water; (2) "gouging" of the surface of the enamel if the vibrating tip is not kept in motion; (3) conduction of the ultrasonic energy through the tooth and into the surrounding bone where it may initiate deleterious changes; (4) damage to the structures of the inner ear. The sensory elements of the cochlea and vestibule are normally protected from high displacement amplitude airborne sounds having frequencies of the order of tens of kilohertz by the relatively high absorption coefficient of air and by the inefficient mechanical conduction system of the eardrum and the ossicles of the middle ear. However, ultrasonic descaling circumvents this inbuilt protective mechanism and couples the ultrasonic energy directly to the bones of the skull via the tooth. Despite the high attenuation coefficient of bone (which is much less at kilohertz frequencies than it is at megahertz frequencies) and the diverging nature of the acoustic beam (i.e. the probe tip acts as a point source of ultrasound), the displacement amplitudes of the acoustic waves transmitted to the inner ear by bone conduction may be high enough to damage the sensory structures (especially while descaling the upper molars and premolars). This potential hazard has not been adequately investigated, but Möller *et al.* (1976) have reported that some patients complained of tinnitus and exhibited temporary threshold hearing shifts (a commonly accepted index of early damage to hearing) after their maxillary teeth had been descaled using an ultrasonic device.

In view of the sensitivities of the hearing mechanisms of the cochlea to ultrasound (Seda *et al.*, 1971) it is recommended that more studies be performed to quantify the magnitude of the potential auditory damage resulting from direct coupling through bone as well as from the airborne ultrasound (Arentsschild and Eichner, 1966; Hayakawa, 1972) during ultrasonic descaling.

One final potential hazard associated with ultrasonic descaling which has received scant attention is the damage to the soft tissues at the gingival margins. Ultrasonic cavitation within the cooling water in the vicinity of the gingivae could produce petechial lesions (Carson and Fishman, 1976) and also drive small particles of dislodged calculus (and bacteria) into the gingival tissues. Also, accidental contact with the probe tip results in areas of necrosis which may become infected. It is not known if the gingival damage associated with careless descaling results in an increased incidence of gingival infection.

6.5.5 Ultrasonic Cleaning Devices

These are usually metal baths having a low frequency (15–40 kHz) transducer attached to the underside of the bath by means of an epoxy resin. The bath is filled with an aqueous medium which usually contains a detergent or a caustic material to aid cleaning, and the mixture is driven to cavitate. The potential hazards of these devices fall into two distinct categories; one is the damage to the sensory structures of the inner ear resulting from airborne acoustic waves, and the other is the debridement of skin by the cavitation-induced streaming.

The airborne ultrasound emitted at the fundamental frequency of the device is not likely to pose a significant threat to hearing because of the relatively high attenuation coefficient of the air to the ultrasound and the inefficient mechanical transmission characteristics of the middle ear. However, the transient cavitation re-radiates a substantial amount of energy as white noise and also an intense "peak" at the first sub-harmonic of the driving frequency. This intense sub-harmonic signal is usually within or close to the audible range and consequently poses a significant threat to hearing unless the bath is located within a soundproofed cabinet or some form of ear protection is worn.

Operators of ultrasonic cleaning baths ought not to immerse their bare hands in the bath while it is cavitating. The acoustic energy is efficiently coupled from water into soft tissues and so cavitational activity may be generated within the tissues of the hand even though this was not detected in the hazardous experiments reported by Fishman (1968). Cavitational activity within the bath may produce petechial lesions on bare skin (Carson and Fishman, 1976) and will also remove the outermost layers of the epidermis exposing the more permeable cells which enables molecules of detergent or caustic material to penetrate the skin and initiate an allergic dermatitis. It is therefore recommended that rubber gloves must be worn if the hands have to be immersed in the bath while it is cavitating. The layer of air trapped between the glove and the fingers acts as an efficient barrier to acoustic transmission as well as the glove itself preventing debridement of the outer epithelial layers.

REFERENCES

Akamatsu, N. and Sekiba, K. (1977). Ultrasound irradiation effect on pre-implanted embryos. *Jap. J. Med. Ultrason.* **4**, 274–278.

Anderson, D. W. and Barrett, J. T. (1979). Ultrasound: a new immunosuppressant. *Clin. Immunol. Immunopathol.* **14**, 18–29.

Anderson, D. W. and Barrett, J. T. (1981). Depression of phagocytosis by ultrasound. *Ultrasound Med. Biol.* **7**, 267–273.

Arentsschild, O. and Eichner, K. (1966). Studies on the problems of hearing damage caused by the use of high speed dental turbines. *Zahnaerztl Rundsch.* **75**, 217–223.

Armour, E., Corry, P. and McGinness, J. (1977). Preferential cytotoxicity of cultured melanoma cells by ultrasound and melanin binding drugs. *Radiat. Res.* **70**, 690.

Au, W. W., Obergoenner, N., Goldenthal, K. L., Corry, P. M. and Willingham, V. (1982). Sister-chromatid exchanges in mouse embryos after exposure to ultrasound *in utero*. *Mutation Res.* **103**, 315–320.

Barnett, S. B. and Kossoff, G. (1977). Negative effect of long duration pulsed ultrasonic irradiation on the mitotic activity in regenerating rat liver. *In* "Ultrasound in Medicine" (Eds D. N. White and R. E. Brown), Vol. 3, pp. 2033–2044. Plenum Press, New York.

Barrass, N., ter Haar, G. and Casey, G. (1982). The effect of ultrasound and hyperthermia on sister chromatid exchange and division kinetics of BHK 21 C13/A3 cells. *Brit. J. Cancer* **45**, 187–191.

Bernstine, R. L. (1969). Safety studies with ultrasonic Doppler technique—a clinical follow-up of patients and tissue culture study. *Obstet. Gynecol.* **34**, 707–709.

Brodie, D. A. and Hanson, H. M. (1960). A study of the factors involved in the production of gastric ulcers by the restraint technique. *Gastroenterology* **38**, 353–360.

Brown, N. T., Galloway, W. D. and Henton, W. W. (1981). Reflex development following *in utero* exposure to ultrasound. Paper No. 1301 at the San Francisco Meeting of the AIUM, Aug 17–21, 1981, p. 119.

Burr, J. G., Wald, N., Pan, S. and Preston, K. Jr (1978). The synergistic effect of ultrasound and ionizing radiation on human lymphocytes. *In* "Mutagen-Induced Chromosome Damage in Man" (Eds H. J. Evans and D. C. Lloyd), pp. 120–128. Yale University Press, New Haven, Connecticut.

Carrano, A. V., Thompson, L. H., Lindl, P. A. and Minkler, J. L. (1978). Sister chromatid exchange as an indicator of mutagenesis. *Nature* **271**, 551–553.

Carson, T. E. and Fishman, S. S. (1976). Biological effects of ultrasound: skin and cutaneous blood vessels. Proc. *West. Pharmacol. Soc.* **19**, 36–39.

Cataldo, F. L., Miller, M. W., Gregory, W. D. and Carstensen, E. L. (1973). A description of ultrasonically induced chromosomal anomalies in *vicia faba*. *Rad. Botany* **13**, 211–213.

Cataldo, F. L., Miller, M. W. and Kaufman, G. E. (1976). Agglomerated nuclei in ultrasonicated *vicia faba* roots: a partial elucidation of their derivation. *Environ. Exp. Botany* **16**, 69.

Chater, B. V. and Williams, A. R. (1982). Absence of platelet damage following the exposure of non-turbulent blood to therapeutic ultrasound. *Ultrasound Med. Biol.* **8**, 85–87.

Chiba, Y., Suehara, N., Sakumoto, T. and Kurachi, K. (1980). The studies on the effect of pulsed ultrasound on the physiology of the mother and the fetus. *In* "The 1979 Report of the Research Grant of the Prevention of Physical and Mental

Disabilities, The Ministry of Health and Welfare, Japanese Government", pp. 179–187.

Child, S. Z., Carstensen, E. L. and Lam, S. K. (1981a). Effects of ultrasound on Drosophila. III. Exposure of larvae to low-temporal-average-intensity, pulsed irradiation. *Ultrasound Med. Biol.* **7**, 167–173.

Child, S. Z., Hare, J. D., Carstensen, E. L., Vives, B., Davis, J., Adler, A. and Davis, H. T. (1981b). Test for the effects of diagnostic levels of ultrasound on the immune response of mice. *Clin. Immunol. Immunopathol.* **18**, 299–302.

Church, C. C., Flynn, H. G., Miller, M. W. and Sacks, P. G. (1981). Rotation of exposure tubes as a requirement for ultrasonically induced cell lysis. *J. Acoust. Soc. Amer.* Suppl. 1, **69**, S27.

Clarke, P. R. and Hill, C. R. (1970). Physical and chemical aspects of ultrasonic disruption of cells. *J. Acoust. Soc. Amer.* **47**, 649–653.

Dauksha, K. K., Sayauskas, S. I. and Yaronene, G. V. (1975). Results of experimental investigation of long-term action of low-intensity ultrasonic waves on the organism. *In* "Abstracts of II All-Union Conference on Ultrasound in Physiology and Medicine", pp. 78–80. Ulyanovsk.

David, H., Weaver, J. B. and Pearson, J. F. (1965). Doppler ultrasound and fetal activity. *Brit. Med. J.* **2**, 62–64.

Docker, M. F. and Petoussi, N. (1982). Does pulsed ultrasound stimulate the foetus? Fifth world congress of ultrasound in medicine and biology. *Ultrasound Med. Biol.* **8**, Suppl. 1, p. 45.

Dohnalek, J., Hrazdira, I., Cecava, Novak and Svoboda, (1965). Penetration of radioiodine through the skin promoted by ultrasound. *Co. Desm.* **40**, 173–176.

Edmonds, P. D. (1972). Interactions of ultrasound with biological structures—a survey of data. *In* "Interaction of Ultrasound and Biological Tissue" (Eds J. M. Reid and M. R. Sikov), pp. 299–317. DHEW Publ. (FDA) 73-8008.

Fahim, M. S. (1978). A transcript of his presentation to the Federation of American Societies for Experimental Biology entitled "Does ointment plus ultrasound clear genital herpes?" appeared in *Medical World News*, May 1, pp. 14–15.

Falus, M., Koranyi, G., Sobel, M., Pesti, E. and Trink, V. B. (1972). Follow-up studies on infants examined by ultrasound during the fetal age. *Orv. Hetil.* **13**, 2119–2121.

Fishman, S. S. (1968). Biological effects of ultrasound: *in vivo* and *in vitro* haemolysis. *Proc. West. Pharmacol. Soc.* **11**, 149–150.

Fleischer, A. C. (1975). Biological basis for ultrasonic enhancement of radiosensitivity. *In* "Ultrasound in Medicine" (Ed. D. N. White), Vol. 1, p. 567. Plenum Press, New York.

Forrest, J. O. (1967). Ultrasonic scaling—a 5-year assessment. *Brit. Dent. J.* **122**, 9–14.

Frost, H. (1977). Heating under dental scaling conditions. *In* "Symposium on Biological Effects and Characterizations of Ultrasound Sources" (Eds D.-W. G. Hazzard and M. L. Litz), pp. 64–76. HEW Publication (FDA) 78-8048.

Fry, F. J., Johnson, L. K. and Erdmann, W. A. (1978). Interaction of ultrasound with solid tumours *in vivo* and tumour cell suspensions *in vitro*. *In* "Ultrasound in Medicine" (Eds D. N. White and E. A. Lyons), Vol. 4, pp. 587–588. Plenum Press, New York.

Fulton, J. F. (1951). "Decompression Sickness: Caisson Sickness, Diver's and Flier's Bends and Related Syndromes". W. B. Saunders, Philadelphia.

Fung, H., Cheung, K., Lyons, E. A. and Kay, N. E. (1978). The effect of low dose ultrasound on human peripheral lymphocyte function *in vitro*. *In* "Ultrasound in

Medicine" (Eds D. N. White and E. A. Lyons), Vol. 4, pp. 583–586. Plenum Press, New York.

Gavrilov, L. R., Kalenlo, G. S. and Ryabukhin, V. V. (1975a). Ultrasonic enhancement of the gamma radiation of malignant tumours. *Sov. Phys. Acoust.* **21**, 119.

Gavrilov, L. R., Tsirulnikov, E. M. and Shchekanov, E. E. (1975b). Responses of the auditory centers of the frog midbrain to labrinth stimulation by focused ultrasound. *Fiziol. Zh. S.S.S.R.* **61**, 213–221.

Gebhart, E. (1981). Sister chromatid exchange (SCE) and structural chromosome aberration in mutagenicity testing (Review Article). *Human Genetics* **58**, 235–254.

Goldman, H. M. (1961). Histologic assay of healing following ultrasonic curettage versus hand-instrument curettage. *Oral Surg.* **14**, 925–928.

Goralchuk, M. V. and Koshik, T. F. (1976). The effects of ultrasound on histological and histochemical changes in the healing process of suppurative ulcers of the cornea. *Oftalmol. Zh.* **31**, 533–555.

Harkanyi, Z., Szollar, J. and Vigvari, Z. (1978). A search for an effect of ultrasound alone and in combination with X-rays on chromosomes *in vivo*. *Brit. J. Radiol.* **51**, 46–49.

Haupt, M., Martin, A. O., Simpson, J. L. Iqbal, M. A., Elias, S., Dyer, A. and Sabbagha, R. E. (1981). Ultrasonic induction of sister chromatid exchanges in human lymphocytes. *Human Genetics* **59**, 221–226.

Hayakawa, W. (1972). Effects of ultrasonics on the cochlea of bats and guinea pigs. *J. Otolaryngol. Japan* **75**, 627–637.

Hellman, L. M., Duffus, G. M., Donald, I. and Sunden, B. (1970). Safety of diagnostic ultrasound in obstetrics. *Lancet* **1**, 1133–1134.

Hering, E. R. and Shepstone, B. J. (1977). Observations on the combined effect of ultrasound and certain cytotoxics on the growth of the roots of *Zea mays*. *Brit. J. Radiol.* **50**, 32–37.

Hertz, R. H., Timor-Tritsch, I., Dierker, L. J., Chik, L. and Rosen, M. C. (1979). Continuous ultrasound and fetal movement. *Amer. J. Obstet. Gynecol.* **135**, 152–154.

Higashiizumi, T., Tashiro, H., Sakamoto, K. and Kanai, F. (1979). Proc. 2nd Meeting of the World Federation of *Ultrasound Med. Biol.* Japan, p. 372.

Hill, C. R. (1967). Changes in tissue permeability produced by ultrasound. *Brit. J. Radiol.* **40**, 317.

Hill, C. R. (1972). Ultrasonic exposure thresholds for changes in cells and tissues. *J. Acoust. Soc. Amer.* **52**, 667–672.

Howkins, S. D. and Weinstock, A. (1970). The effect of focussed ultrasound on human blood. *Ultrasonics* **8**, 174–176.

Hu, J. H. and Ulrich, W. D. (1976). Effects of low-intensity ultrasound on the central nervous system of primates. *Aviat., Space, and Environ. Med.* June, pp. 640–643.

Jones, S. J., Lozdan, J. and Boyde, A. (1972). Tooth surfaces treated *in situ* with peridontal instruments. *Brit. Dent. J.* **132**, 57–64.

Kaufmann, J. S. and Kremkau, F. W. (1978). Influence of ultrasound on mouse leukemia cell DNA synthesis, membrane integrity, and uptake of anticancer drugs *in vitro*. *In* "Ultrasound in Medicine" (Eds D. N. White and E. A. Lyons), Vol. 4, pp. 589–590. Plenum Press, New York.

Kavetsky, R. E., Balitsky, K. P. and Baran, L. A. (1977). Use of ultrasound energy of various intensity in therapy of malignant neoplasm. *In* "Ultrasound in Medicine" (Eds D. N. White and R. E. Brown). Vol. 3B, p. 2111. Plenum Press, New York.

Khokhar, M. T. and Oliver, R. (1975). An investigation of chromosome damage in *vicia faba* root tips after exposure to 1·5 MHz ultrasonic radiation. *Int. J. Radiat. Biol.* **28**, 373.

Koh, S. (1981). The safety of diagnostic continuous wave ultrasonic irradiation—a clinical study. Serum haemoglobin level and scanning electron microscopic finding of maternal and cord blood *in vitro*. *Acta Obstet. Gynaec. Jpn.* **33**, 469–478.

Koranyi, G., Falus, M., Sobel, M., Pesti, E. and van Bao, T. (1972). Follow-up examination of children exposed to ultrasound *in utero*. *Acta Paediatr. Acad. Sci. Hung.* **13**, 231–234.

Kremkau, F. W. (1979). Cancer therapy with ultrasound: a historical review. *J. Clin. Ultrasound* **7**, 287–300.

Kremkau, F. W. (1982). Ultrasonic treatment of experimental animal tumours. *Brit. J. Cancer* **45**, 226–232.

Kremkau, F. W. and Witcofski, R. L. (1974). Mitotic reduction in rat livers exposed to ultrasound. *J. Clin. Ultrasound* **2**, 123–126.

Kunze-Mühl, E. (1975). Chromosome damage in human lymphocytes after different combinations of X-ray and ultrasonic treatment. *In* "Proceedings Second European Conference on Ultrasonics in Medicine" (Eds E. Kazner *et al.*), pp. 3–9. Excerpta Medica, Amsterdam.

Kurachi, K., Chiba, Y., Suehara, N. and Sakumoto, T. (1981). Studies on the effect of pulsed ultrasound on chromosome and erythrocyte, and optimal utility of ultrasound diagnosis in early pregnancy. *Jap. J. Ultrasound Med.* Special Edition, "Studies on the Safety of Pulsed Ultrasound in the Diagnosis of the Fetus during Pregnancy" **8** (4), 271–273.

Lehmann, J. F. and de Lateur, B. J. (1982). Therapeutic heat. Chapter 10 in "Therapeutic Heat and Cold—Third Edition". Williams and Wilkins, Baltimore.

Lele, P. P. (1975). Ultrasonic teratology in mouse and man. Proc. Second Europ. *Congress on Ultrasonics in Med.* Munich, May 12–16.

Lele, P. P. (1977). Thresholds and mechanisms of ultrasonic damage to "organised" animal tissues. *In* "Symposium on Biological Effects and Characterizations of Ultrasound Sources" (Eds D.-W. G. Hazzard, and M. L. Litz), pp. 224–239. DHEW Publication (FDA) 78-8048.

Liebeskind, D., Bases, R., Mendez, F., Elequin, F. and Koenigsberg, M. (1979a). Sister chromatid exchanges in human lymphocytes after exposure to diagnostic ultrasound. *Science* **205**, 1273–1275.

Liebeskind, D., Bases, R., Elequin, F., Neubort, S., Leifer, R., Goldberg, R. and Koenigsberg, M. (1979b). Diagnostic ultrasound: Effects on the DNA and growth patterns of animal cells. *Radiology* **131**, 177–184.

Liebeskind, D., Bases, R., Koenigsberg, M., Koss, L. and Raventos, C. (1981a). Morphological changes in the surface characteristics of cultured cells after exposure to diagnostic ultrasound. *Radiology* **138**, 419–423.

Liebeskind, D., Padawer, K., Wooley, R. and Bases, R. (1981b). Diagnostic ultrasound: Time-lapse and transmission electron microscopic studies of cells insonated *in vitro*. Proc. 10th L.H. Gray Conf., Oxford, England, July 13–16.

Love, L. A. and Kremkau, F. W. (1980). Intracellular temperature distribution produced by ultrasound. *J. Acoust. Soc. Amer.* **67**, 1045–1050.

Lundberg, M., Jerominski, L., Livingston, G., Kochenour, N., Lee, T. and Fineman, R. (1982). Failure to demonstrate an effect of *in vivo* diagnostic ultrasound on sister chromatid exchange frequency in amniotic fluid cells. *Amer. J. Med. Genetics* **11**, 31–35.

Lyons, E. A. and Coggraves, M. (1979). Follow-up study in children exposed to ultrasound *in utero*—an interim report. Paper presented at the Montreal Meeting of the A.I.U.M.

Mackintosh, I. J. C. and Davey, D. A. (1970). Chromosome aberrations induced by an ultrasonic foetal pulse detector. *Brit. Med. J.* **4**, 92–93.

Mackintosh, I. J. C., Brown, R. C. and Coakley, W. T. (1975). Ultrasound and "in vitro" chromosome aberrations. *Brit. J. Radiol.* **48**, 230–232.

McClain, R. M., Hoar, R. M. and Saltzman, M. B. (1972). Teratologic study of rats exposed to ultrasound. *Amer. J. Obstet. Gynecol.* **114**, 39–42.

Meltzer, R. S. and Roelandt, J. (1982). Transmission of echocardiographic contrast through the lungs. *In* "Recent Advances in Ultrasound Diagnosis—3" (Eds A. Kurjak and A. Kratochwil), pp. 439–442. Excerpta Medica, Congress Series 553.

Miller, D. L., Nyborg, W. L. and Whitcomb, C. C. (1979). Platelet aggregation induced by ultrasound under specialised conditions *in vitro*. *Science* **205**, 505–507.

Miller, M. W. (1978). Comparison of micronuclei induction for X-ray and ultrasound exposures of *vicia faba* root meristern cells. *Ultrasound Med. Biol.* **4**, 263–267.

Miller, M. W., Kaufman, G. E., Cataldo, F. L. and Carstensen, E. L. (1976). Absence of mitotic reduction in regenerating rat livers exposed to ultrasound. *J. Clin. Ultrasound* **4**, 169–172.

Möller, P., Grevstad, A. O. and Kristoffersen, T. (1976). Ultrasonic scaling of maxillary teeth causing tinnitus and temporary hearing shifts. *J. Clin. Periodont.* **3**, 123–127.

Morohashi, T. and Iizuka, R. (1977). Symposium on recent studies in the safety of diagnostic ultrasound. The development of low power ultrasonic instruments. *Jap. J. Med. Ultrasound* **4**, 271–273.

Morris, S. M., Palmer, C. G., Fry, F. J. and Johnson, L. K. (1978). Effect of ultrasound on human leucocytes. Sister chromatid exchange analysis. *Ultrasound Med. Biol.* **4**, 253–258.

Mukubo, M., Okai, T., Uezuma, S., Baba, K., Minoura, S., Kumagai, K., Hara, K. and Sakamoto, S. (1981). The safety and irradiation effect of ultrasonic real time scanner on fetal development. *Nippon Choompa Igakkai Koen-Rombunshu* **38**, 555–556.

Mummery, C. L. (1978). The effect of ultrasound on fibroblasts *in vitro*. Ph.D. Thesis, University of London.

Murai, N., Hoshi, K. and Nakamura, T. (1975a). Effects of diagnostic ultrasound irradiated during fetal stage on development of orienting behaviour and reflex ontogeny in rats. *Tohoku J. Exp. Med.* **116**, 17–24.

Murai, N., Hoshi, K., Kang, C-H. and Suzuki, M. (1975b). Effects of diagnostic ultrasound irradiated during foetal stage on emotional and cognitive behaviour in rats. *Tohoku J. Exp. Med.* **117**, 223–233.

Nagy, I. I. and Mihalas, G. I. (1974). Bioelectric potential modification under low intensity ultrasonic field. Proc. Eighth Intern. Congr. Acoustics, London, 1974, p. 363.

Nyborg, W. L. (1977). "Physical Mechanisms for Biological Effects of Ultrasound", pp. 1–59. DHEW Publ. (FDA) 78-8062.

Oosterbaan, W. A. (1981). A potentially harmful aspect of ultrasonic therapy transducers. *In* "Recent Advances in Ultrasonic Diagnosis 3" (Eds A. Kurjak and A. Kratochwil), pp. 27–28. Excerpta Media, Amsterdam.

Pameijer, C. H., Stallard, R. E. and Hiep, N. (1972). Surface characteristics of teeth following peridontal instrumentation: a scanning electron microscope study. *J. Periodont.* **43**, 628–633.

Pizzarello, D. J., Wolsky, A., Becker, M. H. and Keegan, A. F. (1975). A new approach to testing the effect of ultrasound on tissue growth and differentiation. *Oncology* **31**, 226–232.

Pizzarello, D. J., Vivino, A., Madden, B., Wolsky, A., Keegan, A. F. and Becker, M. (1978a). Effect of pulsed low-power ultrasound on growing tissues. I. Developing mammalian and insect tissue. *Exp. Cell Biol.* **46**, 179–181.

Pizzarello, D. J., Vivino, A., Newall, J. and Wolsky, A. (1978b). Effect of pulsed low-power ultrasound on growing tissues. II. Malignant tissues. *Exp. Cell Biol.* **46**, 240–245.

Powell-Phillips, W. D. and Towell, M. E. (1979). Doppler ultrasound and subjective assessment of fetal activity. *Brit. Med. J.* **2**, 101–102.

RED Act (1981). Radiation Emitting Devices Act—Ultrasound Therapy Devices Regulations, pp. 1121–1126. Canada Gazette Part II, Vol. 115, No. 8, 24 April 1981.

Repacholi, M. H. and Benwell, D. A. (1979). Using surveys of ultrasound therapy devices to draft performance standards. *Health Phys.* **36**, 679–686.

Repacholi, M. H., Woodcock, J. P., Newman, D. L. and Taylor, K. J. W. (1971). Interactions of low-intensity ultrasound and ionizing radiation with the tumor cell surface. *Phys. Med. Biol.* **16**, 221–227.

Saad, A. H. and Williams, A. R. (1981). Effects of therapeutic ultrasound on clearance rate of blood-borne colloidal particles *in vivo*. Proc. 10th L. H. Gray Conference, Oxford, England, July 13–16.

Sakamoto, S., Mukuboh, M., Okai, T. and Hara, K. (1981). Safety of pulsed ultrasound—Experimental and Epidemiological studies. *Jap. J. Ultrasound Med.* Special Edition, "Studies on the Safety of Pulsed Ultrasound in the Diagnosis of the Fetus during Pregnancy" **8** (4), 283–285.

Sanada, M., Hattori, A., Watanabe, T., Shu, T., Kasahara, T., Ohn, M. and Tamura, K. (1977). The *in vivo* effect of ultrasound upon human blood platelets. Nikon Choompa Igakukai, *Koen rombunshu* Nov., 149–150.

Sarvazyan, A. P., Belousov, L. V., Petropavlovskaya, M. N. and Ostroumova, T. V. (1980). The interaction of low intensity ultrasound with developing embryos. Paper C18 at the Ultrasound Interaction in Biol. and Med. Symposium, Reinhardsbrunn, East Germany, Nov. 10–14.

Scheidt, P. C., Stanley, F. and Bryla, D. A. (1978). One year follow-up of infants exposed to ultrasound *in utero*. *Amer. J. Obstet. Gynecol.* **131**, 743–748.

Seda, H. J., Suga, F. and Snow, J. B. Jr (1971). Ultrasonic irradiation on the inner ear. *Laryngoscope* **81**, 510–523.

Sekiba, K., Kawai, J., Akamatsu, N., Obata, A., Niwa, K. and Utsumi, K. (1980). Ultrasound irradiation effects on embryos (9). Effects of continuous wave on rat embryo (2). *Nippon Choompa Igakkai Koen-Rombunshu* **37**, 157–158.

Sheldon, T. A. (1978). Assessment of fetal movement. *Brit. Med. J.* **2**, 505.

Shimizu, T. and Tanaka, K. (1980). Experimental teratology of ultrasound exposure in animals. *In* "The 1979 Report of the Research Grant of the Prevention of Physical and Mental Disabilities", the Ministry of Health and Welfare, Japanese Government, pp. 171–176.

Shimizu, T. and Tanaka, K. (1981). Experimental safety-study on sonication of pulsed ultrasound. *Jap. J. Ultrasound Med.*, Special Edition, "Studies on the Safety of Pulsed Ultrasound in the Diagnosis of the Fetus during pregnancy" **8** (4), 289–292.

Shoji, R., Momma, E., Shimizu, T. and Matsuda, S. (1971). An experimental study on the effects of low-intensity ultrasound on developing mouse embryos. J. Faculty of Science I. *Lokkaido Univ.*, Ser. VI, **18**, 51–56.

Siegel, E., Goddard, J., James, A. E. Jr and Siegel, M. S. (1979). Cellular attachment as a sensitive indicator of the effects of diagnostic ultrasound exposure on cultured human cells. *Radiology* **133**, 175–179.

Spring, E. (1972). Radiosensitivity—simultaneous ultrasonic and ionizing radiation. "Modern Trends in Radiotherapy" Vol. 2, pp. 51–58. Butterworths, London.

Stewart, G. J. (1978). The role of hypercoagulability in thrombosis. *Brit. J. Haematol* **40**, 359–362.

Stewart, H. F., Harris, G. R., Herman, B. A., Robinson, R. A., Haran, M. E., McCall, G. R., Carless, G. and Rees, D. (1973). Survey of use and performance of ultrasonic therapy equipment in Pinellas County, Florida, pp. 1–47. DHEW Publication No. (FDA) 73-8039.

Takabayashi, T., Abe, Y., Sato, A. and Suzuki, M. (1979). The effect of intense pulsed ultrasound exposure on fetal mouse. *Acta Obstet. Gynaec. Jap.* **31**, 895–896.

Takabayashi, T., Abe, Y., Sato, S., Sato, A. and Suzuki, M. (1981). Effects of pulse-wave ultrasonic irradiation on Mouse Embryos. *Jap. J. Ultrasound Med.*, Special Edition, "Studies on the Safety of Pulsed Ultrasound in the Diagnosis of the Fetus during Pregnancy" **8** (4), 286–288.

Takemura, H. and Moriyaki, S. (1977). Study on the hemolytic effect of clinical diagnostic ultrasound and the growth rate of cultured cells using a calibrated ultrasound generating system. *Supersonic Med.* **4**, 32–36.

Takeuchi, H., Arima, M. and Mizuno, S. (1966). Studies on the ultrasonic irradiation on rat embryos (The 2nd report)—Pulsed ultrasound for diagnostic use. *Med. Ultrasonics (Jap. Soc. Ultrason. in Med.)* **4**, 20–21.

Taylor, K. J. W. and Pond, J. B. (1972). A study of the production of haemorrhagic injury and paraplegia in rat spinal cord by pulsed ultrasound at low megahertz frequencies in the context of the safety for clinical usage. *Brit. J. Radiol.* **45**, 343–353.

Thompson, W. R. (1957). Influence of prenatal maternal anxiety on emotionality in young rats. *Science* **125**, 698–699.

Timmermans, L. (1973). Complications dues aux radiation A-ultrasons. *J. Urol. Nephrol. (Paris)* **79**, 504–511.

Todd, P. and Schray, C. B. (1974). X-ray inactivation of cultured mammalian cells enhanced by ultrasound. *Rad. Biol.* **113**, 445–447.

Tsutsumi, Y., Sano, K., Kuwabara, T., Takakura, K., Hayakawa, I., Suzuki, T. and Katanuma, M. (1964). A new portable echo-encephalograph, using ultrasonic transducers; and its clinical applications. *Med. Electron. Biol. Engin.* **2**, 21–29.

Ulrich, W. D. (1974). Ultrasound dosage for nontherapeutic use on human beings—extrapolations from a literature survey. *I.E.E.E. Trans. Biomed. Engng* BME-**21**, 48–51.

Wegner, R. D., Obe, G. and Meyenburg, M. (1980). Has diagnostic ultrasound mutagenic effects? *Human Genetics* **56**, 95–98.

Wells, P. N. T. (1974). The possibility of harmful biological effects in ultrasonic diagnosis. *In* "Cardiovascular Applications of Ultrasound" (Ed. R. S. Reneman), pp. 1–17. North-Holland, Amsterdam.

Williams, A. R. (1981). The design of experimental trials and the use of model systems to evaluate the efficacy of therapeutic ultrasound. Proc. Int. Symp. on Therap. Ultrasound, Winnipeg, Manitoba, Sept. 10–12.

Williams, A. R. (1982). Absence of meaningful thresholds for bioeffect studies on cell suspensions *in vitro*. *Brit. J. Cancer* **45**, 192–195.

Williams, A. R. and Chater, B. V. (1980). Mammalian platelet damage *in vitro* by an ultrasonic therapeutic device. *Arch. Oral Biol.* **25**, 175–179.

Williams, A. R., Chater, B. V., Allen, K. A. and Sanderson, J. H. (1981). The use of β Thromboglobulin to detect platelet damage by therapeutic ultrasound *in vivo*. *J. Clin. Ultrasound* **9**, 145–151.

Witcofski, R. L. and Kremkau, F. W. (1978). Ultrasonic enhancement of cancer radiotherapy. *Radiology* **127**, 793.

Woeber, K. (1965). The effect of ultrasound in the treatment of cancer. *In* "Ultrasonic Energy" (Ed. E. Kelley), pp. 237–249. Univ. of Illinois Press, Illinois.

Wong, Y. S. and Watmough, D. J. (1980). Haemolysis of red blood cells *in vitro* and *in vivo* caused by therapeutic ultrasound at 0·75 MHz. Paper C14 at the Ultrasound Interaction in Biology and Medicine Symposium, Reinhardsbrunn, East Germany, Nov. 10–14.

Yaroniene, G. V. (1978). Response of biological systems to low-intensity ultrasonic waves. Proc. FASE-78, Warzawa, pp. 13–16.

Zach, L., Morrison, A. and Cohen, G. (1959). Ultrasonic cavity preparation: histopathologic survey of effects on mature and developing dental tissues. *J. Amer. Dent. Assoc.* **59**, 45–55.

Ziskin, M. C., Bonakdarpour, A., Weinstein, D. P. and Lynch, P. R. (1972). Contrast agents for diagnostic ultrasound. *Investigative Radiol.* **7**, 500–505.

AUTHOR INDEX

A

Abagyan, G. V., 152
Abdulla, U., 55, 204
Abramovich, A., 223
Abramson, D. I., 120
Adler, J., 181
Akamatsu, N., 270
Anderson, D. W., 235–236, 272–273
Anderson, T. P., 225
Andreae, J. H., 89
Arentsschild, O., 301
Armour, E. P., 188, 283
Arslan, M., 117
Au, W. W., 277

B

Baba, M., 228
Bailey, K. I., 239
Baker, N. V., 204
Balamuth, L., 48
Bamber, J. C., 93
Barnett, S. B., 206, 221, 227–228, 272
Barrass, N., 206, 276
Barrett, J. T., 235–236, 273–274
Barth, G., 222
Basauri, L., 115
Beleva-Staikova, R., 215, 238
Benwell, D. A., 298
Bernstine, R. L., 212, 288
Bertenyi, A., 228
Beyer, R. T., 103
Bickford, R. H., 120
Biegler, R., 182
Bierman, W., 220
Blackstock, D. T., 101
Bleany, B. I., 187
Blitz, J., 38
Bobrow, M., 204
Bolt, R. H., 38
Bom, N., 56–57
Borovyagin, V. L., 213
Borrelli, M. J., 225
Boyd, E., 204
Braeman, J., 204
Brendel, K., 74
Brock, R. D., 204
Brodie, D. A., 269
Brookes, M., 223
Brown, G., 40
Brown, N. T., 268
Brug, E., 223
Buchan, J. F., 120
Bucher, N. L. R., 221
Buchtala, V., 222
Buckton, K. E., 204
Bugnon, C., 204
Bundy, M. L., 181
Burgudzhieva, T., 120
Burr, J. G., 285
Burton, R. M., 180
Busse, L. J., 93
Butyagin, P. Yu., 152

C

Campbell, P. N., 183
Carney, S. A., 217
Carrano, A. V., 275
Carson, P., 59, 77
Carson, T. E., 301–302
Carstensen, E. L., 76, 86–89, 91, 93, 103–105
Cataldo, F. L., 266
Cattaneo, V., 223
Cervenka, J., 238
Chapman, I. V., 200

Charm, S. E., 139
Chater, B. V., 150, 156, 165–167, 170, 195–197, 232, 278, 300
Chiba, Y., 279
Child, S. Z., 236, 271–272
Chivers, R. C., 93, 116, 128
Church, C. C., 260
Ciaravino, V., 161, 188
Clarke, G. R., 118
Clarke, P. R., 157–158, 188, 285
Clifton, H., 119
Coakley, W. T., 143, 153, 155, 158–159, 166, 178, 180, 187–188, 204
Coble, A. J., 182
Coggraves, M., 289
Combes, R. D., 179, 203
Cook, B. D., 69
Corry, P. M., 188
Cotterell, T. L., 150
Crick, F. H. C., 178
Crowell, J. A., 138, 194
Crum, L. A., 207
Curie, J. and P., 43

D

Daniels, S., 167, 168
Darling, R. C., 181
Dauksha, K. K., 278
Davey, D. A., 204–205, 265
David, H., 226, 267
Del Duca, M., 151
Denk, J., 120
Dewitz, T. S., 139
Dickson, L. G., 212
Dmitrieva, N. P., 212
Docker, M. F., 267
Dohnalek, J., 284
Dolgopolov, N. N., 200
Donald, I., 55
Doynon, D., 159
Drastichova, V., 221
Duarte, L. R., 222
Duff, R. S., 120
Dunn, F., 74, 86–87, 90–91, 93, 103, 116, 126, 159, 178–180, 182, 201, 209, 225
Dvorak, M., 210, 213
Dyer, H. J., 140, 202–203
Dyson, M., 118, 120, 127, 207–208, 213, 219–220, 223

E

Edmonds, P. D., 86, 209, 280–281
Eichner, K., 301
Elder, S. A., 152
Eller, A., 207
El'Piner, I. E., 140, 151, 212–213
Engel, D., 216
Esche, R., 160
Esmat, N., 225
Evans, E. F., 39
Evans, M. W., 89

F

Fahim, M. S., 117, 239, 284
Falus, M., 288
Filipczynski, L., 74
Fill, E. E., 155
Fischman, H., 204
Fishman, S. S., 166–167, 301–302
Fleisher, A. C., 286
Flynn, H. G., 151, 155
Fofanov, S. I., 120
Ford, W. L., 237
Forrest, J. O., 300
Frank, H. S., 89
Frenkel, J., 178
Frenkel, Ya. I., 151
Frey, P., 38, 89
Fridd, C. W., 209–210
Frizzell, L. A., 115
Frost, H., 301
Fry, F. J., 115–116, 156, 166, 182, 209, 225, 239, 283
Fry, R. B., 68
Fry, W. J., 68, 93
Fulton, J. F., 149, 171, 261
Fung, H., 195, 199, 272

G

Galitsky, A. B., 220
Galperin-Lemaitre, H., 178, 205
Galton, F., 40
Garrison, B. M., 240
Gavrilov, L. R., 228, 267, 284
Gebhart, E., 276

Gershoy, A., 140, 148
Gersten, J. W., 225
Gieder, A., 118
Glick, D., 200, 214
Goldblat, V. I., 222
Goldman, H. M., 301
Goldman, O. E., 208
Goliamina, I. P., 167
Golob, E., 204
Gooberman, G. L., 38
Goralchuk, M. V., 284
Gordon, D., 118
Gornia, F. I., 223
Goss, S. A., 86, 93–97
Graham, E., 149, 179
Gramiak, R., 150
Grandesso, R., 238
Gregory, D. W., 166, 210
Greguss, P., 228
Griffin, D. R., 40, 60
Griffing, V., 151
Guy, A. W., 107–108, 110

H

Haar, G. R. ter, 100, 167, 168, 207–208, 257
Hall, L., 86
Halliwell, M., 118, 223
Hammer, C. E., 60
Hanson, H. M., 269
Hara, K., 205, 218
Harkanyi, Z., 284
Harvey, E. N., 149, 166
Hasegawa, T., 74
Hattori, A., 234
Haupt, M., 276
Hawley, S. A., 86–87, 156, 201
Hayakawa, W., 301
Hellman, L. M., 287
Henneman, E., 225
Hering, E. R., 283
Herrick, J. F., 216, 225
Hertz, R. H., 267
Higashiizumi, T., 264
Hildebrand, B. P., 218
Hill, C. R., 59, 74, 77, 85, 93, 149, 157–158, 160, 178, 188, 205, 260, 283, 285
Holtzmark, J., 133–134
Hopewell, J. W., 100
Howkins, S. D., 143, 154, 278
Hrazdira, I., 181, 188–189, 210, 213, 236, 275
Hu, J. H., 216, 226, 230, 267
Hueter, T. F., 38, 93
Hughes, D. E., 154, 182–183, 188
Hussey, M., 38, 78, 86, 209

I

Iernetti, G., 159, 164
Iizuka, R., 290
Ikeuchi, T., 205
Istomina, O., 200

J

Jackson, F. J., 143
Jaffe, B., 45
James, J. A., 118, 223, 228
Jankowiak, J., 213–214
Johnston, R. L., 116, 209
Johnson, S. J., 118
Jones, S. J., 300
Joshi, G. P., 181
Joule, J. P., 43

K

Kadyrova, R. Kh., 235
Kahlert, K. von, 234
Kashkooli, H. A., 139, 179
Kaufman, G. E., 147, 188
Kaufmann, J. S., 199, 283
Kavetsky, R. E., 283
Keller, M., 216
Kernot, B. A., 183
Kessler, L. W., 86–87
Khan, A. E., 118
Khokhar, M. T., 266
Kinsler, L. E., 38, 89
Klemenkova, I. G., 120
Knaut, K., 223
Knokh, G., 223
Koh, S., 280, 289–290
Koifman, M. M., 235
Kolsky, H., 74
Koranyi, G., 288
Koshik, T. F., 284

Kossoff, G., 55, 73–74, 118, 221, 227, 272
Kratochwil, A., 49
Kremkau, F. W., 86–89, 92, 182, 185, 199, 221, 258, 272, 283–284
Krenztlin, C., 223
Krivitskii, A. K., 120
Krizan, J. A., 113–114
Krusen, F. H., 182
Krylova, T. I., 120
Kunze-Mühl, E., 204, 285
Kurachi, K., 277
Kurjak, A., 49

Lakshminarayanaiah, N., 180
Lamb, H., 152
Lamb, J., 89
Lateur, B. J. de, 120, 297
Latt, S. A., 206
Lauterborn, W., 155
Lehmann, J. F., 107–110, 120, 182, 297
Lele, P. P., 69, 93, 112–113, 115, 117, 169, 209, 257
Lepreshkin, W. W., 208
Levina, S. I., 220
Lewandowska, L., 120
Li, G., 188
Liebeskind, D., 193, 198–199, 201, 205–206, 237, 274–277
Lizzi, F. L., 115, 226–227, 230
Loch, E. G., 218
Loomis, A. L., 201
Lota, M. J., 181
Love, L. A., 185, 258
Ludlam, C. A., 196
Ludwig, G., 74
Lunan, K. D., 233–234
Lundberg, M., 277
Lyon, M. F., 239
Lyons, E. A., 289
Lypacewicz, G., 74

M

Macintosh, I. J. C., 204–205, 265–266
Macleod, R. M., 179
Madsen, P. W., 225
Majewski, C., 200, 213–214
Mannor, S. M., 239
Marcus, P. W., 93
Markham, D. E., 220
Marmor, J. B., 117
Marmur, R. K., 215
Martin, C. J., 141, 147, 166, 210
Mason, W. P., 136–137
McClain, R. M., 217, 269
McNally, N., 200
Meltzer, R. S., 264
Mendez, J., 180
Mermut, S., 204
Messino, D., 161
Meyers, R., 83
Michael, P. L., 229
Middleman, S., 179
Mihalas, G. I., 267
Miller, D. L., 143–147, 165, 198, 264
Miller, M. W., 147, 188, 193, 221, 266, 272
Minnaert, M., 143
Möhres, F. P., 60
Molinari, G. A., 228
Möller, P., 229, 301
Moore, J. L., 187–188
Moriyaki, S., 279
Morohashi, T., 290
Morris, D. R., 91, 129
Morris, S. M., 206, 276
Mortimer, A. J., 230
Morton, K. I., 188
Mountford, R. A., 92
Mueller, P., 180
Muggeo, M., 238
Muir, T. G., 103–104
Mukubo, M., 289–290
Mummery, C. L., 201, 275
Murai, N., 226, 268–269
Myenberg, M., 206

N

Nagy, I. I., 267
Neppiras, E. A., 151, 155, 157–158, 188
Neumann, A., 223
Newman, D. L., 181, 236
Nicholas, D., 85
Novak, A., 40
Nover, A., 120
Noltingk, B. E., 151, 155, 157
Nyborg, W. L., 86, 97–99, 114, 133, 135,

140, 142–143, 145, 148, 153–155, 158–159, 164, 166, 185, 210, 264, 280–281, 291

O

Oakley, E. M., 224
O'Brien, W. D. Jr., 90–91, 210, 218, 239
Ochs, A. L., 180
Oka, M., 69
Oliver, R., 266
Oosterbaan, W. A., 300
Ostrovskij, E., 200

P

Paaske, W. P., 120
Pameijer, C. H., 300
Parry, J. S., 137
Patrick, M. K., 58, 207, 220
Pasechnik, V. I., 180
Pashovkin, T. N., 217
Paul, B. J., 220
Pauly, H., 92
Petoussi, N., 267
Petrov, V. I., 223
Pierce, G. W., 40
Pinamonti, S., 236–237, 275
Pinchuk, V. G., 213
Pizzarello, D. J., 217–219, 222, 271
Plevinskis, V. P., 215
Poliakov, V. A., 223
Pond, J., 115, 126, 284
Pospisilova, J., 214
Powell-Phillips, W. D., 267
Prasad, N., 199
Pritchard, N. J., 143
Prudhomme, R. D., 151
Pye, J. D., 39–40

R

Lord Rayleigh, 155
Repacholi, M. H., 285, 298
Robinson, H. P., 211, 215–216
Rodgers, A., 154
Roelandt, J., 264
Rohr, K. R., 180
Roman, M. P., 118
Rooney, J. A., 59, 74, 77, 139, 143, 180
Rott, H.-D., 204
Rottova, A., 214
Ruban, E. L., 200
Rudin, D. O., 180
Ryabchenko, N. I., 178

S

Saad, A. H., 237–238, 273
Saggio, M. A., 194
Sakamoto, S., 270
Saksena, T. K., 155
Samosudova, N. V., 212
Sanada, M., 234, 278
Sanders, M. F., 138
Sarvazyan, A. P., 217, 282
Scheidt, P. C., 288–289
Schmid-Schönbein, H., 129
Schmitt, F. O., 140
Schneider, F., 90
Schneider, H., 60
Schreck, R. R., 206
Schray, C. B., 285
Schwan, H. P., 89, 91–92, 128
Seda, H. J., 301
Sekiba, K., 270
Senapati, N., 93
Serr, D. M., 204, 218
Sewell, G. D., 39–40
Shah, P. M., 150
Sheldon, T. A., 267
Shepstone, B. J., 283
Shimizu, T., 269–270
Shoji, R., 217, 269
Siddiqi, F., 180
Siegel, E., 199, 237, 274–275
Sikov, M. R., 117, 218
Simont, Y., 159
Simpson, G. M., 239
Sjoberg, A., 70–71
Slade, J. S., 137
Smith, A., 92
Smyth, M. G., 239
Sokolov, V. S., 180
Sommer, C. V., 194
Spencer, J. L., 202
Spring, E., 284
Stefanovic, V., 180
Stenner, L., 118
Stephens, R. J., 211–212

Stereva, S., 215, 238
Stewart, C. R., 204
Stewart, G. J., 235, 264
Stewart, H. F., 298
Stolzenberg, S. J., 212, 218, 225
Storm, D. L., 154
Straburzynski, G., 214
Stratmeyer, M. E., 209, 218, 236
Stuhlfauth, K., 225
Stumpff, U., 49
Summer, W., 58, 208, 220
Suppipat, N., 48

T

Takabayashi, T., 269–270
Takagi, S. F., 225
Takemura, H., 279
Talbert, D. G., 229
Tanaka, K., 269–270
Tanka, D., 216
Taylor, K. J. W., 181, 236, 284
Temple, P. R., 154
ter Haar, G. R., see Haar, G. R. ter
Thacker, J., 152, 178–179, 203, 205
Thompson, W. R., 269
Timmermans, L., 288
Tippe, A., 216
Tirrell, M., 179
Todd, P., 285
Towell, M. E., 267
Treton, J. A., 179
Tsokh, R. M., 228
Tsutsumi, Y., 215–216, 269
Tuchman, L. S., 220
Tucker, S. H., 200

U

Ulrich, W. D., 226, 267, 280–281

V

Valtonen, E. J., 118, 210, 213
van der Decken, A., 183

Wachsmann, F., 222
Wang, S. Y., 151
Warwick, R., 217
Watmough, D. J., 156, 188–191, 232, 263
Watson, J. D., 178, 187
Watts, P. L., 204
Webster, D. F., 199, 211, 221
Wegner, R.-D., 206, 276
Wehner, W., 49
Weinstock, A., 278
Weissler, A., 157
Wells, P. N. T., 38, 49, 54–55, 67, 73–74, 83, 89–90, 93, 100, 116, 118, 205, 209, 215, 228, 280–281
Wells, R. E., 129
Went, J. M. van, 120, 235
Wercham, R. E., 69
Whitlow, A. L., 60
Willard, G. W., 154, 159
Williams, A. R., 91, 113–114, 119, 129–131, 135, 137–141, 144, 146–147, 150, 152, 156, 161–167, 170, 178, 189, 191–192, 195–198, 230–234, 237–238, 259, 263–264, 273, 278, 298, 300
Willis, J. N., 166
Witcofski, R. L., 221, 272, 284
Wittenzellner, R., 166
Woeber, K., 284
Wong, B. L., 139
Wong, Y. S., 232, 263
Wood, M. R., 220
Wood, R. W., 201
Wyard, S. J., 207

Y

Yamamoto, Y., 228
Yaroniene, G. V., 279
Yoshioka, K., 69, 74
Young, R. R., 225

Z

Zach, L., 301
Zana, R., 90
Zaretskii, A. A., 88
Ziskin, M. C., 171, 264
Zorina, O. M., 88

SUBJECT INDEX

A

"A" scan, 50–51
Absorbing target, 72
Absorption, 7, 16, 39, 83–97
  effects of cross-linking, 92
  frequency dependence, 96
  of liver, 92
  of polymers, 88
  of proteins, 88–89
Acoustic impedance, 22
  of common materials, 5
Acoustic microstreaming, 133–134
Acoustic saturation, 105
Advantages of using cells in suspension, 184
Amplitude of a wave, 2
Angle of incidence, 18
Angle of reflection, 18
Anoxia from blood cell stasis, 262
Antinodes, 25
Aperture, 8
Applications of ultrasound, 46–60
Artificial membranes, 180–181
Atomisation of liquids, 154–155
Attenuation, 83–86, 93–97
  frequency dependence, 94–97
Audible sound, 33–36

B

"B" scan, 52–56
B/A, 103
Backbone scission, 178
Bacteriocidal capacity, 194
Beam geometry, 61
Beam plots, 64
Behavioural effects, 201
Biochemical changes, 199–201, 211, 214–217
Bioeffects approach, 255, 265–286
Biological membranes, 181–183
Blood cell stasis, 207–209
Blood coagulation system, 230–235, 268–280
Bone, 109, 129
Bone welding, 223
Bony fractures, 222–224
"Boundary curves", 280–283
"Breathing pulsation" of bubbles, 142
Burglar alarms, 47

C

Calorimeter, 68
Capacitance techniques, 74
Cataract production, 226–227
Caustic curve, 27
Cavitation, 126 et seq.
  detection, 156–158
  enhancement by stirring, 161–164, 259–260
  enhancement in standing waves, 207–209, 262
  in vivo, 148–150, 165–171, 232–233
  models of, 133–137
Cell cycle, 185–187
Cell surface changes, 274–275
Chemical techniques, 75, 150–152, 157
Cholesteric liquid crystals, 69
Chromosome abnormalities, 204–206, 265–266
Clinical trials, 118–120
Compound "B" scan, 55
Compressional waves, 11–12
Converging lenses, 28
Curie temperature, 46

D

Dauermodifiers, 203
Decibels (dB), 33
Densities of common materials, 5
Design of clinical trials, 297–299
Detection of cavitation, 156–158
Detection of ultrasound, 60
Dextrans, absorption of, 87–88
Diffraction pattern, 61–65
Diffusion of heat, 98
Disadvantages of using cells in suspension, 184–185
Displacement amplitude, 14
Displacement antinodes, 25
Displacement nodes, 25
DNA, 177–179
Doppler devices, 49
"Dose" of ultrasound, 77–78, 254
Dosimetry of ultrasound, 60–78
Drosophila, 270–271

E

Ear, 228–229
Echolocation in animals, 60
Eddy currents, 42
Electron micrographic studies, 189–193, 211–213
Electrophoretic mobility changes, 181
Embryonic growth rate effects, 217–219, 269–272
Energy of a wave, 4
Enzyme, 179–180
Epidemiological approach, 256, 286–291
Eye, 226–228

F

Far field, 61–65
"Fatigue" of biological structures, 126, 129
Finite amplitude effects, 101–106
Fletcher–Munsen curves, 34–35
Flexural waves, 15
Focal length, 29
Focal lesions, 111–112
Focusing sound, 26
Foetal activity, 226, 267
Foetal growth and development, 217–219, 269–272
Formation of microdroplets, 154–155
Fountain effect, 70
Fractures in bone, 222–224
Fraunhoffer zone, 61–65
Free radicals, 132, 150–152, 157, 179
Frequency dependence
  of absorption, 96
  of attenuation, 95
Frequency of sound waves, 7
Frequency weighting networks, 35
Freznel zone, 61–65
Fruit flies, 270–271
Functional changes
  in cells, 194–199
  in tissues, 217–240
Fundamental frequency, 44
Fusion of bone, 223

G

Galton whistle, 40
Generation of ultrasound, 40–46
Genetic effects, 202–206
Giant cells, 193–194
Grey scale displays, 56

H

Harmonics, 16, 44
Hartman generator, 41
Healing of fractures, 222–224
  of ulcers, 219–220
  of wounds, 219–221
Heating of tissues, 97
Histological assessment of damage, 111, 116
Hormesis, 188, 200
Hormonal system, 238
"Hot spots", 65
Huygens principle, 16
Hydration zone, 89
Hydrodynamic nucleation, 161–164, 259–260
Hydrophones, 68
Hyperaemic reflex, 257

I

Immunological system, 235–238, 272–274
Implosion of bubbles and cavities, 155
Industrial applications, 47–48
Intensity, 23, 32, 66–67
Interface (energy partition at), 23
Interface, 10, 61–65
Interphase (of cell), 186
Intravascular haemolysis, 232–233
Ionizing radiation "dose", 78
Isolated organelles, 183

J

Joints-treatment, 110

L

Large aperture, 9
Lens, 28
Lesion boundary temperature, 112
Limb bud regeneration, 221–222
Liquid crystals, 69
Liver regeneration, 221, 272
Longitudinal wave, 12
Loudness of sounds, 34
Luciferin/Luciferase, 146
Lymphocytes, 204–205

M

"M" mode displays, 51–52
Magnetostriction, 41–43
Mason horn, 136–137
Mechanistic approach, 255–264
Medical applications, 48–60
Meniére's disease, 83, 228
Membranes
  artificial, 180–181
  biological, 181–183
Microbubble formation, 154–155
Microdroplet formation, 154–155
Micronuclei *in vivo*, 149–150
Microorganisms, 203
Microsomes, 183
Microstreaming, 133–134
Mode conversion, 16
Models simulating stable cavitation, 133–137
Morphological changes, 189–194, 209–211
Moss, 202
Motility of cells, 201
Muscular system, 229–230
Mutants, 202

N

"N" wave, 102
Near field of transducer, 61–65
Negative bioeffects, 200
Nervous system, 224–226, 267–269
Nodes, 25
Nonlinear effects, 101–106
Nonlinearity parameter (B/A), 103
Nuclear damage, 192–193, 265–266, 275–278
Nucleation of cavitation
  by γ and X-rays, 161
  by muscular activity, 261
  by stirring, 161–164, 259–260
Nuclepore membranes, 144–146

O

Optical techniques, 74–75
Organelles-isolated, 183
Organs of special sense, 226–229
Oscillation of bubbles, 155
Overtones, 16

P

Parabolic reflector, 27
Partial standing wave, 26
Partition of energy at an interface, 23
Peas, 202
Perceived loudness of sound, 33
Periosteum, 16
Phase, 15
Phon, 34
Phonophoresis, 181–182, 284

Physiological effects, 209–240
Piezoelectricity, 43–46
Placebo effect, 119
Plane progressive wave, 8
Platelets, 137–138, 195–198, 231–235, 278
Point source, 8
Polyethylene glycol, absorption, 87–88
Potassium leakage, 181
Power, 2
Power density, 32
Pressure in an acoustic field, 14–15
Pressure antinode, 26
Pressure node, 26
Propagation of a wave, 4

Q

Q factor, 43
Quartz wind streaming, 70

R

Radiation pressure, 69–74, 262
Ray, 20–21
Real time displays, 56–57
Recommendations, 294–302
dental exposures, 300–302
diagnostic exposures, 295–296
general, 294–295
therapeutic exposures, 297–300
ultrasonic cleaning baths, 302
Reciprocity techniques, 76
Red blood cells, 129, 139
Reflecting target, 73
Reflection, 16–18
Refraction, 16, 19
Regenerating liver, 221, 272
Relative loudness of audible sounds, 33–35
Relaxation processes, 90–91
Reproductive system, 238–240
Resonance, 30–31
Resonant frequency, 44, 46
of gas bubbles, 142–143
Resonant length, 30, 42
Reticuloendothelial system, 237–238
Ribosomes, 183
Ring patterns, 65–66
Root mean square (rms), 32

S

"Safety levels" for diagnostic exposure, 291–293
Saturation (acoustic), 105
Scattering, 83–86
Schlieren techniques, 75–76
Shear waves, 11
Simple harmonic motion (SHM), 3–5
Sine wave, 4, 12
Siren, 41
Small aperture, 9
Snell's law, 21
SOAB rubber, 71
Sonochemical effects, 150–152
Sonoluminescence, 155, 157
Spherical aberration, 29
Stabilized gas bubbles, 142–146
Standing waves, 23–26, 207–209, 262
Standing wave ratio, 26
Stasis of blood cells, 207–209, 262
Strain in biological materials, 128
Streaming around solid bodies, 142
Stringed instruments, 16
Supersonic implosion, 155
Surface characteristics of cells, 236, 274–275
Surface waves, 152–156
Surface wave analogy, 2
Surgical applications, 117–118, 225
Synergistic effects, 283–286

T

Temperature elevation in tissues, 97–99
Teratogenic effects, 202
Thermal damage, 110–113
Thermistors, 69
Thermocouples, 69
"Threshold" curves, 280–283
Tinnitus, 229
Tissue heating, 97–99
Transport across membranes, 181
Transverse waves, 11, 15

U

Ulcer treatment, 219–220
Uterine contraction, 229, 230

V

Velocity gradient, 135
Velocity of sound
  in common materials, 5
  in gases, 6
  in liquids, 6
  in solids, 6
Velocity transformers, 30–31, 42, 137
Viability of cells, 187–189

W

Wavelength, 2
Weber-Fechner law, 33
Wedge resonators, 41
Weight reduction in foeti, 217–219
Welding of bones, 223
Whispering galleries, 27–28
White noise, 48
"Willard" events, 159
Wound healing, 219–224